Ecosystem Services from Agriculture and Agroforestry

Ecosystem Services from Agriculture and Agroforestry

Measurement and Payment

EDITED BY

Bruno Rapidel, Fabrice DeClerck, Jean-François Le Coq and John Beer

earthscan
from Routledge

First published by Earthscan in the UK and USA in 2011

For a full list of publications please contact:
Earthscan
2 Park Square, Milton Park, Abingdon, Oxon OX14 4RN
711 Third Avenue, New York, NY 10017

Earthscan is an imprint of the Taylor & Francis Group, an informa business

ISBN 978-1-84971-147-0 (hbk)

Typeset by MapSet Ltd, Gateshead, UK
Cover design by Benjamin Youd

A catalogue record for this book is available from the British Library

Library of Congress Cataloging-in-Publication Data

Ecosystem services from agriculture and agroforestry : measurement and payment / Bruno Rapidel ... [et al.].
p. cm.
Includes bibliographical references and index.
ISBN 978-1-84971-147-0 (hardback)
1. Agriculture--Environmental aspects. 2. Agroforestry--Environmental aspects. 3. Agricultural ecology. 4. Ecosystem services. I. Rapidel, Bruno.
S589.75.E28 2011
630.208'6–dc22

2010048663

Contents

Part II Marketing Ecosystem Services

Part III From Theory to Practice: Tales of Success and Lessons Learned

List of Figures, Tables and Boxes

Figures

Tables

Boxes

List of Contributors

Jacques Avelino is a plant pathologist at the Centre de Coopération Internationale en Recherche Agronomique pour le Développement (CIRAD), based at the Centro Agronómico Tropical de Investigación y Enseñanza (CATIE) in Costa Rica. As a coffee disease specialist, he works on the effects of plant species diversity on pests and diseases at plot and landscape scales.

John Beer is the research and development director of CATIE, Costa Rica, where he has worked for the last 30 years on the value of agroforestry systems as a systemic response to the complex problems faced by rural agricultural- and forestry-based communities.

Liliana Bravo-Monroy is a PhD student from the Centre for Agri-Environmental Research (CAER) at the University of Reading, UK. She is currently conducting her fieldwork in Colombia.

Francisco Casasola is a silvopastoral systems specialist at CATIE, Costa Rica. As part of his activity within the livestock and environmental management programme, his interest is focused on the payment of environmental services in agricultural landscapes.

René Castro is full professor and director of the Sustainable Development Programme at Instituto Centroamericano de Administración de Empresas (INCAE) Business School, Costa Rica, where he teaches sustainability and environment-related courses. He is the former minister of environment and energy and currently is Costa Rica's chancellor at the Ministry of Foreign Affairs.

Fabrice DeClerck is a landscape and community ecologist at CATIE, Costa Rica, with a special interest in the mechanisms that drive the relationship between biodiversity and ecosystem functioning, and their application to the sustainable management of agricultural landscapes.

Jan Fehse heads the forest carbon team at EcoSecurities (Oxford, UK). His responsibilities include carbon offset certification, commercializing offsets in the voluntary carbon market and providing consultancy services on most aspects related to forest carbon projects, markets and policy development.

Lucas Alejandro Garibaldi is an ecologist and post-doc at the Laboratorio Ecotono (CONICET-CRUB) and lecturer at the University of Buenos Aires in Argentina.

Federico Gómez-Delgado is an expert in hydrological modelling and statistics. He has wide experience in these subjects after working for the Instituto Costarricense de Electricidad, as well as for CATIE and CIRAD in Costa Rica and the Swedish Cooperation.

Cliserio González Hernández is a PhD student from the joint doctorate programme between CATIE, Costa Rica, and Wales University. His PhD research is related to the economic analysis and environmental services provided by agroforestry systems with certified coffee in Costa Rica and Nicaragua.

Beatriz González-Rodrigo is an expert in rural development engineering. She works as scientific assessor for Consejo Superior de Investigaciones Científicas (CSIC; Spanish National Research Council) and is an associate professor at the Polytechnic University of Madrid, Spain.

Kristell Hergoualc'h is a researcher in ecosystems' functioning, with a focus on climate change mitigation through the reduction of greenhouse gas emissions from land-use change, at the Center for International Forestry Research (CIFOR). She completed her PhD research at CIRAD in Costa Rica on soil greenhouse gas emissions and carbon storage in coffee plantations.

G. Martijn ten Hoopen is an epidemiologist at CIRAD, Costa Rica. A specialist of cocoa diseases, he works on the control (especially biological control) of pests and diseases.

Muhammad Ibrahim is the leader of the Livestock and Environmental Management Programme at CATIE, Costa Rica. He has developed and led several research and development projects in Latin America, with the aim of improving the livelihood of rural communities, the conservation of natural resources and adaptation to climate change.

Iván Islas Cortés is director of environmental economics for the National Institute of Ecology in Mexico, where he is currently working on economic valuation, economic instruments and energy efficiency.

Bernard Kilian is an associate professor at INCAE Business School in Alajuela, Costa Rica, where he teaches operations management and microeconomics. He is also the director of research at the Centre for Sustainable Market Intelligence (CIMS) at INCAE, where he conducts research in agribusiness.

Alexandra-Maria Klein is an ecologist and head of the Ecosystem Functioning Group at the Institute of Ecology and Environmental Chemistry in the Leuphana University of Lüneburg, Germany.

Jean-François Le Coq is an agro-economist working with CIRAD, Costa Rica, and posted at Universidad Nacional, Heredia, Costa Rica, where he analyses the political framework and governance of the instruments of agro-environmental policies, payment for environmental services programmes and green certifications schemes.

Josué León is an agronomist at CATIE, Costa Rica. He is the watershed programme coordinator for the CATIE Copán River Watershed FOCUENCAS programme in Copán Ruinas, Honduras.

Róger Madrigal-Ballestro is a research fellow at Environment for Development Center (EfD/CATIE) and the academic coordinator of the masters programme in socio-economics at CATIE. He has a masters degree in environmental economics and he is a PhD candidate in economics at Freiburg University in Germany. His scholarly interests are in the areas of community-based management of common pool resources and institutional analysis of water policies. He also has extensive experience with the design and implementation of payments for ecosystem services schemes in Latin America.

Miguel Marchamalo, a hydrology and geomatics specialist, is an associate professor at the Polytechnic University of Madrid in Spain and manages scientific cooperation projects about communities, water and forests in Central America.

Alejandra Martínez Salinas is a Nicaraguan ecologist who leads CATIE's permanent bird monitoring station in Costa Rica. She specializes in the conservation of bird biodiversity in human-dominated landscapes and the management of biological corridors.

Colin Moore is a senior ecosystem services manager at EcoSecurities, Oxford, UK, where he is responsible for conducting due diligence and providing project development services for carbon forestry projects.

Iris Motzke is an ecologist currently working towards her PhD in pollination services at the Agroecology Group at the University of Göttingen in Germany.

Nathan Muchhala is a post-doctoral fellow with the University of Nebraska-Lincoln. His research focuses on the evolutionary ecology of bat pollination in the Neotropics.

Rodrigo Murillo is an independent consultant and researcher under contract basis for the INCAE Business School, Costa Rica. He is an associate professor at the University of Costa Rica's Mechanical Engineering Department where he teaches economic engineering.

Till Neeff is senior commercialization manager for carbon credit origination at EcoSecurities, Oxford, UK. He carries out project appraisals, develops business plans and negotiates commercial deals for carbon forestry projects.

Roland Olschewski is an environmental and forest economist and the current head of the research group Environmental and Resource Economics at the Swiss Federal Research Institute WSL, Switzerland.

Alexander Pfaff is an associate professor of public policy, economics and environment at Duke University (Durham, NC, USA), and an environmental and natural resources economist focused on how economic development, the environment and natural resources affect each other.

Cornelis Prins is a sociologist at CATIE, Costa Rica, where he leads research and development activities on rural development, action research in rural innovation and institution building.

Bruno Rapidel is an agronomist at CIRAD, based at CATIE, Costa Rica. As the coordinator of the Pôle de Compétences en Partenariat (PCP), an interdisciplinary research group on agroforestry systems with perennial crops, he specializes in the development of methods to design cropping systems, using modelling tools and participatory approaches.

Juan Robalino is director of the Latin American and Caribbean Environmental Economics Programme and a research fellow at the EfD Initiative at CATIE, Costa Rica. He is an economist whose research has focused on the evaluation of conservation and development policies.

Carlos Manuel Rodríguez is vice president of conservation policy at Conservation International, based in Costa Rica. A lawyer, politician and, above all, a conservationist, he held various political posts in Costa Rica, including minister of environment and energy, where he was a pioneer in the development of payment for ecosystem services.

José Eduardo Rolón Sánchez is director of public policy in Community and Biodiversity A.C., where he works in designing, analysing and evaluating the political and institutional feasibility of public policies that use economic incentives for environmental conservation in coastal and marine areas of Mexico.

Ina Salas Boucher is an environmental economist. She currently works at the Food and Agriculture Organization of the United Nations (FAO) as a national consultant specialist in environmental concerns in rural development in Mexico.

Claudia Sepúlveda is an agroecologist at CATIE, Costa Rica. As part of the Livestock and Environmental Management Programme, she has focused her activities on sustainable certification mechanisms as a means of adding value to cattle farm products through enhanced adoption of silvopastoral technologies.

Gabriela Soto, soil ecologist and agroecologist at CATIE, Costa Rica, is an organic coffee production specialist. She is currently working on the promotion of sustainable coffee production and evaluating and comparing the impacts of certification labels upon the coffee sector.

Raffaele Vignola is an expert in climate change adaptation policies and environmental services. He works as a scientific assessor in the Climate Change Programme, CATIE, Costa Rica, and project manager in these subjects in tropical regions.

Laura Villalobos is an environmental economist currently working for the EfD Initiative at CATIE, Costa Rica. Her research focuses on evaluating the impacts of different conservation policies, such as payments for environmental services and protected areas in Costa Rica and Mexico.

Cristóbal Villanueva is a researcher at CATIE, Costa Rica, specializing in livestock farming systems. His interests focus on the design of cattle farm systems based on associating trees with pastures in order to balance productivity and ecosystem service provision. He is part of the Livestock and Environmental Management Programme at CATIE.

Foreword I

Sara Scherr

For most of the last century, agricultural landscapes were valued almost exclusively for their role in supplying crop and livestock products for local consumption and national and international markets. The importance of managing lands for ecosystem services – in particular, watershed protection and biodiversity conservation – was widely recognized. But it was assumed that farmlands themselves had little or no ecological value, and that ecosystem services could be provided effectively only *outside* of agricultural lands, mainly through protecting or restoring natural vegetative cover, whether forests, natural grasslands or wetlands. During recent decades, the negative impacts of agricultural intensification and expansion upon ecosystems have become a critical threat in many regions, prompting new policy responses to reduce these effects, both carrots (subsidies for practices reducing agrochemical pollution and erosion, tax incentives for conservation easements) and sticks (regulations, zoning restrictions). The first generation of payments for ecosystem services to farmers – payments made by governments or private actors that are contingent upon producing ecosystem benefits – focused on taking land out of agricultural production, and restoring natural habitat on farms and in the broader landscape. Policy action was framed within a 'trade-off' paradigm, in which policy-makers, businesses and farming communities had to decide how much land and water to keep in production and how much to allocate to conservation.

But a rethinking of this paradigm is under way. New scientific evidence is showing that well-planned and well-managed farmed fields and grazing lands can actually *produce* and even *restore* ecosystem services. By including agrobiodiverse land cover and adjusting management systems, they can provide critical supplemental – in some cases, primary – habitat for wildlife. Farmlands with healthy, high organic-matter agricultural soils; deep-rooted crop, grass and trees producing food and commodities in strategic locations; low-impact tillage practices; year-round vegetative cover; and sensible protections for nearby water resources can protect most important watershed services. Such benefits are especially significant where farmlands are included as part of strategic habitat and watershed planning, coordinated with public and private land conservation in the landscape.

In the Neotropics, the promise of this approach to ecosystem management is especially high. The main agricultural areas of this region, which provide food security and livelihoods to most of the rural population, as well as critical export earnings, are located in or around regionally important watersheds and globally important centres of biodiversity. A rich diversity of indigenous plant species valued for food, medicines, feed and other uses could potentially enrich commercial agriculture. A growing research community is generating new solutions and systems for ecologically friendly farming and ranching that are economically productive and profitable. However, currently, the dominant agricultural and livestock production systems produce relatively low levels of ecosystem services, or undermine the 'natural capital' for producing them. For farmers, shifting to more eco-friendly crop mixes and practices can be costly and risky, and requires new complex knowledge, and there is little public or private funding available for technical assistance or risk management.

Payment for ecosystem services (PES) schemes for farmers and farming communities offer one of the most promising policy and market instruments to accelerate this transition. PES for eco-friendly farming systems needs only to cover the costs of establishment, transition and early risks of local adaptation, and can then be maintained at a low level (or even withdrawn, if the new practice is more profitable than the old). Financing will often still be needed for nature conservation and restoration on private lands; but as these are more expensive (they must compensate farmers for a stream of lost revenues from land taken out of production), they can be targeted at those parts of the agricultural landscape where eco-friendly production is clearly not enough. There are numerous beneficiaries of the ecosystem services produced in these agricultural regions who derive concrete financial benefits from them, so that potential 'buyers' include not only taxpayers financing 'public goods', but also municipal utilities, bottling companies, tourism operators, fishers, agencies managing flood and other risks, as well as the food industry and consumers of agricultural goods who want to establish 'green' credentials.

This is the promise; we are still in the early stages of realizing that promise. This volume provides a synthesis of experience to date. Drawing together the findings of a decade of research by CATIE, CIRAD and many national and international research and development organizations active in the Neotropics, this book presents numerous lessons learned for targeting, designing, managing and monitoring agricultural PES. The contributors apply a rigorous and sceptical scientific eye to the body of experience, describing failures as well as successes, and highlight challenges that have not yet been solved. They also touch on fundamental questions of social equity, political philosophy and governance. This volume provides numerous ideas for shaping the next generation of agricultural PES in the Neotropics, and is relevant to innovators around the world.

Sara Scherr
President, EcoAgriculture Partners
January 2011

Foreword II

Patrick Caron

Since the hunger riots in 2008, agricultural and food issues have dramatically returned to the forefront of the global political agenda as one of the 21st century's major concerns. As a consequence, it is evident that world agriculture more broadly lies at the heart of the main global challenges: food security due to demographic growth and evolution in food consumption patterns; environment due to climate change, natural resources and fertility management, pollutions, water scarcity and biodiversity concerns; energy due to the growing scarcity of fossil fuels and the potential for biomass use; and the fight against poverty and inequality. To fulfil the multiple functions that are locally and globally expected from agriculture and to address the unpredictable evolution of human consumption patterns, there is no doubt that agriculture should increase production of newer and better commodities. The biggest challenge will be to feed 9 billion people in 2050 while preserving ecosystems which provide other products and services in a changing environment. The unique and globally complex questions regarding the relationships between agriculture, food security, health, energy, economic and social development, and the environment, call for renewed investments in research.

Business as usual is no longer an option and the need to mobilize all knowledge sources for alternative ways of producing and managing resources and rural space is imperative. CIRAD (a French research centre working with developing countries to tackle international agricultural and development issues, www.cirad.fr) has chosen to invest and promote research activities towards ecological intensification, such as new technologies based on imitating rather than forcing 'nature'. This calls for renewed approaches to look at the interactions between agriculture and environment, for a shift in paradigms that paved the 'modernization' of agriculture during the last century and for renewed links between fragmented epistemic and political communities.

In this context, agroforestry is looked upon as a potential option for the stimulation of interactions between trees and crops. The research investment objective is to better understand ancestral or emerging practices and systems and to assess their performances and impacts by taking into account the entire range of expected functions. From an environmental point of view, this relates to biodiversity preservation, an increase in carbon storage, erosion control, and an increase in soil fertility. Agroforestry also provides farmers with essential alternative sources of income. This effort should contribute to innovative and more efficient practices through the mobilization of both farmers' knowledge and the most advanced academic science.

The integration of environmental concerns into agricultural transformation should not be referred to as the 'ecologization' of agriculture. Agriculture is often looked upon as an enemy of nature; however, there is a strong assumption that agriculture can and should provide environmental goods and could be acknowledged for such services. In recent years a movement towards a greener economy through payment for environmental/ecosystem services that prevent environmental degradation or contribute to site improvement has emerged. Hypothetically, management of natural resources could be conducted through the connection between market and public intervention. However, numerous questions regarding the efficacy of this concept remain and assessment of the initial payments for environmental/ecosystem services programmes is necessary, both from the points of view of the payment of services and the economic and ecological advantages and disadvantages of agroforestry systems in comparison to monocultures. Throughout this book we will learn from experiences and case studies while nourishing current thoughts and innovative processes with a remarkable reflexivity!

This book fulfils the main objective of the Mesoamerican Scientific Partnership Platform for Agroforestry Systems with Perennial Crops that is implemented since 2007 by CATIE, CIRAD, Bioversity International, CABI, INCAE and Promecafé, all six partners related to agricultural research and development in Central America. The objective of this platform is to contribute to maintaining and increasing the competitiveness and sustainability of the agricultural sector of Mesoamerica through quantification, valuing and development of all potential products and environmental services of agroforestry systems with perennial crops, in particular coffee and cocoa. This platform was designed as an institutional tool and engagement for securing long-term joint investment on agroforestry systems regarding:

i) the evaluation of environmental services provided by agroforestry;
ii) the design of competitive, sustainable and diversified management strategies for agroforestry;
iii) the assessment of the impacts of agroforestry systems on farmer's livelihoods and strategies;

iv) the strengthening of farmers' business organizations; and
v) the understanding of the institutional arrangements along value chains for agroforestry products and services.

The publication of this book after only four years of study and analysis is a positive signal of the intellectual and institutional dynamics of this partnership. It not only accounts for the joint innovative production of knowledge, but also paves the way for the next activities and opens up promising research avenues for the future.

Patrick Caron
Director General for Research and Strategy, CIRAD
February 2011

List of Acronyms and Abbreviations

AAU	assigned amount unit
AFS	agroforestry systems
ALM	agricultural land management
APFC	Advance Paid Forest Certificate
A/R	afforestation and reforestation
ARESEP	Autoridad Reguladora de Servicios Públicos (Public Services Regulating Authority Bureau)
ARR	afforestation, reforestation and revegetation
ASL	above sea level
AyA	Acueductos y Alcantarillados (Water Piping and Sewage Company)
B	boron
BCCR	Banco Central de Costa Rica
BCS	biological control services
BMP	Bird Monitoring Project
C	carbon
4C	Common Code for Coffee Community
Ca	calcium
CAER	Centre for Agri-Environmental Research (UK)
CAFE	Coffee and Farmer Equity (Starbucks)
CAM	Microbasin Environmental Committee
CAR	Climate Action Reserve
CATIE	Centro Agronómico Tropical de Investigación y Enseñanza (Tropical Agricultural Research and Higher Education Center, Costa Rica)
CBD	Convention on Biological Diversity
CCB	Climate Community and Biodiversity Standards
CCOF	California Certified Organic Farmers
CCX	Chicago Climate Exchange
CDM	Clean Development Mechanism
CER	Certified Emission Reduction/Certificate of Emissions Reductions

CH_4	methane
CIFOR	Center for International Forestry Research (one of 15 centres within the Consultative Group on International Agricultural Research, with headquarters in Bogor, Indonesia)
CIMS	Centre for Sustainable Market Intelligence
CIRAD	Centre de Coopération Internationale en Recherche Agronomique pour le Développement (French Centre for International Cooperation in Agricultural Research for Development)
Cl	chlorine
CM	crop management
CNFL	Compañía National de Fuerza y Luz (National Power and Illumination Company)
CO_2	carbon dioxide
COHDEFOR	Honduran Forest Development Corporation
CONAFOR	Comisión Nacional Forestal (Mexican National Forestry Commission)
CONAGUA	Comisión Nacional del Agua
CONANP	National Commission of Natural Protected Areas
CONAPO	National Population Council
COOCAFE	Consortium of Cooperatives of Coffee Growers
CPF	Forest Protection Fund
CREAMS	Chemicals, Runoff and Erosion from Agricultural Management Systems model
CSIC	Consejo Superior de Investigaciones Científicas (Spanish National Research Council)
CVC	citrus variegated chlorosis
DANIDA	Danish International Development Agency
DBH	diameter at breast height
DFG	Dirección General Forestal (General Bureau of Forestry)
DM	dead material
DNDC	DeNitrification-DeComposition model
DOC	dissolved organic carbon
ECD	electron capture detector
EfD Initiative	Environment for Development Initiative (capacity-building programme in environmental economics, focusing on research, policy advice and teaching in Central America, China, Ethiopia, Kenya, South Africa, and Tanzania)
ERPA	Emission Reduction Purchase Agreement
ERU	emission reduction unit
ES	ecosystem/environmental services
ESC	Environmental Services Certificate
ESPH	Empresa de Servicios Públicos de Heredia (Water and Power Utility Company, Heredia)

EU	European Union
EU ETS	European Union Emissions Trading Scheme
EUROSEM	European Soil Erosion Model
FACMP	Forest Payment Certificate for Management Purposes
FAO	Food and Agriculture Organization of the United Nations
FCPC	Forest Carbon Partnership Facility
FDF	Forest Development Fund
FID	flame ionization detector
FLO	Fairtrade Labelling Organizations
FMP	farm management plan
FOCUENCAS	Innovation, Learning and Communication for the Adaptive Co-Management of Watersheds
FOG	Florida Organic Growers
FONAFIFO	Fundo Nacional de Financiamiento Forestal (National Forestry Financing Fund)
FPC	Forestry Payment Certificate
FUNDECOR	Fundación de Desarrollo de la Cordillera Volcánica Central (Central Volcanic Mountain Range Development Foundation)
GEF	Global Environment Facility
GEF-Silvopastoral project	Global Environment Facility-funded Integrated Silvopastoral Systems for Ecosystem Management project
GHG	greenhouse gas
GIS	geographic information system
GPP	gross primary production
GPS	global positioning system
GWP	global warming potential
HES	hydrological ecosystem services
IBSA	Biodiversity Index for Environmental Services
ICAFE	Instituto del Café de Costa Rica (Costa Rican Coffee Institute)
ICE	Instituto Costarricense de Electricidad (Costa Rican Institute of Electricity)
ICRAF	International Centre for Research in Agroforestry
IETA	International Emissions Trading Association
IFC	Institute for Forest Conservation
IFM	improved forest management
IFOAM	International Federation of Organic Agricultural Movements
IHCAFE	Instituto Hondureño del Café
INA	National Agrarian Institute
INCAE	Instituto Centroamericano de Administración de Empresas (Central American Institute of Business Administration)
INE	National Institute of Ecology
IPCC	Intergovernmental Panel on Climate Change

IRR	internal rate of return
ISO	International Organization of Standardization
IUCN	International Union for Conservation of Nature
JAS	Japanese Agricultural Standard
K	potassium
KfW	Kreditanstalt für Wiederaufbau (German Development Bank)
lCER	long-term Certified Emission Reduction
LISEM	Limburg Soil Erosion Model
MAG	Ministerio de Agricultura y Ganadería (Ministry of Agriculture and Livestock)
MANCORSARIC	Association of Municipalities of the Maya Route of Copán Ruinas, Santa Rita, Cabañas and San Jeronimo
MBC	Mesoamerican Biological Corridor
MEA	Millennium Ecosystem Assessment
MESAP	Board of Environment and Production
MIDEPLAN	Ministerio de Planificación de Costa Rica
MINAE	Ministerio de Ambiente y Energía (Ministry of Environment and Energy)
MMMF	Modified Morgan-Morgan-Finney model
MMS	metric multidimensional scaling
Mn	manganese
MW	megawatt
N	nitrogen
N_2	di-nitrogen
N_2O	nitrous oxide
NBP	net biome production
NECB	net ecosystem carbon balance
NEP	net ecosystem production
NGO	non-governmental organization
NH_3	ammonia
NOE	Nitrous Oxide Emission model
NOP USDA	National Organic Program of the United States Department of Agriculture
NPP	net primary production
NPV	net present value
NYBOT	New York Board of Trade
OAS	Organic Accreditation System
OCIA	Organic Crop Improvement Association
OGBA	Organic Growers and Buyers Association
OTC	over the counter
OTCO	Oregon Tilth Certified Organic
P	phosphorus
PA	protected area
PAID	photo-acoustic infrared detector

PASTIS	Predicting Agricultural Solute Transport in Soils model
PAP	Protected Areas Project
PCA	principal components analysis
PCP	Pôle de Compétences en Partenariat
PDD	project design document
PES	payment for ecosystem/environmental services
PHES	payment for hydrological ecosystem/environmental services
PRODEFOR	Programa de Desarrollo Forestal (Forestry Development Programme)
PRODEPLAN	Programa de Plantaciones Comerciales (Commercial Plantation Programme)
PRONARE	Programa Nacional de Reforestación (National Programme of Reforestation)
PSM	propensity score matching
R2	coefficient of determination
Ra	autotrophic respiration
RA	Rainforest Alliance
Re	total ecosystem respiration
REDD	reduce/reducing emissions from deforestation and degradation
REDD+	reducing emissions from deforestation and degradation and enhancing carbon stocks
Rh	heterotrophic respiration
RISEMP	Regional Integrated Silvo-pastoral Ecosystem Management Project
RMU	removal unit
Rs	soil respiration
R:S ratio	root–shoot ratio
RUSLE	Revised Universal Soil Loss Equation
S	sulphur
SAN	Sustainable Agriculture Network
SANAA	National Autonomous Service of Aqueducts and Sewers
SEMARNAT	Secretaría de Medio Ambiente y Recursos Naturales (Mexican Ministry of Environment and Natural Resources)
Si	silicon
SINAC	National Conservation Areas System
SOC	soil organic carbon
SWAT	Soil and Water Assessment Tool
TCD	thermal conductivity detector
tCER	temporary Certified Emission Reduction
TroFCCA	Tropical Forests and Climate Change Adaptation
UNFCCC	United Nations Framework Convention on Climate Change
USDA	US Department of Agriculture

USLE	Universal Soil Loss Equation
VCS	Voluntary Carbon Standard
VCU	voluntary carbon unit
VOS	Voluntary Offset Standard
WCA	Wildland Conservation Area
WEF	World Economic Forum
WEPP	Water Erosion Prediction Programme
WFPS	water-filled pore space
WTA	willingness to accept
WTO	World Trade Organization
WTP	willingness to pay

Introduction

Bruno Rapidel, Fabrice A. J. DeClerck, Jean-François Le Coq and John Beer

Payment for ecosystem services (PES) is rapidly gaining attention as a potential mechanism to integrate protection of natural resources within production landscapes. However, we are at a critical crossroads where PES has captured the imagination of conservationists, economists and landowners alike; yet there is little documented evidence that PES either ensures the delivery of the service that is being paid for or that it improves the livelihood of landowners. Furthermore, it is important that this growing attention to PES does not hinder the development and use of other means to secure the provision of ecosystem services, which use either other financial incentives (Jack et al, 2008) or less 'trendy' ways, such as command and control, integrated development conservation approaches and law enforcement (Wunder et al, 2008). Another commonly used approach is to purchase private lands in sensitive areas as a means of protecting them from harmful human activities (e.g. local authorities who secure recharge areas to ensure the provision of high-quality water). The selection and combination of the most efficient means to guarantee the provision of the desired ecosystem services will depend on the local context. In many cases, combinations of mechanisms or policy mixes will be the most effective means of ensuring ecosystem services. For example, a community may develop a communally agreed territorial management plan to maximize the ecosystem services, but may also depend on law enforcement to ensure that the plan is followed.

In this book we review what has been learned regarding PES schemes that have evolved in the Neotropics, the tropical zone of the American continent and the Caribbean, over the past 20 years. We focus on the Neotropics because, as a whole, the region has been a global leader in developing PES schemes that are currently being developed in other regions, and we feel that these experiences will be of value to other tropical as well as temperate areas. We present case studies that demonstrate the successes as well as the limits of PES in the hope that, by learning from these examples, future PES schemes will be strengthened both in their capacity to ensure environmental sustainability in managed landscapes and to become an important component of local, regional and global economies.

The concept of ecosystem services is not new. As recalled by Mooney and Ehrlich (1997), Plato (in his book *Critias*) mentioned the services delivered by forest ecosystems; he hypothesized that the deforestation of Attica in ancient times had led to current (i.e. around 400 BC) soil erosion and the drying up of springs. The human race has obviously been well aware of the links between nature and our livelihoods since before the origins of agriculture. However, the increasing dominance of urban lifestyles has weakened this awareness of our dependence on nature; the formerly obvious links have to be recalled and re-emphasized.

This concept has been stressed relatively recently in the ecological literature, particularly through the work edited by Daily (1997), *Nature's Services: Societal Dependence on Natural Ecosystems*, where ecosystem services were defined as 'the conditions and processes through which natural ecosystems, and the species that make them up, sustain and fulfil human life'. In large part, the objective of Daily's work was to remind the world that human society depends on ecosystems for the provision of these services. Economists took on the concept and developed approaches to value these services. In the famous paper 'The value of the world's ecosystem services and natural capital', Costanza et al (1997) valued the services provided by all the ecosystems of the world at US$33 trillion, almost twice the gross world product (sum of the gross national product of all nations – that is, US$18 trillion), and claimed that the value of the services provided by ecosystems must be incorporated within national accounting systems. The study of these services increased exponentially during the following years and was synthesized in 2005, providing an overview of the state of the world's ecosystems (MEA, 2005). The Millennium Ecosystem Assessment (MEA) definition of ecosystem services remained: 'the benefits [that] people obtain from ecosystems'. However, the MEA expanded the classification of ecosystem services by identifying broad categories of services received:

- provisioning;
- regulating;
- cultural; and
- supporting (see Figure 0.1).

Notably absent from both Daily's (1997) and the MEA (2005) definition is the explicit identification of biodiversity conservation as an ecosystem service; rather, both definitions recognize the value of biodiversity through its effects on other services. This ecological viewpoint is somewhat distinct from socio-

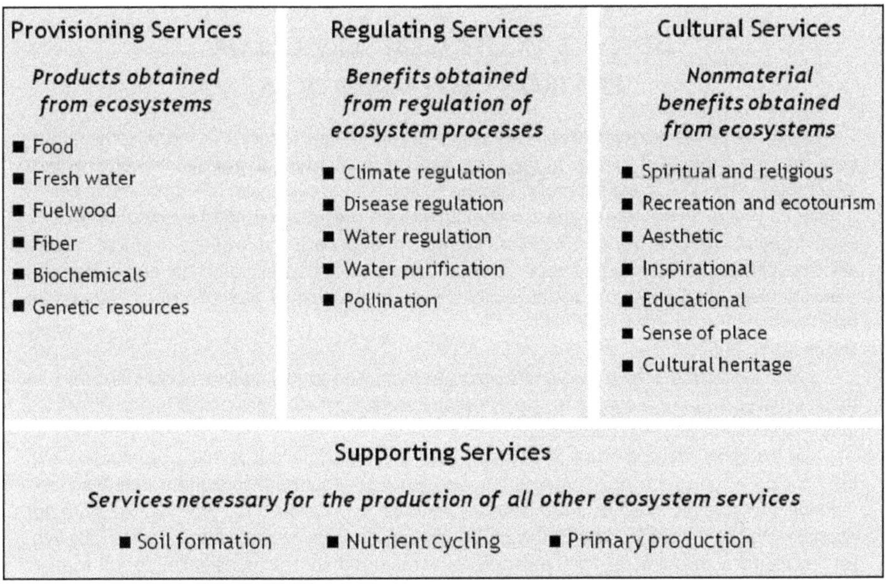

Figure 0.1 *The services provided by ecosystems*

Source: MEA (2003)

economic perspectives that often include biodiversity or habitat conservation as a service.

From an economic point of view, the MEA definition mixes two categories: services whose value is easy to determine on the market (referred to as ecosystem goods in Daily's definition) and which may already be included in accounting systems, and services contributing significantly to human welfare that are not captured in existing welfare accounting. These last services are more narrowly defined by some as environmental services (Boyd and Banzhaf, 2006), which in some definitions excludes the provision of marketed goods. The origins and specificities of both concepts are developed in Box 0.1.

PES concepts and schemes arose as an alternative approach to finance conservation efforts. They aim at including the value of the ecosystem services in local, national or even global accounting systems in order to maintain or even increase their provision. A currently widely accepted definition for PES (Wunder, 2005) is a voluntary transaction where a well-defined environmental service is bought by a buyer from a provider if (and only if) the providers secure the provision of this service. This strict definition has few real applications, as has been demonstrated in previous reviews (Smith, 2006; Wunder et al, 2008) and is also apparent in this book. We will nevertheless use this definition, for lack of a better one, but will detail how the examples we present differ.

Box 0.1 Ecosystem services or environmental services

The use of the terms 'ecosystem services' and 'environmental services' has led to some confusion; they are often used to refer to the same concept in an interchangeable manner. The term 'ecosystem services' is more generally used by ecologists and biological scientists, and places a greater emphasis on the role of species assemblages on the provisioning of ecosystem services. Indeed, the notion of ecosystem services emerged in large part from a series of studies generally classified as biodiversity and ecosystem function studies that explored general patterns in how changes in species composition, richness and abundance altered critical ecosystem functions such as primary productivity, decomposition and pollination (Loreau et al, 2002; Naeem et al, 2009).

Some have suggested that the term ecosystem may be too complex and does not capture those services that are provided at the landscape scale, such as hydrological functions, which may encompass multiple ecosystems.

On the other hand, the use of the term 'environmental services' is mainly associated with the principle of payment for environmental services (PES) – that is, a mechanism to reward land managers for the provision of environmental services, within which saleable products are not considered since it is assumed that this particular 'service' is rewarded by existing markets (e.g. for agricultural products). This principle is embedded in the consideration of market approaches to protect the environment, and the term 'environmental services' is thus more currently used by economists. As such, it was frequently associated with the climate change issue, hydrologic functions of the forest and the role of ecotourism in the funding of conservation programmes.

Apart from these differences, the use of the two terms depends on cultural patterns and languages. In some cultures, environmental services may refer more broadly to waste and secondary resource managements (e.g. rubbish collections and mechanical sewage removal). In the Spanish language, which dominates the Neotropics, '*servicios ambientales*' ('environmental services') is the term most commonly used. Here, we emphasize that both terms can be used interchangeably when they refer to the services provided by nature.

From Forests to Agricultural Landscapes

A common perception is that ecosystem services are only provided by trees or forests. This perception is probably related to the instinctive idea that forests represent the former state of nature, and that this natural state is 'good'. However, forests provide far too little of the 'provisioning services', especially food, to cover the current and growing needs of the world's population. In order to provide an adequate level of ecosystem services, supplying the needs of current generations while maintaining this potential for future generations, our management of agricultural landscapes will have to change. From a singular focus on crop production, agricultural landscapes will have to encompass the production of ecosystem services. This can be achieved in multiple ways. For example, we can envision landscapes where crop production and the provision of ecosystem services are segregated into patches of natural vegetation and crops, using natural vegetation patches in critical zones (e.g. on riverbanks to filter sediments or in corridors to connect larger forests and preserve biodiversity). Or we can modify the way in which crops are grown so that the agricultural land uses continue to provide at least part of the ecosystem services that previously were supplied by natural vegetation. These two strategies are not mutually exclusive. Moreover, some of the services discussed in this book are essential elements of agricultural rather than forested landscapes (e.g. pest control and pollination).

Before we can rationalize the use of such strategies, numerous questions have to be answered: what is the functional role of agroecological landscapes? To what degree can ecosystem services be optimized in agricultural landscapes, in ecological and economic terms; or, in other words, what are the trade-offs between the provision of specific services and in which cases can facilitation occur? How can decision-makers implement institutional and legal frameworks that ensure that stakeholders modify their practices, providing a balance of ecosystem services better adapted to the needs of society? What will be the social consequences of these interventions, particularly in terms of equity between stakeholders?

In this book, we will thus focus on the contribution of agricultural ecosystems, including agroforestry systems, to ecosystem services. Agroforestry systems, associating crops (includes pastures) with trees, are now frequently promoted as a means of balancing the provision of different ecosystem services. They are particularly widespread in the Neotropics, as has been highlighted in a recent study by Zomer et al (2009), who in a global comparison demonstrate that Mesoamerica surpasses all other regions globally in terms of having the greatest integration of trees within agricultural landscapes.

Organization of this Book

The chapters of this book are organized in three parts: Part I: 'Measuring ecosystem services' covers the ecological aspects of agricultural PES; Part II: 'Marketing ecosystem services' covers the economic aspects of PES; and Part III: 'From theory to practice: Tales of success and lessons learned' seeks to draw lessons from the current experiences with PES implementation in the Neotropics, of local or global scope, including aspects of social equity and policy-making.

The ecological side of agricultural PES

In Part I, we review the existing evidence of the provision of services by agricultural ecosystems, and identify the complementary evidence that is needed. We also focus on the methods and tools for monitoring these services, as many ecosystem services either have not been generally recognized and/or quantified. We focus our attention on the services that have received the greatest amount of attention by ecologists, economists and development practitioners: climate regulation, water provision and regulation, pest and disease regulation, pollination, and soil erosion (see Table 0.1). Although we support the view that biodiversity is not a specific service, but is the means through which services are provided, we include a chapter on how biodiversity is measured, focusing

Table 0.1 *Description of the services and the topics covered in Part I: 'Measuring ecosystem services'*

Chapter	*Authors*	*Service considered*	*Main topic*
1	Hergoualc'h	Climate regulation	Relative contribution of different greenhouse gases
2	Marchamalo et al	Water regulation	Methods for targeting and up-scaling
3	DeClerck and Martínez Salinas	Biodiversity conservation	Complementarity of different metrics
4	Avelino et al	Pest and disease regulation	Trade-offs in cropping systems
5	Garibaldi et al	Pollination	Relationships between biodiversity and pollination
6	Villanueva et al	Climate regulation; biodiversity conservation	Bundling, using simple indicators

on measures that are of value for conservation goals, which frequently are included in PES.

In Chapter 1, Hergoualc'h presents the greenhouse gas (GHG) balances of agroforestry systems at the field level in Costa Rica and different methods to estimate them. Since agricultural practices modify not only carbon (C) sequestration, but also methane (CH_4) and nitrous oxide (N_2O) emissions – GHGs that have much greater warming potentials than carbon dioxide (CO_2) – it is essential that we assess the emission of all these gases when estimating the climate regulation service (or disservice) provided by these cropping systems.

The rapid evolution of PES necessitates better mechanisms for demonstrating the causal relationship between land uses and the provisioning of specific ecosystem services. It will also necessitate tools that permit the buyers of ecosystem services to rapidly evaluate which portions of the landscape are the greatest providers of ecosystem services in order to target interventions in those portions of the landscape that provide the greatest amount of service per dollar paid. In Chapter 2, Marchamalo and colleagues present methods used to assess hydrological services (water regulation and soil conservation) at the watershed scale and include methods currently used for up-scaling and targeting priority areas for intervention. Hydrological services are highly spatially dependent (e.g. the place where they are delivered in the watershed determines their importance). As an example, the same reduction in erosion achieved by farmers who increase soil cover may have a far greater effect on river water quality when their land is located close to the river compared to fields distant from the river; in the latter case, the eroded material may be deposited before reaching the river. In the case of contamination from livestock operations, the difference could be even more significant. Therefore, when planning an incentive scheme to foster the provision of hydrological services (e.g. to reduce sedimentation/contamination of a dam), it is important to have methods to identify priority areas.

Although the value of biodiversity is difficult to quantify, there is no doubt that many people appreciate biodiversity and are willing to pay more for products that contribute to its conservation (e.g. for coffee from a farm that has contributed to the conservation of tropical species). However, determining the contribution of crop production to biodiversity conservation can be tricky. Using data from a permanent bird monitoring programme on the Tropical Agricultural Research and Higher Education Center (CATIE) farm in Costa Rica as a guideline, DeClerck and Martínez Salinas (Chapter 3) present different metrics to assess biodiversity conservation, discussing their strengths, weaknesses and complementarities in relation to the species protected. They present and discuss the proxies used by current coffee certification programmes in Central America to assess biodiversity conservation in these agricultural systems, which are the most common examples of conservation payments in agricultural landscapes of the region.

Some ecosystem services are directly received by the provider of the service. Pollination and pest control services are two examples. In the case of crop disease regulation, the service benefits a farmer through a reduction in production costs and eventual threats to his health when pesticide use is reduced. Society can also benefit from reduced water and air contamination. In the tropics, pest and disease pressure on agricultural crops is high. The quest for strategies that result in high yields, as well as limited recourse to pesticides, is at the cutting edge of agricultural development and is based on the assumption that ecological knowledge can replace synthetic inputs. Avelino and colleagues focus on this quest (Chapter 4), reviewing perennial crop disease regulation services provided by agroforestry systems. Abundant literature on this topic is available but relatively poorly integrated. Limited evidence has been able to identify the best overall pest control strategies when targeting the whole pest and disease complex in a particular environment, under farmers' practices and with economical constraints.

Garibaldi and colleagues present the existing knowledge on pollination services in the Neotropics (Chapter 5). Pollination is usually considered as an agricultural service linked to biodiversity that can improve yields, production quality and stability over time. Although bees usually come to mind when we think of pollination, this service is provided by a huge variety of organisms, including vertebrates and insects. The authors present the methods and main results (quantification) in the case of tropical crops, and show how pollination services can be increased by certain management practices, at landscape or farm scales. They present a detailed account of the pollination of coffee in the tropics, its effects on coffee yield and its economic valuation.

From an ecological point of view, it also appears that in many cases interventions that favour a single service favour other related services. Thus, a key line of research focuses on 'bundling' (i.e. a strategy to concurrently evaluate a set of services, allowing for grouped payments and an increased ratio between amounts paid for the service and transaction costs of making the payments). In Chapter 6, Villanueva and colleagues present an example of the successful use of this approach in silvopastures in Central America and Colombia. Their method is based on expert knowledge of the value of different land uses for different ecosystem services and a comparison between actual land management and reference states. It uses the proxy variable of land use, bundling the services of carbon sequestration and biodiversity conservation.

The economic and social aspects of agricultural PES

The principle of PES is simple (see Figure 0.2 for the case of a coffee plantation). In the current situation, within the functioning of the commodity market, the land manager tends to maximize his benefit by choosing intensive systems that generate high production of private goods that can be sold. If other stakeholders desire a different state of the land, that provides more ecosystem services (ES), they establish a PES mechanism that should compensate for any reduction of economic benefits of the land manager, resulting from lower yields of saleable products. If the cost of providing the service – the decrease in private goods production by the land manager – is lower than the benefits

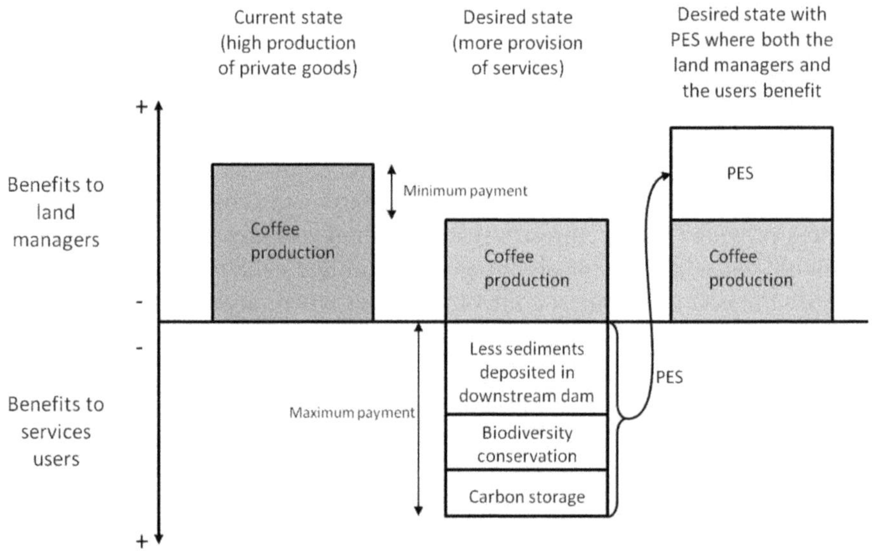

Figure 0.2 *The logic of payment for ecosystem services (PES)*

Note: In this example, the current state is a coffee plantation with little or no shade trees. Intensively managed, it produces high coffee yields and economic benefits for the land manager. The desired state is a coffee plantation with a high density and diversity of shade trees. These trees will provide protection of the soil, habitat for birds and carbon storage in their stems; but their presence will usually lead to a reduction of coffee yields (mostly due to competition for resources). The increased ecosystem services provision has a value for the users; this is the maximum amount that could be available for PES. The decreased income, in this example, due to decreased coffee yield, is the minimum payment to convince the land manager to modify its management. In this idealized and simplified example situation, a PES could be given by the users to the managers in a form that the manager benefits from (his income would increase in relation to the current state), as well as the users. The payment given to the service providers is less than the value of the ES benefit obtained by the users from modifying the management of the plantation.

Source: adapted from Pagiola and Platais (2007)

accruing to users, then negotiation between providers and users of PES is possible. The price will be established during this negotiation at a value that is somewhere in between these two limits: cost and benefit. In this negotiation, the value of the provision of ES plays a fundamental role; it has to be estimated in the rationality of every stakeholder (i.e. benefits from commodity markets are not the only motivation). The determination of this value is the main focus of Part I of this volume.

There is a fundamental discrepancy between the value actually paid in PES schemes and the value of the services at the global level. For example, Costanza et al (1997) established the value of the services at almost twice the value of the goods produced by the world. But, in reality, PES premiums are only a small percentage of the value of the goods produced. Uncertainty regarding the quantity of the services provided by a particular land use is certainly part of the explanation of this discrepancy. But another significant part is explained by the way in which markets have developed, with their cortege of valuing methods, negotiation and intermediaries. In Part II, we try to cast some light on this complex reality.

In order to illustrate the principles and reality of marketing ES, we rely on examples of PES developed in Latin America, particularly for services whose measurement methods were described in the first section. Water regulation, carbon sequestration and biodiversity conservation are the services usually targeted by PES, and we concentrate our attention on them (see Table 0.2).

Hydrological services are local and their location (watershed) can be relatively well defined, although their scale can range from a few square kilometres to tens or hundreds of thousands of square kilometres. Moreover, it is relatively easy to identify the potential beneficiaries of increased hydrological services, whether water for drinking, electricity production, crop irrigation or others such as reduced flood risk. The providers, as detailed in Chapter 2, can also be identified, although this may require complex measurement and modelling, particularly to bridge the scale gap between field activities (provider) and

Table 0.2 *Description of the services and the topics covered in Part II: 'Marketing ecosystem services'*

Chapter	*Authors*	*Service considered*	*Main topic*
7	Madrigal-Ballestero	Hydrological regulation	Commonly used economic valuation methods
8	Moore and Neeff	Climate regulation	Accessing the carbon market
9	Neeff and Fehse	Climate regulation	Roles and functions in the carbon market
10	DeClerck and Le Coq	Biodiversity conservation	Valuing biodiversity or the services that it supports
11	Le Coq et al	Biodiversity conservation; climate and water regulation	Comprehensive comparison between PES and eco-labelling schemes

the watershed level at which the service is enjoyed. Provided the required evidence of the link between the agricultural activities and the provision of services is documented and recognized by stakeholders, it is possible to establish a direct link between a group of users and a group of providers of this service. This makes this service a good example for PES scheme development, in a sense very close to Wunder's strict definition. Madrigal-Ballestero (Chapter 7) presents the valuation methods used in various cases of payment for hydrological ecosystem services (PHES) in Latin America.

Climate regulation services usually refer to the carbon balance of ecosystems, unlike the cautious approach developed in Chapter 1 by Hergoualc'h. The service is usually related to carbon sequestration in ecosystems. The characteristics of this service imply that a very different approach to PES is required: the service is global, in the sense that increased carbon sequestration in any region of the world has a climate regulation effect at the worldwide scale. This independence from locality has a negative consequence: providers and users of this service are spread over the whole world, and communication between them is not direct. Moreover, free-riding can be easy: any users enjoy the service, whether they accepted to pay for it or not.

Due to the widespread concern about climate change and effective international organization, this service has two additional characteristics:

1 the establishment of a huge and well-known scientific body in the form of the Intergovernmental Panel on Climate Change (IPCC; www.ipcc.ch), which gathers evidence, reviews it and makes cautious and (almost) consensual statements and previsions;
2 the existence of an international regulating body, the United Nations Framework Convention on Climate Change (UNFCCC), and an international agreement, the Kyoto Protocol, which entered into force in 2005 under the UNFCCC, whereby at least some developed countries agreed to make a commitment to quantitatively reduce their GHG emissions.

This latter characteristic led to the creation of a carbon market because the Kyoto Protocol allows for carbon credit exchanges between countries that have acquired GHG emission reduction commitments, and also between those countries and other non-committed countries who can voluntarily reduce their GHG emissions and thus emit carbon credits. Moreover, in addition to this 'official' market, a number of public or private initiatives have arisen to develop voluntary markets with varying requirements for traded carbon credits.

Therefore, in the case of the service of climate regulation, the marketing issue is more that of accessing existing markets than developing new ones. Project developers may waste time and money trying unsuccessfully to access existing carbon markets. Such examples of unsuccessful attempts are relatively common, as the Costa Rican history of environmental fundraising shows (see Chapter 14). In order to reduce the risk of such a disappointment, Moore and

Neeff, in Chapter 8, provide practical guidance that can help project developers to understand the various steps required to establish a business plan for forestry- and land use-based carbon projects. We hope that this information will help project developers to determine whether, and under what conditions, pursuing carbon markets is a sensible proposition.

Chapter 9 presents a complementary aspect of carbon markets: Neeff and Fehse explain how carbon markets function as outlets where land management projects can sell carbon ecosystem services as carbon offsets. They describe the different roles of participants to be played in the Clean Development Mechanism (CDM) market, the diverse actors who play these roles, the contract types that are available, and how role allocation changes in the case of voluntary carbon markets. They discuss criteria that can be used when choosing to work with particular types of carbon market intermediaries in a sound commercialization strategy. Finally, they analyse the current limitations to voluntary markets as well as their future developments.

Biodiversity is the main topic of Chapter 10. Biodiversity conservation – or, conversely, the increasing rate of species extinction – is recognized as a pressing challenge at the global scale. However, compared to global climate regulation, it suffers from the lack of unambiguous methods to estimate its value, a scientific body that reviews and discusses scientific evidence, and internationally agreed-upon targets and means to revert the current trends. Biodiversity is a public good whose conservation should be the concern of all humankind, as is climate change regulation.

In Chapter 10, DeClerck and Le Coq analyse the current methods used to value biodiversity and explore various mechanisms that may increase its value. They first argue that biodiversity is not an ecosystem service, but rather a supporting feature of the ecosystem, which may, in turn, provide very different ecosystem services. As a result, specific mechanisms to reward biodiversity conservation efforts, under strict PES schemes, are unlikely to blossom. There are some exceptions, which generally function at a local level, such as ecotourism activities next to a national park. Biodiversity is generally financed through indirect mechanisms established for other services, such as carbon or water. The best-developed transnational mechanism used to reward biodiversity is probably the eco-labelling of agricultural and forestry products.

In Chapter 11, Le Coq and colleagues compare two market-oriented mechanisms that support the provision of ES: PES as defined by Wunder (2005) and eco-labelling. The experience accumulated on both mechanisms in Latin America allows meaningful comparison. This comparison explores the structure and functioning of both approaches, as well as their efficiency and their impacts upon equity. Both schemes have the potential to support ES provision, but can have controversial impacts and inherent limitations in terms of ES provision and poverty alleviation. The authors request a greater use of comparable methodologies to evaluate these two schemes and call for further reflection on the possible synergies between these two instruments in practice.

Presentation of the case studies

In Part III, most of the aspects presented in the preceding two sections are discussed in the context of case studies. Relying on the relatively long experience of the Latin American region with these PES and PES-like mechanisms, six different experiences and points of view were selected. These experiences, listed in Table 0.3, allow us to review some of the key elements of current PES and the lessons that we have learned from them. How can a country secure funding for national PES? How can we increase the efficiency of a PES mechanism through better targeting of priority areas? How can we monitor and improve PES? And, finally, what is the right balance between PES and other means to achieve the desired objective of conserving natural resources?

The first contribution in Part III draws upon the different Costa Rican PES experiences to illustrate and discuss the issue of sustainable funding (Chapter 12). After describing the history of the national PES programme, the oldest at this scale in the Neotropics, Murillo and colleagues detail the successive funding sources accessed and the quest for sustainable sources. They hypothesize that citizen awareness needs to be enhanced to secure funding and develop this thesis using a specific Costa Rican example, developed by the public service enterprises of the city of Heredia. Several useful lessons are drawn from this long and diverse experience in Costa Rica.

Another contrasting experience is the Mexican payment for hydrological environmental services. Also a national state-funded PES scheme, it is somewhat younger than the Costa Rican experience. Its design and its objectives are very different from the Costa Rican programme: it relies mainly on geographical characterization to prioritize land for PES; highly targeted to ensure additionality while pursuing the goal of efficient use of economic resources, it does not allow for bundling. After it started, poverty alleviation was included in its objectives at a similar level of importance as environmental

Table 0.3 *Description of the services and the topics covered in Part III: 'From theory to practice: Tales of success and lessons learned'*

Chapters	*Authors*	*Service considered*	*Main topic*
12	Murillo et al	Climate and water regulation	Sustainable funding of PES in Costa Rica
13	Rolón Sánchez et al	Water and climate regulation	Targeting national PES mechanisms in Mexico
14	Robalino et al	Climate regulation	Monitoring the efficacy of PES in Costa Rica
15	Soto and Le Coq	Biodiversity conservation	Eco-labelling coffee in Central America
16	Prins and León	Biodiversity conservation; climate regulation	Collective action for drinking water management in Honduras
17	Rodríguez	Natural resource management	Institutional setting and political support required to developed PES in Latin America

protection. In Chapter 13, Rolón Sánchez and his colleagues report the progress made to date with the design and operation of this system and identify further drawbacks to be overcome.

The monitoring of a PES scheme is essential to build confidence among users (and funders). This monitoring is far from simple. In Chapter 14, Robalino et al show us how this monitoring has been performed for the Costa Rican national PES scheme. The results of this impact assessment show the effect of the evolution of the system over time, and the authors propose some ways for further improvements.

Other mechanisms have been widely used in this region to improve both the management of natural resources in agricultural systems and the social and economic status of landowners. Eco-labelling began with organic agriculture and a very specific objective; but the concept of eco-labelling was rapidly extended to other types of eco-friendly agriculture. The best and oldest case study, as the basis of an analysis of the features and efficacy of eco-labelling, is the coffee sector: the age and diversity of this eco-labelling experience is unequalled. In Chapter 15, Soto and Le Coq review the highs and lows of the Central American experience over a period of 20 years. They present the various existing coffee eco-labels, their origins, objectives and structures, and discuss their effects on ES provision, especially on biodiversity, as well as their socio-economic impacts. Concluding that this mechanism is a tool to support the provision of ES and to increase the welfare of farmers, they identify key bottlenecks whose resolution would contribute to improving the efficiency of eco-labelling.

Prins and Léon (Chapter 16) present the experience with the creation of a new local payment for hydrological ecosystem services (PHES) scheme. They argue that PHES can promote good water management practices; but to be effective, PHES must be embedded in a broad framework that includes the promotion of collective actions and other regulations that go hand in hand with PES development and at a pace determined by particular local conditions. They stress that the objective of their intervention in Copan, Honduras, is water conservation, not PHES, and the importance of introducing PHES cautiously. In their thesis, PES should be seen as a secondary means. In order to secure the provision of water for this tourist area, the main focus of their intervention is on the conditions and temporal aspects that have to be considered when seeking to improve the dialogue between stakeholders of the watershed. They detail the different means to give momentum to collective action and their contribution of each to the final objective.

In Chapter 17, Rodríguez uses his experience as former minister of the environment of Costa Rica during a period of strong PES development to further broaden our understanding of the requirements for successful PES schemes. Focusing his contribution on the national government level, he argues that a very high level of coordination among state institutions is needed to ensure successful national PES schemes. Environmental protection cannot be fostered with a sectorial approach as economic issues have overwhelming importance for governments. If a sectorial approach is chosen, the environment will be dealt with by the Ministry of Environment, with very limited power over its companion ministries who deal with the economy. Therefore, a cross-cutting approach has to be chosen, where the environment is embedded in other sectors so that the environmental consequences of every political decision are weighted and taken into consideration. Within this framework, Rodríguez further draws lessons about the institutional features required for PES to be effective, and also about the conditions that pilot PES projects must fulfil to provide useful insights in order to build a national framework.

Finally, Rapidel and colleagues, in Chapter 18, synthesize new knowledge presented in this book, particularly regarding the specific characteristics of Neotropical agricultural and agroforestry systems as providers of ecosystem services, and the means to increase this provision. They present a quick overview of published information on ecosystem services and payments, expanding this with the new knowledge presented in this volume.

References

Boyd, J. and Banzhaf, S. (2006) *What Are Ecosystem Services? The Need for Standardized Environmental Accounting Units*, Discussion paper RFF DP 06-02, Resources for the Future, Washington, DC

Costanza, R., d'Arge, R., de Groot, R., Farber, S., Grasso, M., Hannon, B., Limburg, K., Naeem, S., O'Neill, R. V., Paruelo, J., Raskin, R. G., Sutton, P. and van den Belt, M. (1997) 'The value of the world's ecosystem services and natural capital', *Nature*, vol 387, no 6630, pp253–260

Daily, G. (ed) (1997) *Nature's Services: Societal Dependence on Natural Ecosystems*, Island Press, Washington, DC

Jack, B. K., Kousky, C. and Sims, K. R. E. (2008) 'Designing payments for ecosystem services: Lessons from previous experience with incentive-based mechanisms', *Proceedings of the National Academy of Sciences of the United States of America*, vol 105, no 28, pp9465–9470

Loreau, M., Naeem, S. and Inchausti, P. (eds) (2002) *Biodiversity and Ecosystem Functioning, Synthesis and Perspectives*, Oxford Biology, Oxford University Press, Oxford, UK

MEA (Millennium Ecosystem Assessment) (2003) *Ecosystems and Human Well-Being: A Framework for Assessment*, World Resources Institute and Island Press, Washington, DC

MEA (2005) *Ecosystems and Human Well-Being: Synthesis*, Island Press, Washington, DC

Mooney, H. A. and Ehrlich, P. R. (1997) 'Ecosystem services: A fragmentary history', in G. C. Daily (ed) *Nature's Services: Societal Dependence on Natural Ecosystems*, Island Press, Washington, DC, and Covelo, CA, pp11–19

Naeem, S., Bunker, D. E., Hector, A., Loreau, M. and Perrings, C. (2009) *Biodiversity, Ecosystem Functioning, and Human Well-Being: An Ecological and Economic Perspective*, Oxford Biology, Oxford University Press, Oxford, UK

Pagiola, S. and Platais, G. (2007) *Payments for Environmental Services: From Theory to Practice*, World Bank, Washington, DC

Smith, K. R. (2006) 'Public payments for environmental services from agriculture: Precedents and possibilities', *American Journal of Agricultural Economics*, vol 88, no 5, pp1167–1173

Wunder, S. (2005) *Payments for Environmental Services: Some Nuts and Bolts*, Occasional Paper no 42, CIFOR, Jakarta, Indonesia

Wunder, S., Engel, S. and Pagiola, S. (2008) 'Taking stock: A comparative analysis of payments for environmental services programs in developed and developing countries', *Ecological Economics*, vol 65, no 4, pp834–852

Zomer, R., Trabucco, A., Coe, R. and Place, F. (2009) *Trees on Farm: Analysis of Global Extent and Geographical Patterns of Agroforestry*, World Agroforestry Centre, Nairobi, Kenya

Part I

Measuring Ecosystem Services

1

Principles and Methods for Assessing Climate Change Mitigation as an Ecosystem Service in Agroecosystems

Kristell Hergoualc'h

Introduction

The concentration of carbon dioxide (CO_2) and other greenhouse gases (GHGs) in the atmosphere has increased considerably over the last four decades. This increase primarily results from the burning of fossil fuels and the conversion of tropical forests to agriculture, with concomitant negative impacts upon the global climate. Agricultural activities account for about 13.5 per cent of total anthropogenic GHG emissions (Rogner et al, 2007) and release mainly nitrous oxide (N_2O) and methane (CH_4) (about 45 per cent of agricultural GHG emissions each), with CO_2 accounting for the remaining share (Baumert et al, 2005). Agricultural N_2O and CH_4 emissions are expected to increase by 35 to 60 per cent in 2030 due to increased nitrogen fertilizer use, animal manure production and livestock numbers. In contrast, CO_2 emissions are likely to remain at the same level due to stable or declining deforestation rates, and increased adoption of conservation tillage practices (Smith et al, 2007). Mitigating agricultural GHGs can be achieved by reducing emissions through more efficient management of carbon (C) and nitrogen (N) flows and by enhancing C storage in soil and vegetation (Smith et al, 2007). Agroforestry systems (AFS) are one means by which the impacts of climate change can be mitigated. The role that AFS play can be increased through payment for ecosystem services (PES) systems that reduce agricultural emissions and increase the quantity of carbon stored. Carbon sequestration (or atmospheric CO_2 removal) as an ecosystem service of AFS is generally quantified as the amount of C stored in trees. Nevertheless, increasing tree density in agroforests may also modify soil fluxes of N_2O or CH_4, which have a global warming potential (GWP) 298 and 25 times higher than CO_2, respectively (Forster et al, 2007). Nitrogen-fixing species used as shade trees (e.g. in coffee plantations) may increase soil emissions of N_2O (Hergoualc'h et al, 2008) and reduce the soil CH_4 sink (Palm et al, 2002).

This chapter reviews the biogeochemical processes controlling the fluxes of CO_2, N_2O and CH_4 between agroecosystems and the atmosphere. It also presents the principles and methods for measuring these fluxes and assessing the net balance of GHGs in tropical agroecosystems.

Biogeochemical Processes Involved in GHG Exchanges between the Agroecosystem and the Atmosphere

Carbon dioxide (CO_2)

The amount of CO_2 that is 'fixed' from the atmosphere by plants (i.e. oxidized into organic compounds during photosynthesis) is called gross primary production (GPP) (see Figure 1.1). About half of the GPP is incorporated within new plant tissues (foliage, wood and roots) and the other half is converted back to atmospheric CO_2 by autotrophic respiration (Ra) (respiration by plant tissues) (IPCC, 2001). Annual plant growth is defined as net primary production (NPP) and relates to GPP and Ra as:

$$NPP = GPP - Ra. \quad [1.1]$$

Each year part of the standing biomass is transferred to litter and soil layer carbon pools, where it is decomposed by soil microorganisms and fauna through heterotrophic respiration (Rh) (Ryan and Law, 2005). Heterotrophic respiration includes the decomposition of the recently deposited biomass, but also contains decomposition of organic matter that accumulated in the ecosystem during the past decades, centuries or millennia (Luyssaert et al, 2007). The rate of decomposition depends on the chemical composition of the dead tissues and on environmental conditions such as temperature and moisture (IPCC, 2001). The difference between NPP and Rh determines how much C is lost or gained by the ecosystem in the absence of disturbances that remove C from the ecosystem, such as harvesting or fire (IPCC, 2001). This C balance is the net ecosystem production (NEP):

$$NEP = NPP - Rh. \quad [1.2]$$

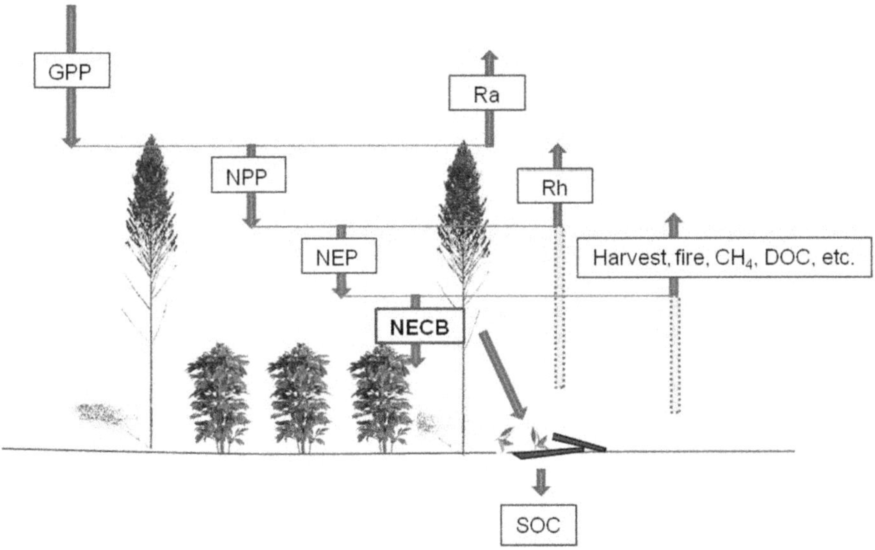

Figure 1.1 *The carbon cycle in agroecosystems*

Notes: CH_4 = methane; DOC = dissolved organic carbon; GPP = gross primary production; NECB = net ecosystem carbon balance; NEP = net ecosystem production; NPP = net primary production; Ra = autotrophic respiration; Rh = heterotrophic respiration; SOC = soil organic carbon.
Source: chapter author

The sum of Rh and Ra represents the total ecosystem respiration (Re) and the sum of the below-ground fraction of Ra and Rh is the soil respiration (Rs).

When other losses of carbon are accounted for, including fires, harvesting/removals (eventually combusted or decomposed), erosion and export of dissolved or suspended organic carbon (DOC) by rivers to the ocean and microbially produced CH_4, what remains is the net biome production (NBP) (IPCC, 2001), also called net ecosystem carbon balance (NECB) (Luyssaert et al, 2007). By definition, for an ecosystem in steady state, Rh and other carbon losses would just balance NPP, and NECB would be zero (IPCC, 2001).

Nitrous oxide (N_2O)

In the soil, nitrification and denitrification are the main processes that form N_2O (Hergoualc'h et al, 2008). Nitrification is the aerobic oxidation of NH_4^+ or NH_3 to NO_2^- and NO_3^-, with N_2O production under either oxic or anoxic (nitrifier denitrification) conditions (Wrage et al, 2001). Denitrification is the anaerobic reduction of NO_3^- and NO_2^- to N_2O and N_2 (see Figure 1.2).

The main drivers of N_2O emissions from soil are substrate supply, as N additions and mineralization of organic N in soil, soil water content and temperature (Skiba and Smith, 2000). Increased N supply to the soil through either N fertilization or N transfers from litter in systems with N_2-fixing trees is known to substantially increase soil emissions of N_2O (Rochette and Janzen,

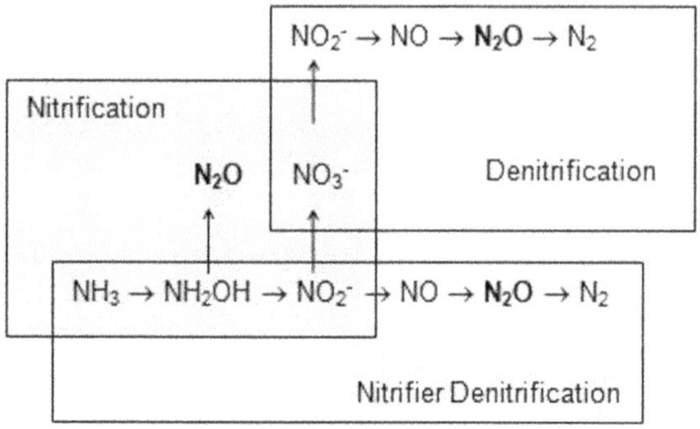

Figure 1.2 *Pathways for N_2O production*

Source: adapted from Wrage et al (2001)

2005; Stehfest and Bouwman, 2006). The effect of soil water content on soil N_2O effluxes has been described as a monotonic curve with maximum emissions around field capacity, which is often around 60 per cent of water-filled pore space (WFPS) (Davidson, 1991). At higher water contents, N_2O is primarily reduced to N_2. Increased soil temperature generally causes exponential increases in N_2O emissions (Smith et al, 2003).

Methane (CH_4)

The microbial metabolism of CH_4 is complex since soils can both produce and consume CH_4, even simultaneously (Schimel and Holland, 2005). Methane production (Methanogenesis) is achieved by a group of anaerobic Archaea known as methanogens and occurs through either CO_2 reduction:

$$CO_2 + H_2 \rightarrow CH_4 \qquad [1.3]$$

or acetate fermentation:

$$CH_3COOH \rightarrow CH_4 + CO_2. \qquad [1.4]$$

Methane oxidation (*Methanotrophy*) is achieved by a group of aerobic bacteria called methanotrophs that use CH_4 as their sole source of C and energy:

$$CH_4 + 2O_2 \rightarrow CO_2 + 2H_2O. \qquad [1.5]$$

Several studies (Mosier et al, 1991; Knowles, 1993; Schnell and King, 1994) have demonstrated a link between methanotrophy and the nitrogen cycle. Because of the molecular similarity of CH_4 and ammonia (NH_3), autotrophic nitrifiers are able to oxidize CH_4 and methanotrophs can oxidize NH_3. Accordingly, CH_4 oxidation can be inhibited by high soil ammonium concentrations, which occur after fertilization. On the other hand, in soil with a high cation exchange capacity, NH_4^+ resulting from fertilization would tend to be adsorbed on the argilo-humic complex of the soil rather than inhibiting methanotrophy (Mosier et al, 1991). Thus, while the overall increase in nitrogen fertilizer use will probably decrease global soil CH_4 consumption, leading to greater atmospheric CH_4 concentrations, the patterns are complex, with potential surprises in store (Schimel and Holland, 2005).

The total CH_4 consumption rate is controlled by the potential biological demand and diffusion into the soil profile, notably by the WFPS (Mosier et al, 2004). Del Grosso et al (2000) observed that CH_4 oxidation was optimal at WFPS of about 7.5 per cent for coarse-textured soils and 13 per cent for fine-textured soils.

Upland agricultural soils tend to consume CH_4 rather than produce it; but the net flux (i.e. the balance of production and consumption) will vary depending on the soil, its properties and the vegetation. Vegetation is critical because it can affect the rate of CH_4 production by supplying easily decomposable carbon through root exudates or recently dead roots (Schimel and Holland, 2005). Wetland soils (natural and paddies) emit CH_4, although CH_4 oxidizers can consume up to 90 per cent of the CH_4 produced in the anaerobic zone before it reaches the atmosphere. Many wetland plants have air channels in their stems called *aerenchyma* that allow CH_4 to diffuse upwards through the plant. In many wetland systems, 90 per cent or more of the CH_4 that makes it out to the atmosphere is transported through the plants (Schimel and Holland, 2005).

Methods for Assessing the Net GHG Balance in Agroecosystems

Principles and methods for measuring the net ecosystem carbon balance (NECB)

There are two general methods for estimating the net C balance in an agroecosystem: the gain–loss approach and the stock-difference approach. The first method is a process-based approach that requires the annual rates of biomass (above and below ground) accumulation from growth and losses associated with harvest and burning, and the annual transfer into and out of dead organic matter and soil stocks. The second approach requires estimates of C stocks in biomass (above and below ground), dead organic matter (deadwood and litter) and soil (soil organic matter) at different points in time (IPCC, 2006). The first approach provides a net C flux without considering the C stock, whereas the net C flux obtained from the second approach integrates the stocks of the agroecosystem in at least two different points in time. The times when the fluxes or stocks are measured should be carefully chosen with respect to the expected growth of the agroecosystem. The evolution of C stocks in the agroecosystem can be assessed diachronically (measurement of the stocks at two points in time, at least, at one site) or synchronically (measurement of the stocks at the same time in at least two sites, which have the same initial state and have been identically cultivated but for a different period of time). For measurements achieved synchronically, also called chronosequences (IPCC, 2006), efforts are made to control between-site differences, such as soil type and topography. The evolution of fluxes in the agroecosystem may also be followed, diachronically or synchronically.

In a gain–loss approach, the NECB can be assessed from measurements of NEP and C loss from fires, harvests, soluble and physical removal, and CH_4 flux. The NEP can be measured using meteorological techniques and calculated as the annual integral of the measured CO_2 exchange between the agroecosystem and the atmosphere. Meteorological techniques estimate CO_2 fluxes by measuring concentrations and concentration gradients. Tower-based micrometeorological methods fall into two general categories: eddy correlation and concentration gradient techniques. Even though meteorological methods can yield integrated flux information for large areas and are useful for measuring seasonal variations, they require expensive and sophisticated instrumentation and are dependent on a uniform large surface and constant atmospheric conditions (Van Cleemput and Boeckx, 2006).

The stock-difference approach, covered in the following section, is more commonly used than the gain–loss approach for assessing the NECB.

Measurement of carbon stocks

In terrestrial ecosystems, there are four fundamental C pools:

1 above-ground biomass,
2 below-ground biomass,
3 necromass (or dead organic matter), and
4 soils (mineral and organic horizons),

although they may be further sub-divided depending on the ecosystem under investigation (Hamburg, 2000).

The IPCC (2006), for instance, subdivides the necromass pool into deadwood and litter in croplands. For each of these specific pools, there are several methods available for evaluating stocks. The most commonly used methods for measuring C stocks of these four pools are presented below.

Above-ground living biomass

All measurements of above-ground biomass in trees are dependent on the reliability of allometric relationships (Hamburg, 2000). Trees, like almost all organisms, have a highly deterministic form. Thus, by knowing one or two key attributes, many other attributes can be predicted. For trees, the total height of the tree and diameter of the trunk at 1.35m above the soil surface (also known as diameter at breast height, or DBH) allow the biomass of a tree to be predicted with a high degree of accuracy. Allometric equations can be locally developed by destructive sampling, derived from literature for supposedly comparable ecosystem types, or estimated from fractal branching analysis (Hairiah et al, 2001). Table 1.1 presents several allometric relationships from literature for various components of AFS in Central America and Indonesia. Once the allometric equation for a species has been established or chosen, one only needs to measure DBH of the tree and possibly some other parameter used in the equation, such as height, in order to estimate the biomass of individual trees. The weight of dry matter in trees is converted into a weight of carbon by multiplying the total weight with the C fraction of the dry matter. This is obtained by direct measurement or by using the default factor of 0.5 proposed by the IPCC (2003). The total carbon stored in the trees is the sum of the carbon estimates for all trees found within a given area.

The biomass of the understorey living vegetation is measured by destructive sampling. All vegetation within a 1 square metre quadrat is cut and weighed. Subsamples of the vegetation are weighed before and after drying at 80°C and the C fraction of the dry matter is analysed by dry combustion.

Table 1.1 *Allometric relationships for components of agroforestry systems in Central America and Indonesia*

Species	*Diameter (D) range (cm)*	*Equation*	*Coefficient of determination (R2)*	*Reference*
Bambusa (bamboo)	3–7	$W = 0.131D^{2.28}$	0.95	Priyadarsini (1998)*
Coffea robusta	1–10	$W = 0.281D^{2.06}$	0.95	Arifin (2001)*
Coffea arabica	0.3–7.5	$W = 10^{(-1.113 + 1.578 \text{Log}_{10} D15 + 0.581 \text{Log}_{10} H)}$	0.94	Segura et al (2006)
Coffea arabica		$W = 0.67 + 0.0007\exp(H/0.4)$	0.94	Powell and Delaney (1998)
Cordia alliodora	5–44	$W = 10^{(-0.755 + 2.072 \text{Log}_{10} D)}$	0.95	Segura et al (2006)
Inga densiflora	8.5–18.5	$W = 0.34D^{1.8}$	0.92	Siles et al (2010)
Inga punctata	5–44	$W = 10^{(-0.559 + 2.067 \text{Log}_{10} D)}$	0.97	Segura et al (2006)
Inga tonduzzi	5–44	$W = 10^{(-0.936 + 2.348 \text{Log}_{10} D)}$	0.95	Segura et al (2006)
Junglans olanchana	5–44	$W = 10^{(-1.417 + 2.755 \text{Log}_{10} D)}$	0.97	Segura et al (2006)
Eucalyptus deglupta	6–22	$W = \exp[-2.76 + 2.61\text{Ln}(D)]$	0.98	De Miguel Magaña et al (2004)
Eucalyptus deglupta	11–32	$W = \exp[-2.31 + 2.47\text{Ln}(D)]$	0.99	De Miguel Magaña et al (2004)
Paraserianthes falcatari	8– 18	W = 0.027D2.83	0.82	Sugiharto (2002)*
Pinus caribbea	5–28	W = 0.042D2.66	0.91	Waterloo (1995)*
Musa (banana)	7–27	W = 0.03D2.13	0.99	Arifin (2001)*
Terminalia ivorensis	9–23	$W = \exp[-2.28 + 2.41\text{Ln}(D)]$	0.94	De Miguel Magaña et al (2004)

Notes: * Cited by Hairiah et al (2001).
W = tree biomass (kilograms dry matter per tree); D = diameter at breast height (cm) (1.35m from the soil surface); H = height (m); D15: diameter at 15cm from the soil (cm).

Below-ground living biomass

Root plasticity and variability (spatial and temporal) together with below-ground sampling challenges make it very difficult to accurately measure root biomass (Johnson et al, 2006). All root biomass-sampling techniques (e.g. soil cores, monoliths, minirhizotron, etc.) are hampered by high variability, loss of fine root biomass and high labour requirements (Johnson et al, 2006). Nevertheless, below-ground living biomass can be estimated with some accuracy, though with lower precision than above-ground tree components using allometric equations (Hamburg, 2000). Root structures are genotypically determined, except for the proportion of root mass relative to above-ground biomass (root–shoot ratio, or R:S ratio), which is phenotypically more plastic – in other words, it responds more to changes in the environment. Site characteristics and tree age all influence the R:S ratio (Hamburg, 2000). In a recent literature review, Mokany et al (2006) found that R:S ratios were negatively related to shoot biomass, mean annual precipitation and temperature, forest stand age and height. Their results for tropical and subtropical sites are presented in Table 1.2. Nevertheless, the use of a site-specific R:S ratio of a vegetation type provides a more accurate root biomass prediction than that obtained using a general R:S ratio (Hamburg, 2000; Mokany et al, 2006).

Necromass (deadwood and litter)

Sampling the necromass pool involves stratifying the types of dead material by where it is in the ecosystem and how fast it is likely to decompose (Hamburg, 2000). The most accurate method for estimating these pools is by direct measurement. The litter (tree necromass, undecomposed plant materials or

Table 1.2 *Root–shoot ratios for major biomes in the tropics and subtropics*

Vegetation category	*Shoot biomass (Mg dry matter ha^{-1})*	*Median*	*Standard error (SE)*	*Low extreme*	*High extreme*	*Number of data points (n)*
Tropical/subtropical	< 125	0.205	0.036	0.092	0.253	4
moist forest/plantation	> 125	0.235	0.011	0.220	0.327	10
Tropical/subtropical	< 20	0.563	0.086	0.281	0.684	4
dry forest/plantation	> 20	0.275	0.003	0.271	0.278	2
Tropical/subtropical moist woodland		0.420	0.032	0.292	0.548	7
Tropical/subtropical dry woodland		0.322	0.085	0.259	0.710	6
Shrubland		1.837	0.589	0.335	4.250	8
Savanna		0.642	0.111	0.397	1.076	5
Tropical/subtropical grassland		1.887	0.304	0.380	4.917	15
Tropical/subtropical arid shrubland/desert		1.063	–	1.063	1.063	1

Source: based on Mokany et al (2006)

crop residues, leaves and branches) is collected in quadrats whose size and number are defined according to the heterogeneity of the litter at the plot scale. The litter is carefully rinsed, dried, sieved (2mm) and weighed. Its moisture and C fraction are analysed.

Soil

Soil consists of a mineral fraction (fragments of parent rock, colloidal elements and mineral ions) and an organic fraction (decomposing plant and animal residues, humus, root deposits and microbial biomass). The most common method for measuring soil C stocks consists of simultaneously determining the C content and bulk density of systematically defined soil layers (every 10cm, in general) up to a certain soil depth. In mineral soils, although the organic matter is typically distributed along the first 1m to 1.5m of soil, it decreases quickly with increasing depth. Some methods specify a default soil depth of 30cm (Hairiah et al, 2001; IPCC, 2003) and do not include mineral C (IPCC, 2003); others consider that both mineral and organic C should be accounted for and that soil should be sampled to a depth of at least 1m (Hamburg, 2000). In organic soils, C content does not necessarily decrease with depth, in which case an assessment of the soil C stock requires measurements over the entire depth of the soil profile. The soil C stock is generally calculated on a volumetric basis (as the product of the soil C content, bulk density and layer thickness); however, to compare situations with different bulk densities, a calculation based on an equivalent mass basis is more appropriate (Roscoe and Buurman, 2003; Blanchart et al, 2006). For this, the soil mass in each depth interval is calculated for the least compacted system. Taking the same soil mass, the equivalent depths and C stocks are calculated for the more compacted systems.

Measurement of GHG fluxes

There are two general approaches for measuring GHG fluxes *in situ*. The first approach involves meteorological techniques, described earlier in 'Principles and methods for measuring the net ecosystem carbon balance (NECB)'. The second approach uses closed (static) or open (dynamic) chambers installed on the soil surface. In closed chambers, the gas is sampled manually or automatically from the headspace at different time intervals and the change in concentration in the chamber over time is used to calculate the gas flux. The calculation of the flux goes through a linear regression analysis of the gas concentration increase with time and a calculation of the chamber volume and area. For manual sampling, the gas samples are transferred to the laboratory in sealed pre-evacuated vials for gas chromatography analysis. The gas chromatograph requires an electron capture detector (ECD) for the analysis of N_2O, a flame ionization detector (FID) for the analysis of CH_4, and a FID or a thermal conductivity detector (TCD) for the analysis of CO_2. With automatic sampling, a gas flow system transfers periodic samples to a detector. In the field, photoacoustic infrared detectors (PAIDs) are generally used. In open chambers, outside air flows into the chamber via an inlet and is forced to flow over the enclosed soil surface before leaving the chamber via an outlet. In most cases, sampling is automatic. The concentration of the respective gas is measured at the inlet and outlet sides. The soil surface gas flux is calculated from the inlet and outlet concentration differences, the gas flow rate, and the volume and area covered by the chamber (Van Cleemput and Boeckx, 2006). The chamber approach is relatively simple, low cost, sensitive and portable, but can cover only small areas.

In the laboratory, soil cores, intact or not, can be incubated in airtight containers (glass flasks and cylinders) to measure soil flux. The soil is placed under controlled temperature, moisture and substrate conditions. The atmosphere in the container is regularly sampled for subsequent analysis by gas chromatography, or chemisorption into NaOH in the case of CO_2, if no gas chromatograph is available. The gas flux is calculated through a linear regression analysis of the gas concentration increase with time, the volume of headspace and the dry weight of soil.

Modelling constitutes another way of assessing soil GHG emissions. A number of different approaches have been used to develop models, all of which draw to varying degrees on field and laboratory data. In general, there are three types of models (Parton et al, 1996):

1 microbial growth models that simulate C and N dynamics by explicitly representing the dynamics of the microbes and other organisms responsible for the processes involved (mineralization, nitrification, denitrification, etc.) – for instance, DeNitrification-DeComposition (DNDC) (Li, 2000);
2 process-oriented models which represent C and N dynamics assuming that the different C and N cycling processes can be expressed as a function of the soil water content, temperature, pH, etc., such as NGAS (Parton et al, 1996), nitrous oxide emission (NOE) (Hénault et al, 2005) and CENTURY (Parton, 1996);
3 soil structural models that simulate soil physical processes, such as the diffusion of gases and solutes into soil aggregates, and explicitly represent the distribution of soil aggregates – for instance, Predicting Agricultural Solute Transport in Soils (PASTIS) (Lafolie, 1991).

These models are of varying complexity, from the relatively simple IPCC (2006) emission factor approach, to very complex models such as DNDC. Some focus on simulating soil fluxes of N_2O (NGAS and NOE), CO_2 (Del Grosso et al, 2005) or CH_4 (Del Grosso et al, 2000), exclusively. Other models called 'soil–plant models' consider the ecosystem C and N cycles in some detail (mineralization, assimilation by plants and microbes, leaching, microbially driven transformations, interactions between the C and N cycles, and biophysical drivers of ecosystems) (Frokling et al, 1998). For those models, soil emissions of GHG are but one small component of these complex cycles. The existing soil–plant models have all grown from different projects with different objectives and therefore they all make different decisions when faced with inevitable trade-offs between detail and simplicity, site specificity and general portability (Frokling et al, 1998). Both DNDC and ExpertN (Stenger et al, 1999) have been developed specifically to look at N and C biogeochemistry in agroecosystems and have therefore made much effort to incorporate and include a variety of agricultural management activities (manure fertilization, planting, harvesting, weeding, tillage, irrigation, etc.). Finally, some models were designed to simulate processes at the field scale (CENTURY, DNDC and ExpertN), while others simulate processes at a regional scale (NASA CASA) (Potter et al, 1994, 1996).

Calculation of the Net GHG Balance in Agroecosystems

Principles

Each greenhouse gas has different heat-trapping properties, so to compare the impacts of emissions and reductions of different gases, GHGs are indexed according to their global warming potential (GWP). The GWP is the ability of a GHG to trap heat in the atmosphere relative to an equal amount of CO_2 and is calculated as the ratio of the radiative forcing of 1kg of the GHG emitted to the atmosphere to that of 1kg of CO_2 over a period of time, generally 100 years. The GWPs of N_2O and CH_4 are 298 and 25, respectively, over a time horizon of 100 years (Forster et al, 2007). Fluxes of N_2O and CH_4 are converted into CO_2-equivalent as follow:

$$N_2O\ (\text{Mg } CO_2\text{-equivalent ha}^{-1}\text{ y}^{-1}) = \frac{M_{N_2O}}{2M_N} GWP_{N_2O} \times \quad N_2O\ (\text{Mg N-}N_2O\text{ ha}^{-1}\text{ y}^{-1}) \tag{1.6}$$

$$CH_4\ (\text{Mg } CO_2\text{-equivalent ha}^{-1}\text{ y}^{-1}) = \frac{M_{CH_4}}{M_C} GWP_{CH_4} \times \quad CH_4\ (\text{Mg C-}CH_4\text{ ha}^{-1}\text{ y}^{-1}) \tag{1.7}$$

with M_{N_2O}: molar mass of N_2O (44 g mol^{-1}); M_N: molar mass of N (14 g mol^{-1}); and M_{CH_4}: molar mass of CH_4 (16 g mol^{-1}); M_C: molar mass of C (12 g mol^{-1}).

The net GHG balance is generally expressed in CO_2-equivalent and is made up of two terms: emissions of N_2O and net ecosystem carbon balance (NECB). By definition (see the earlier section on 'Carbon dioxide (CO_2)'), CH_4 fluxes are part of NECB. When using a gain–loss approach for assessing NECB, CH_4 fluxes expressed in CO_2-equivalent will be deduced from NEP. CH_4 fluxes expressed in CO_2-equivalent should additionally be accounted for when NECB is assessed using the stock-difference method.

Finally, indirect GHG emissions arising from the production of agricultural inputs (fertilizers, pesticides and lime), fuel combustion and use of machinery on the farm may also be taken into account. Indirect emissions of GHGs may contribute as much as half of the total GHG budget of agricultural crops (Robertson et al, 2000; Mosier et al, 2005).

Coffee monocultures and agroforestry systems in Costa Rica: A practical example

To conclude the chapter, we present an example case study carried out at the Costa Rican Coffee Institute's (ICAFE's) experimental station in the Central Valley at San Pedro de Barva. The station is located in a medium altitude zone (1180m) with a wet climate (mean annual temperature of 21°C and average annual rainfall of 2300mm). The soil is a fine-textured Andosol derived from volcanic ashes. Carbon storage and soil N_2O emissions were studied in coffee monocultures and coffee AFS with nitrogen-fixing shade trees, under two agricultural management regimes. Two of the coffee plantations (monoculture and AFS with *Inga densiflora*) were conventionally managed (250kg mineral N ha^{-1} y^{-1}); the two others (monoculture and AFS with *Erythrina poeppigiana*) were organic with applications of 150kg organic N per hectare per year (ha^{-1} y^{-1}) supplied as coffee pulp. The conventionally managed plantations were established in 1997 after long-term conventional coffee cultivation, with a planting density of 5000 coffee plants ha^{-1} and 278 trees ha^{-1} in the AFS. The organic coffee plantations were installed in nearby plots in June 1999 after clearing a seven-year-old fallow vegetation, with a planting density of 5000 coffee plants ha^{-1} and 420 trees ha^{-1} in the AFS. For all coffee plots, biomass in the litter and coffee plants was measured in litter traps and by destructive sampling of eight coffee plants per plot, respectively. The above-ground biomass of the *Inga* trees was quantified by developing allometric relationships between tree DBH and the above-ground biomass contained in the stem, branches and leaves. We then applied this allometric equation to an inventory of the trees made in July 2004 (Siles et al, 2010). The above-ground biomass of the *Erythrina* trees was assessed after a severe pruning of all branches. The branches were weighed to calculate their biomass, which was added to the biomass of the remaining trunks estimated from DBH and height measurements (Harmand et al, 2007). The components of the above-ground biomass (plant biomass + litter) were converted into a C-stock by multiplying by a C fraction of 0.48. We then calculated mean annual C accumulation rates (Mg CO_2 ha^{-1} y^{-1}) by dividing the mean C stock value of the coffee systems by the age of the plantation.

Soil N_2O emissions were measured monthly during one year (4 October to 5 September) using the static chamber method in the two conventionally fertilized systems. In the organic systems, soil N_2O emissions were measured at three different dates in 2005: during the dry season, at the beginning and in the middle of the rainy season. Annual N_2O emission rates were calculated, first, by linear integration between the measurement dates and divided by 365 days in the conventional systems; and, second, as the mean daily N_2O fluxes extrapolated to one year in the organic systems. The net GHG balances of the systems

Table 1.3 *Carbon stocks in the above-ground biomass (plant biomass + litter) in two pairs of coffee monoculture and coffee agroforestry systems (AFS) conventionally fertilized or organically managed in the Central Valley of Costa Rica*

Fertilization Treatment	*Coffee system*	*C stocks in above-ground biomass (Mg C ha^{-1})*			
		Trees: mean (standard error)	*Coffee: mean (standard error)*	*Litter: mean (standard error)*	*Total: mean (standard error)*
Mineral fertilizer 250kg N ha^{-1} y^{-1}	AFS with *Inga densiflora*	13.5 (0.0)	9.3 (0.5)	2.4 (0.1)	25.1 (0.6)
	Monoculture		9.0 (1.1)	1.5 (0.1)	10.5 (1.1)
Organic fertilizer 150kg N ha^{-1} y^{-1}	AFS with *Erythrina poeppigiana*	9.8 (0.5)	5.7 (0.3)	5.3 (0.6)	20.8 (0.8)
	Monoculture		4.5 (0.2)	0.5 (0.1)	5.0 (0.3)

were calculated in CO_2-equivalent as the difference between C sequestrated in the above-ground biomass (plant biomass + litter) and N_2O emissions, applying the N_2O global warming potential of 298 for a 100-year period.

Carbon stocks in the above-ground biomass were 2.4 and 4 times greater in the coffee AFS than in the coffee monoculture, for conventional and organic management, respectively (see Table 1.3), confirming the potential of AFS of enhancing C storage.

Soil emissions of N_2O were calculated at 5.8kg (0.5) and 4.3kg (0.3) N ha^{-1} y^{-1} for the conventional management, 3.7kg (1.5) and 1.8kg (0.8) N ha^{-1} y^{-1} for the organic management in the AFS and the monocultures, respectively. Annual N_2O emissions were approximately 1.4 and 2 times greater under the AFS than under the monocultures, respectively, for the

Table 1.4 *Carbon sequestration rate in above-ground biomass (plant biomass + litter) [1]; soil N_2O emissions expressed in CO_2-equivalent [2]; and net GHG balance [1] – [2] in two pairs of coffee monoculture and AFS conventionally fertilized or organically managed in the Central Valley of Costa Rica*

Fertilization treatment	*Coffee system*	*C sequestration rate in above-ground phytomass Mg CO_2 ha^{-1} y^{-1}: mean (standard error)*	*Soil N_2O emissions Mg CO_2 ha^{-1} y^{-1}: mean (standard error)*	*Net GHG balance Mg CO_2 ha^{-1} y^{-1}: mean (standard error)*
Mineral fertilizer 250kg N ha^{-1} y^{-1}	AFS with *Inga densiflora*	13.2 (0.3)	2.7 (0.2)	10.5 (0.4)
	Monoculture	5.5 (0.6)	2.0 (0.0)	3.4 (0.6)
Organic fertilizer 150kg N ha^{-1} y^{-1}	AFS with *Erythrina poeppigiana*	12.7 (0.5)	1.7 (0.7)	11.0 (0.9)
	Monoculture	3.1 (0.2)	0.9 (0.4)	2.2 (0.4)

conventional and organic managements (see Table 1.4). These results confirm that N_2-fixing legumes can increase soil N_2O emissions. In spite of greater soil N_2O emissions under the AFS than under the monoculture, the AFS showed net GHG balances three and five times greater than the monoculture for conventional and organic management, respectively (see Table 1.4).

Neither soil C stock changes nor CH_4 fluxes were accounted for in the GHG balance. We assumed that changes in soil C stocks are low in these agroecosystems. CH_4 fluxes were monitored in the conventional systems and appeared to be very low: –0.001Mg and –0.003Mg C ha^{-1} y^{-1} in the AFS and the monoculture, respectively (Hergoualc'h et al, 2008). Finally, if we assume that fertilizer processing, transport and application releases 1.4 mole of CO_2 per mole of N applied, as proposed by Robertson and Grace (2004), the net GHG balance in the conventionally fertilized coffee plantations would be reduced by 1.1Mg CO_2 ha^{-1} y^{-1}.

Conclusions

By planting trees in agroecosystems, agroforestry significantly contributes to the reduction of atmospheric CO_2 concentrations. Carbon sequestration can be quantified proportionally to the amount of additional C stocked in the AFS when compared to the same system without trees. Methods for quantifying the additional C stored are of different complexity and accuracy. The simplest is to assign a mean C stock value for the planted species. A more accurate method would be to quantify the C stored in the trees by using an allometric relationship specific to the planted species. Ideally, the net ecosystem C balance should be measured in the agroecosystem of reference (without trees) and in the AFS in order to quantify the additional C stored.

While some doubt exists regarding a potential contribution of other components to the net greenhouse gas budget, quantification of the changes in flux values due to the inclusion of the trees should be evaluated. Suspicion may arise when the trees are nitrogen-fixing species or when increased N fertilization rates are required for the growth of the trees. Again, the methods are of different complexity and accuracy. Indirect GHG emissions should also be accounted for when they are modified due to the presence of trees.

References

Arifin, J. (2001) *Estimasi cadangan C pada berbagai sistem penggunaan lahan di Kecamatan Ngantang, Malang*, Skripsi-S1, Unibraw, Malang, Indonesia

Baumert, K., Herzog, T. and Pershing, J. (2005) *Navigating the Numbers: Greenhouse Gas Data and International Climate Policy*, World Resources Institute (WRI), Washington, DC

Blanchart, E., Roose, E. and Khamsouk, B. (2006) 'Soil carbon dynamics and losses by erosion and leaching in banana cropping systems with different practices (Nitisol, Martinique, West Indies)', in E. Roose and C. Feller (eds) *Soil Erosion and Carbon Dynamics*, Advances in Soil Science, vol 15, CRC Press, Boca Raton, FL, pp87–102

Davidson, E. A. (1991) 'Fluxes of nitrous oxide and nitric oxide from terrestrial ecosystems', in J. E. Rogers and W. B. Whitman (eds) *Microbial Production and Consumption of Greenhouse Gases: Methane, Nitrogen oxides and Halomethanes*, American Society of Microbiology Press, Washington, DC, pp219–235

De Miguel Magaña, S., Harmand, J.-M. and Hergoualc'h, K. (2004) 'Cuantificación del carbono almacenado en la biomasa aérea y el mantillo en sistemas agroforestales de café en el Sur Oeste de Costa Rica', *Agroforestería en las Américas*, vol 41–42, pp98–104

Del Grosso, S. J., Parton, W. J., Mosier, A. R., Ojima, D. S., Potter, C. S., Borken, W., Brumme, R., Butterbasch-Bahl, K., Criss, P. M., Dobbie, K. and Smith, K. A. (2000) 'General CH_4 oxidation model and comparisons of CH_4 oxidation in natural and managed systems', *Global Biogeochemical Cycles*, vol 14, pp999–1019

Del Grosso, S. J., Parton, W. J., Mosier, A. R., Holland, E. A., Pendall, E., Schimel, D. S. and Ojima, D. S. (2005) 'Modeling soil CO_2 emissions from ecosystems', *Biogeochemistry*, vol 73, pp71–91

Forster, P., Ramaswamy, V., Artaxo, P., Berntsen, T., Betts, R., Fahey, D. W., Haywood, J., Lean, J., Lowe, D. C., Myhre, G., Nganga, J., Prinn, R., Raga, G., Schulz, M. and van Dorland, R. (2007) 'Changes in atmospheric constituents and in radiative forcing', in S. Solomon, D. Qin, M. Manning, Z. Chen, M. Marquis, K. B. Averyt, M. Tignor and H. L. Miller (eds) *Climate Change 2007: The Physical Science Basis. Contribution of Working Group I to the Fourth Assessment Report of the Intergovernmental Panel on Climate Change*, Cambridge University Press, Cambridge and New York, pp129–234

Frokling, S., Mosier, A., Ojima, D., Parton, W., Potter, C., Priesack, E., Stenger, R., Haberbosch, C., Dörsch, P., Flessa, H. and Smith, K. (1998) 'Comparison of N_2O emissions from soils at three temperate agricultural sites: Simulations of year-round measurements by four models', *Nutrient Cycling in Agroecosystems*, vol 52, pp77–105

Hairiah, K., Sitompul, S. M., van Noordwijk, M. and Palm, C. A. (2001) 'Methods for sampling carbon stocks above and below ground: ASB_LN 4B', in M. van Noordwijk, S. E. Williams and B. Verbist (eds) *Towards Integrated Natural Resource Management in Forest Margins of the Humid Tropics: Local Action and Global Concerns*, ASB Lecture Notes 1–12, International Centre for Research in Agroforestry (ICRAF), Bogor, Indonesia

Hamburg, S. P. (2000) 'Simple rules for measuring changes in ecosystem carbon in forestry-offset projects', *Mitigation and Adaptation Strategies for Global Change*, vol 5, pp25–37

Harmand, J.-M., Hergoualc'h, K., Dzib, B. et al (2007) 'Carbon sequestration in biomass and litter of coffee agroforestry plantations in Central America', Paper presented at the 2nd International Symposium on Multi-Strata Agroforestry Systems with Perennial Crops, CATIE, Turrialba, Costa Rica, 17–21 September 2007

Hénault, C., Bizouard, F., Laville, P., Gabrielle, B., Nicoullaud, B., Germon, J. C. and Cellier, P. (2005) 'Predicting in situ soil N_2O emission using an NOE algorithm and soil database', *Global Change Biology*, vol 11, pp115–127

Hergoualc'h, K., Skiba, U., Harmand, J.-M. and Hénault, C. (2008) 'Fluxes of greenhouse gases from Andosols under coffee in monoculture or shaded by *Inga densiflora* in Costa Rica', *Biogeochemistry*, vol 89, pp329–345

IPCC (Intergovernmental Panel on Climate Change) (2001) *Climate Change 2001: The Scientific Basis. Contribution of Working Group I to the Third Assessment Report of the Intergovernmental Panel on Climate Change*, Cambridge University Press, Cambridge and New York

IPCC (2003) *Good Practice Guidance for Land Use, Land-Use Change and Forestry*, Institute for Global Environmental Strategies (IGES), Japan

IPCC (2006) *2006 IPCC Guidelines for National Greenhouse Gas Inventories*, Institute for Global Environmental Strategies (IGES), Hayama, Japan

Johnson, J. M.-F., Allmaras, R. R. and Reicosky, D. C. (2006) 'Estimating source carbon from crop residues, roots and rhizodeposits using the national grain-yield database', *Agronomy Journal*, vol 98, pp622–636

Knowles, R. (1993) 'Methane: Processes of production and consumption', in L. A. Harper, A. R. Mosier, J. M. Duxbury and D. E. Rolston (eds) *Agricultural Ecosystem Effects on Trace Gases and Global Climate Change*, vol 55, ASA, Madison, WI, pp145–156

Lafolie, F. (1991) 'Modeling water flow, nitrogen transport and root uptake including physical non-equilibrium and optimization of the root water potential', *Fertilizer Research*, vol 27, pp215–231

Li, C. (2000) 'Modeling trace gas emissions from agricultural ecosystems', *Nutrient Cycling in Agroecosystems*, vol 58, pp259–276

Luyssaert, S., Inglima, I., Jung, M., Reichstein, M., Papale, D., Piao, S., Schulze, E.-D., Wingate, L., Matteucci, G., Aubinet, M., Beer, C., Bernhofer, C., Black, K. G., Bonal, D., Chambers, J., Ciais, P., Davis, K. J., Delucia, E. H., Dolman, A. J., Don, A., Gielen, B., Grace, J., Granier, A., Grelle, A., Griffis, T., Grünwald, T., Guidolotti, G., Hanson, P. J., Harding, R., Hollinger, D., Kolari, P., Kruijt, B., Kutsch, W., Lagergren, F., Laurila, T., Law, B., Le Maire, G., Lindroth, A., Magnani, F., Marek, M., Mateus, J., Migliavacca, M., Misson, L., Montagnani, L., Moncrieff, J., Moors, E., Munger, J. W., Nikinmaa, E., Loustau, D., Pita, G., Rebmann, C., Richardson, A. D., Roupsard, O., Saigusa, N., Sanz, M. J., Seufert, G., Soerensen, L., Tang, J., Valentini, R., Vesala, T. and Janssens, I. A. (2007) 'Global patterns in forest CO_2-balance: An analysis based on a new global database', *Global Change Biology*, vol 13, pp2684–2697

Mokany, K., Raison, R. J. and Prokushkin, A. S. (2006) 'Critical analysis of root:shoot ratios in terrestrial biomes', *Global Change Biology*, vol 12, pp84–96

Mosier, A. R., Schimel, D. S., Valentine, D. W., Bronson, K. F. and Parton, W. J. (1991) 'Methane and nitrous oxide fluxes in native, fertilized, and cultivated grasslands', *Nature*, vol 350, pp330–332

Mosier, A. R., Wassmann, R., Verchot, L., King, J. and Palm, C. (2004) 'Methane and nitrogen oxide fluxes in tropical agricultural soils: Sources, sinks and mechanisms', *Environment, Development and Sustainability*, vol 6, pp11–49

Mosier, A. R., Halvorson, A. D., Peterson, G. A., Robertson, G. P. and Sherrod, L. (2005) 'Measurement of net global warming potential in three agroecosystems', *Nutrient Cycling in Agroecosystems*, vol 72, pp67–76

Palm, C., Alegre, J., Arevalo, L., Mutuo, P., Mosier, A. and Coe, R. (2002) 'Nitrous oxide and methane fluxes in six different land use systems in the Peruvian Amazon', *Global Biochemical Cycles*, vol 16, p1073, doi:10.1029/2001GB001855

Parton, W. J. (1996) 'The CENTURY model', in D. S. Powlson, P. Smith and J. U. Smith (eds) *Evaluation of Soil Organic Matter Models Using Existing Long-Term Datasets*, Springer, Berlin, Germany, pp283–293

Parton, W. J., Mosier, A. R., Ojima, D. S., Valentine, D. W., Schimel, D. S., Weier, K. and Kulmala, A. E. (1996) 'Generalized model for N_2 and N_2O production from nitrification and denitrification', *Global Biogeochemical Cycles*, vol 10, pp401–412

Potter, C. S., Matson, P. A. and Vitousek, P. M. (1994) 'Evaluation of soil database attributes in a global carbon cycle model: Implications for global change research', in W. Michener (ed) *Environmental Information Management and Analysis: Ecosystem to Global Scales*, Taylor and Francis, London, pp281–302

Potter, C. S., Matson, P. A., Vitousek, P. M. and Davidson, E. A. (1996) 'Process modeling of controls on nitrogen trace gas emissions from soils worldwide', *Journal of Geophysical Research*, vol 101, pp1361–1377

Powell, M. H. and Delaney, M. (1998) *Carbon Sequestration and Sustainable Coffee in Guatemala: Final Report*, Winrock International, Arlington, VA

Priyadarsini, R. (1998) *Studi cadangan C dan populasi cacing tanah pada berbagai macam system pola tanam berbasis pohon*, Thesis S2, Unibraw, Malang, Indonesia

Robertson, G. P. and Grace, P. R. (2004) 'Greenhouse gas fluxes in tropical and temperate agriculture: The need for a full-cost accounting of global warming potentials', *Environment, Development and Sustainability*, vol 6, pp51–63

Robertson, G. P., Paul, E. A. and Harwood, R. R. (2000) 'Greenhouse gases in intensive agriculture: Contributions of individual gases to the radiative forcing of the atmosphere', *Sciences*, vol 289, pp1922–1925

Rochette, P. and Janzen, H. (2005) 'Towards a revised coefficient for estimating N_2O emissions from legumes', *Nutrient Cycling in Agroecosystems*, vol 73, pp171–179

Rogner, H. H., Zhou, D., Bradley, R., Crabbé, P., Edenhofer, O., Hare, B., Kuijpers, L. and Yamaguchi, M. (2007) 'Introduction', in B. Metz, O. R. Davidson, P. R. Bosch, R. Dave and L. A. Meyer (eds) *Climate Change 2007: Mitigation. Contribution of Working Group III to the Fourth Assessment Report of the Intergovernmental Panel on Climate Change*, Cambridge University Press, Cambridge and New York, pp95–116

Roscoe, R. and Buurman, P. (2003) 'Tillage effects on soil organic matter in density fractions of a Cerrado Oxisol', *Soil & Tillage Research*, vol 70, pp107–119

Ryan, M. G. and Law, B. E. (2005) 'Interpreting, measuring, and modeling soil respiration', *Biogeochemistry*, vol 73, pp3–27

Schimel, J. and Holland, E. (2005) 'Global gases', in D. Sylvia, J. Fuhrmann, P. Hartel and D. Zuberer (eds) *Principles and Applications of Soils Microbiology*, Upper Saddle River, NJ, pp491–509

Schnell, S. and King, G. M. (1994) 'Mechanistic analysis of ammonium inhibition of atmospheric methane consumption in forest soils', *Applied and Environmental Microbiology*, vol 60, pp3514–3521

Segura, M., Kanninen, M. and Suarez, D. (2006) 'Allometric models for estimating aboveground biomass of shade trees and coffee bushes grown together', *Agroforestry Systems*, vol 68, pp143–150

Siles, P., Harmand, J.-M., and Vaast, P. (2010) 'Effects of *Inga densiflora* on the microclimate of coffee (*Coffea arabica* L.) and overall biomass under optimal growing conditions in Costa Rica', *Agroforestry Systems*, vol 78, pp269–286

Skiba, U. and Smith, K. A. (2000) 'The control of nitrous oxide emissions from agricultural and natural soils', *Chemosphere: Global Change Science*, vol 2, pp379–386

Smith, K. A., Ball, T., Conen, F., Dobbie, K. E., Massheder, J. and Rey, A. (2003) 'Exchange of greenhouse gases between soil and atmosphere: Interactions of soil physical factors and biological processes', *European Journal of Soil Science*, vol 54, pp779–791

Smith, P., Martino, D., Cai, Z., Gwary, D., Janzen, H., Kumar, P., McCarl, B., Ogle, S., O'Mara, F., Rice, C., Scholes, B. and Sirotenko, O. (2007) 'Agriculture', in B. Metz, O. R. Davidson, P. R. Bosch, R. Dave and L. A. Meyer (eds) *Climate Change 2007: Mitigation. Contribution of Working Group III to the Fourth Assessment Report of the Intergovernmental Panel on Climate Change*, Cambridge University Press, Cambridge and New York, pp498–540

Stehfest, E. and Bouwman, L. (2006) 'N_2O and NO emission from agricultural fields and soils under natural vegetation: Summarizing available measurement data and modeling of global annual emissions', *Nutrient Cycling in Agroecosystems*, vol 74, pp207–228

Stenger, R., Priesack, E., Barkle, G. and Sperr, C. (1999) 'Expert-N, a tool for simulating nitrogen and carbon dynamics in the soil–plant–atmosphere system', in M. Tomer, M. Robinson and G. Gielen (eds) *NZ Land Treatment Collective, Technical Session 20: Modelling of Land Treatment Systems*, New Plymouth, pp19–28

Sugiharto C. (2002) *Kajian Aluminium sebagai factor penghambat pertumbuhan pohon sengon (*Paraserianthes falcataria *L. Nielsen)*, Skripsi S1, Unibraw, Malang, Indonesia

Van Cleemput, O. and Boeckx, P. (2006) 'Greenhouse gas fluxes: Measurement', in R. Lal (ed) *Encyclopedia of Soil Science*, second edition, Taylor & Francis, Columbus, OH, pp787–790

Waterloo, M. J. (1995) *Water and Nutrient Dynamics of* Pinus caribaea *Plantation Forests on Former Grassland Soils in Southwest Viti Levu, Fiji*, PhD thesis, Vrije Universiteit, Amsterdam, The Netherlands

Wrage, N., Velthof, G. L., Van Beusichem, M. L. and Oenema, O. (2001) 'Role of nitrifier denitrification in the production of nitrous oxide', *Soil Biology & Biochemistry*, vol 33, pp1723–1732

2

Quantifying Services and Identifying Watershed Priority Areas for Soil and Water Conservation Programmes

Miguel Marchamalo, Raffaele Vignola, Federico Gómez-Delgado and Beatriz González-Rodrigo

Introduction

Regional payment for ecosystem services (PES) schemes targeting hydrological ecosystem services are rapidly gaining in popularity and dominance in Central America. In some biological corridors, they have become the driving force behind forest conservation initiatives (DeClerck et al, 2010). Soil conservation and water regulation are strongly interrelated ecosystem processes that provide important services. Indeed, soil conservation influences the provision of water services in terms of quality, quantity and continuity of water supply. These relationships make soil conservation and water regulation fully integrated topics, with particular relevance to managing for increasingly needed climate change adaptation.

Soil and water services include the provision of water in quantity (total yield), continuity (inter- and intra-annual stability) and quality (regulation of sediment load and of biological and chemical pollutions), conservation of the soil to support ecosystem production, reduction of flood risks, and maintenance of local climate. They are mainly provided by agricultural and forest landscapes. On-site service users, such as farmers, make soil management decisions at the farm scale, which affect the amount and quality of off-site services received downstream at watershed scales (e.g. the amount of sediment load in fresh water). As such, soil conservation programmes that target improvement in water quality have to take these two scales into consideration in order to estimate the amount of service provided (e.g. reductions in sediment load) and to define priority areas where interventions should be targeted. The sustainability of hydrological services is of concern to national policies, particularly in countries that are strongly dependent on agriculture. Indeed, many regions around the world are experimenting with local or national PES programmes that provide monetary rewards to farmers for implementing soil conservation practices aimed at reducing the loss of fertile topsoil and sediment loads in freshwater streams. Stanley Heckadon of the Smithsonian Tropical Research Institute in Panama half-jokingly commented that Panama's greatest export product is soil.

Despite the increasing number of these experiences and other innovative initiatives at national and local scales, designers and managers of soil conservation programmes are usually faced with the same questions:

- How much do land use and management affect soil and water conservation?
- How can we measure and quantify soil and water conservation/degradation?
- How can we improve the efficiency of soil conservation and water harvesting programmes?
- What is the effect of spatial and temporal scales?
- How can we prioritize measures and optimally assign funds for the production of soil and water services?

Empirically based answers are needed to respond to these questions; however, research in the Neotropics is faced with several challenges that must first be overcome. These include:

- a challenging physical environment, with high variability and complexity of natural systems;
- low data availability; and
- important trade-offs with urgent social needs, which must be fulfilled first.

In this chapter we provide an overview of a range of methods and tools that can be applied to quantify soil conservation services in the Neotropics. The selection of these tools and methods will vary depending on the objectives of the user and data availability. We present three main blocks of information: the basics of soil and water services; measurement and quantification tools; and the integration of methodologies – all of them illustrated with a case study on the Birrís River watershed in Costa Rica. We use this example to demonstrate how priority areas can be identified where soil conservation programmes are targeted.

Soil and Water Conservation: A Key Issue in National and Local Land-Use Planning Policies

PES schemes offer a tool for national and local land-use planning in tropical countries. Previous soil conservation policies were based on enforced conservation of vulnerable and degraded lands in high-priority watersheds or ecosystems. Governments declared watersheds as high priority based on whether important social, economic or environmental systems were at risk because of land degradation. New approaches tend to offer financial compensation to owners of high-priority vulnerable lands in return for valuable hydrological services (Goldman et al, 2007).

However, despite the renewed focus on compensation for hydrological services, there remains a very critical issue of scale. Watersheds are functional units for the study of the influence of climate, soil cover, soil type, slope, geology and human activities on the quality and quantity of available water (Swisher and Todd-Bockarie, 1996). They are also a natural landscape unit for managing hydrological services, in particular. Nevertheless, whereas land-use planning decisions for hydrological services are made at the watershed scale, management decisions are made by landowners at the farm or property scale. Consequently, the analysis of soil conservation practices available and potentially improvable under a soil conservation programme must begin at the farm level (Müller, 1997); but impacts must be scaled to the landscape level. The management level, which corresponds to farm or property levels in Western economies, generates goods and services that are generally demanded at the watershed level by society.

The hydrological cycle: Linking soil and water services

Let's begin by reviewing the basics. The hydrological and hydro-social cycles determine the distribution and availability of water (Falkenmark and Folke, 2002). The hydrological cycle refers to the physical phenomena associated with water movement through the geosphere, biosphere and atmosphere, and energy exchange through physical states (vapour, liquid and ice). The hydro-social cycle frames the processes for water use in society (supply, demand, sanitation, rules and markets) after which water returns, though often altered, to the hydrological cycle. Human activities affect most of the natural processes of the hydrological cycle through irrigation of agricultural and urban lands, water capture and storage in dams, water movement through canals and pipelines, and even by changing evapotranspiration rates through land-use change.

Land-use change is among the most significant anthropogenic effects on the water cycle, affecting the topsoil layer, which is the interface between the atmospheric and the underground phases of the hydrological cycle (see Figure

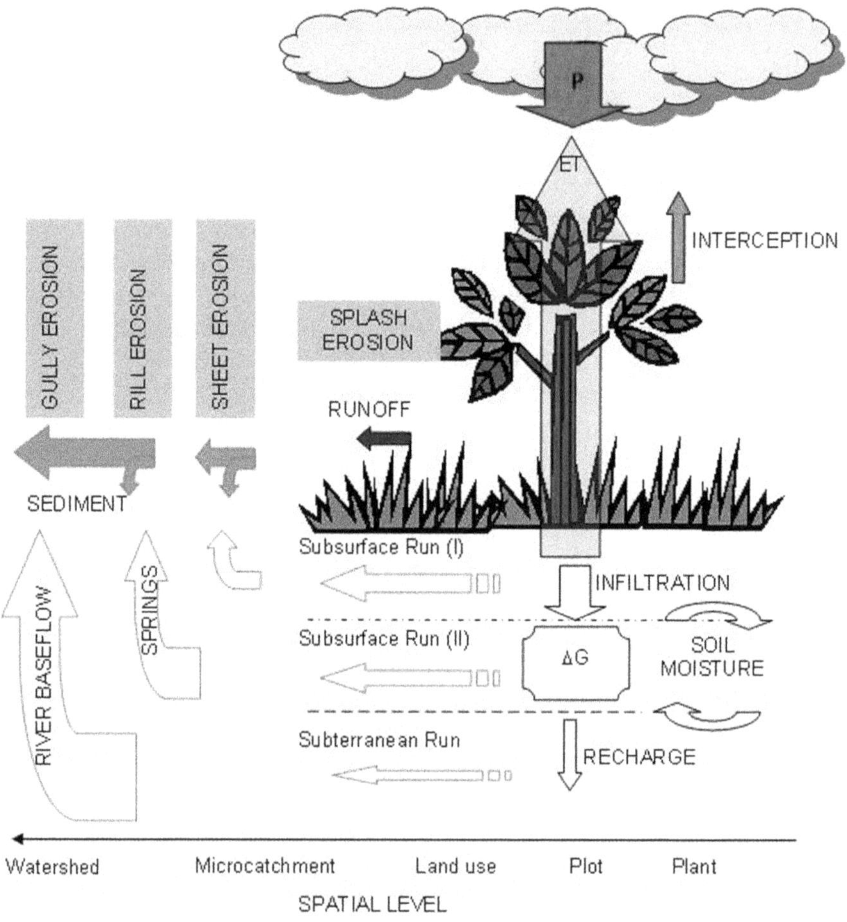

Figure 2.1 *Hydrological cycle: Linking soil and water conservation*

Notes: P = precipitation; ET = evapotranspiration; ΔG = soil moisture variation.
Source: chapter authors

2.1). Slight alterations of this soil layer can alter the cycle as a whole, threatening the availability of water, particularly underground water. Soil and water services are therefore tightly linked and dependent. Surface soil layers play an important role in reducing raindrop impact and runoff generation. Any modification of this surface layer can alter these phenomena, leading to land degradation and water loss.

Figure 2.1 presents a schematic view of the hydrologic cycle, from precipitation (P) to the processes that occur when raindrops reach land surface and vegetation cover, to the dynamics of water in soil. The vertical dimension represents elevation, which is linked to the potential energy of water. On the horizontal axis, we represent the way in which water moves on the Earth's surface, in increasing spatial domains, from plot scale on the right to watershed scale on the left of the figure. This movement involves water and sediment fluxes that are highlighted in white and grey, respectively.

Understanding impacts upon soil and water services: On-site and off-site effects

Erosion and sedimentation affect both sites where these processes occur (on-site effects) as well as downstream locations where sediment in movement causes problems to existing economic or social activities (off-site effects). Erosion and sedimentation are natural processes that are crucial to some ecosystems, especially in lowland areas, deltas and coastlands. However, when anthropogenic activities accelerate on-site erosion and trigger off-site impacts, they may cause severe alterations of downstream activities that depend on soil and water services. The intensity of the impacts of on-site erosion upon off-site services and activities depends on different factors – for example, distance from the fluvial network (distance from the nearest stream or river), topography, land use, and spatial configuration or arrangement of these land uses within the landscape. It is important to note that because of the impacts of these factors, not all portions of the landscape are equally important to (or capable of providing) soil and water services. For example, soil conservation is more critical in areas directly adjacent to rivers or plots situated on steep slopes (Goldman et al, 2007).

The spatial nature of hydrological services highlights the important need for PES schemes that target portions of the landscape most critically needed to provide hydrological services, or where gaps in the provisioning of these services are most evident. The main sources of sediments can be identified using modern modelling tools and geographic information systems (GIS); however, the identification of critical priority areas must be based on the quantification of ecosystem services provided and must take into account the effects of land use and human activity at different scales. This prioritization depends on the perspective of decision-makers and the relevant spatial scale of interest (Fisher et al, 2009).

The effects of land use on erosion, pollution and landslides can be classified into five categories:

1. on-plot effects;
2. on-farm effects;
3. neighbouring farm effects;
4. direct downstream effects; and
5. indirect downstream effects (De Graaff, 2000).

On-plot and on-farm effects can be considered on-site effects, while the last three categories can be considered off-site effects. Direct downstream effects are those that directly affect land and infrastructure located downstream (e.g. sedimentation and flooding), whereas indirect downstream effects include the reduction of hydropower production, impairment of drinking water systems, and effects on human health.

Quantification and prioritization requirements for the implementation of soil and water PES

Quantification and prioritization are critical tasks in the design of a PES programme. Quantification provides an account of services, backing the compensation scheme, while prioritization allows optimizing resources allocation, targeting areas for the programme. The implementation of soil and water PES requires a green accountancy of goods, services and production. This accounting can be divided into three main, but integrated, categories: sociological, biophysical and economic. In order to conduct such accounting, we first need a spatially explicit quantification of the actual and potential production of ecosystem services (litres of water produced, tonnes of soil conserved, etc.). In market terms, this determines and measures the product being sold. In a second phase it is essential to identify priority areas for PES investment in order to make it optimum and efficient. Which portions of the landscape are the most vulnerable to erosion? Or in market terms, which portion of the landscape is going to be able to provide the greatest amount of service per dollar invested? Third, there must be a determination of the amount of the compensation payment made to support desired land-use cover in priority areas. How much should the landowner be paid? Finally, a sensitivity analysis is needed to evaluate and quantify the ecosystem services produced under actual and potential land-use scenarios.

As mentioned above, soil and water services depend strongly on topography, soil properties, land use and distance to rivers and streams. Spatial analysis is essential to quantify and prioritize soil and water environmental services. In this chapter, we focus on this aspect of biophysical quantification and targeting and on the tools available for this. It is important to point out that sociological and economic components of PES are essential and should not be overlooked (see, in particular, Chapters 7 and 16 of this volume). In order to demonstrate how targeting can work, we present the example of the Birrís watershed in Costa Rica.

Soil Conservation and Hydropower Production in the Birrís Watershed, Costa Rica: Setting the Stage

Costa Rica relies on hydropower production to support its current and future development targets. Many of Costa Rica's hydroelectric dams are adversely affected by sedimentation due to severe erosion in adjacent slopes and headwaters. Sedimentation in hydroelectric dams causes a reduction in electricity generation potential and increased operational costs due to an increase in frequency and duration of sediment management operations, and the consequences of stopping generation and buying additional energy from other sources. Thus, sustainable electricity planning in Costa Rica depends greatly on a rational land-use planning strategy, especially in energy-producing river basins (Marchamalo and Romero, 2007).

The Reventazón River Basin is the primary electricity-producing basin in Costa Rica, accounting for 32 per cent of total hydropower production (Jaubert, 2001). The lifespan of its hydropower dams critically depends on the sediment inputs from adjacent slopes, which affects the overall quality of water in the dam. Each year, up to 1.5 million tonnes of sediments are removed from the dams to ensure the largest possible lifespan. More than US$2 million are spent each year in removing sediments and providing alternative energy sources during sediment removal (Rodríguez, 2001).

The Birrís River Basin (over 4800ha) encompasses a land mass with elevations ranging from the top of the Irazú Volcano (3400m above sea level, or ASL) down to the Reventazón River Valley (1250m ASL) that drains into the Atlantic. The state-owned JASEC Birrís Hydropower System (SHB) manages the Birrís River Basin, which supplies Cartago with 20MW of power. The predominant land use in the basin comprises pastures and intensive agricultural farming on deep volcanic soils (Andosols). The cool weather found at these high elevations is well suited to the production of crops such as potatoes and carrots, as well as dairy products. As such, the area is the source of more than half the vegetables produced nationwide. The intense production supports a dense population of 161 inhabitants per square kilometre, which is significantly greater than the national average of 78 inhabitants per square kilometre (INEC, 2001). In this area, 61 per cent of the population is employed in this intensive farming.

The combination of steep slopes, heavy rains and intensive horticultural production has resulted in the Birrís watershed producing some of the largest sediment loads in the country (Sanchez-Azofeifa et al, 2002) to the detriment of the more than five hydroelectric plants located downstream of this basin. The quantity and quality of electricity generation in the Birrís Hydropower System has fallen recently due to high sedimentation rates in dams and the impact of spates upon structures. This is related to the extension of landslides, the intensive agricultural production in steep slopes with erodible soils, and the decrease of forest cover in tributaries flowing into the Birrís Hydropower System. As a result, Cartago has increased its reliance on external energy sources and raised its energy prices. For these reasons, the Birrís River Basin was declared a high-priority basin in the *Reventazón River Land-Use Plan* (Costa Rican Law 8023).

In addition to the importance of the Reventazón River for energy production, the watershed has also been noted for its vulnerability to extreme weather events. This is especially critical for soil and water services in the Birrís watershed. The intensive land use combined with observed (Aguilar et al, 2005) and projected (Magrin et al, 2007) increases in extreme precipitation events threatens the future sustainability of local hydropower production. A recent project (the European Union-funded Tropical Forests and Adaptation to Climate

Change; see www.cifor.cgiar.org/trofcca/_ref/home/index.htm) focusing on ecosystem-based adaptation (UNCBD, 2009; Vignola et al, 2009) has developed and evaluated on- and off-site soil and water services provided under alternative land-use scenarios and extreme precipitation incidences (Vignola et al, 2010) with local stakeholders. Findings indicated that land use has a dominant effect over climate extremes in terms of affecting the amount of sediments produced. Moreover, consultations with stakeholders (i.e. upstream farmers and downstream hydropower facility) showed that focusing on high-priority areas can be both environmentally (Goldman et al, 2007) and socially acceptable strategies.

The local hydropower utility JASEC commissioned the *River Birrís Land-Use Plan* (CATIE, 2003) to provide guidelines in implementing a soil conservation programme, which includes the establishment of an environmental tariff in the electric bill to finance direct incentives to upstream farmers to promote soil conservation practices. The plan's implementation calls for specific attention to quantifying the benefits derived from off-site upstream soil conservation and developing a prioritization method to target areas with the highest return on environmental investments. The critical first step is determining where interventions should be targeted.

Tools for the Quantification and Estimation of Soil Conservation in the Neotropics

This section presents quantification tools, distinguishing and combining direct measurements and modelling approaches. There is a vast range of tools available to quantify the amount of soil and water services provided by soil conservation interventions. The driver of soil and water services, however, is directly tied to on-site interventions that reduce erosion at the farm scale, as well as increase the water-holding capacity of these landscapes in an attempt to mitigate the impacts of floods. For the Birrís watershed, while the primary goal is to reduce sedimentation of reservoirs behind hydroelectric dams, most of the interventions needed to provide this service will be achieved by implementing targeted soil conservation practices. Note, however, that while the methods presented here include some water balance estimates, the quantification of parameters such as evapotranspiration, soil water flux and storage, which are necessary when assessing the provision of water and its dynamics (e.g. supply of drinking water during the year and mitigation of floods) are not discussed. A good presentation of these methods and tools can be found in technical material, such as United Nations Food and Agriculture Organization's (FAO's) irrigation and drainage paper (Allen et al, 1998). In this description of quantification tools, we focus on measuring and modelling soil erosion and sedimentation, land-use conflicts analysis and the connectivity approach. We show how these tools were applied to the Birrís watershed and the results obtained, and discuss how the land-use conflict analysis and the modelling tools were combined in the Birrís watershed to target areas for intervention.

Measurement and modelling

Quantification and measurement methods

Quantification methods provide direct measurements of soil loss at different scales. These require an experimental design in order to provide adequate and reliable data. The methods selected usually depend on the objectives and scope of the quantification process. Table 2.1 presents a range of methods for direct measurement. A more detailed description of these methods can be found in *Soil Erosion Research Methods* (Lal et al, 1994).

A monitoring programme may combine and integrate different quantification methods at different scales with the objective of assessing soil and water service provision. Some methods are based on periodic point measures, such as land degradation measurements based on frequent surveys of land level referred to a given elevation, or a distinct geomorphologic feature, such us gully heads, rock outcrops and bridge pillars. These methods result in punctual data that can be used to evaluate 'on-site' effects. On the other hand, methods such as instrumented catchments, bathymetry and remote sensing may be used for evaluating 'off-site' effects at different scales.

Table 2.1 *Direct measurement methods for soil erosion quantification*

Method	*Type*	*Scale*	*Evaluation*	*Description*
Pin	Experimental	Point	On-site	Steel pin dug into soil. The initial level of the pin head is recorded and monitored periodically to estimate land degradation (erosion) or aggradation (sedimentation).
Pin transects	Experimental	Cross-section	On-site	A line of pins across land.
Cut/fill measurement	Observation	Point	On-site	Topographic measurement and calculation of sediment volume at a specific location (m^3).
Gully measurement	Observation	Point plot	On-site	Measurement of the progress of gullies (m, m^2, m^3).
Erosion plot	Experimental	Plot	On-site	Plot of land delimited in order to collect all sediment produced within its limits.
Rainfall simulator	Experimental	Plot	On-site	Rainfall-generating devices that allow simulating storms over small plots; sediments are collected.
Instrumented catchments	Experimental	Catchment	On-site/ off-site	Catchment that is instrumented at a given pour point to measure liquid and solid flows.
Digital terrain model/ topography/ bathymetry	Observation	Plot catchment	On-site/ off-site	Direct measurement of sediment storage in water bodies.
Remote sensing	Observation	Plot catchment	On-site/ off-site	Analysis of remotely sensed images to classify erosion and sedimentation areas.

Soil erosion models

Soil erosion models are simplified representations of complex systems and can be categorized into three types according to fundamental processes, algorithms used and data demand (Saavedra, 2005):

1 empirical models;
2 physical models; and
3 conceptual models.

Empirical models are primarily based on analysis of field experiments or data plots using statistical inference to extract empirical patterns. These models have lower data requirements than physical and conceptual models. Normally, they require data at a higher level of aggregation in space and time and are often useful in identifying critical areas of high erosion. Physical models are based on knowledge and simulation of physical erosion processes, transport and sedimentation, to ensure conservation of mass and energy over time. Conceptual models are based on the representation of physical erosion processes by empirical equations. Conceptual models are conglomerates of empirical and physical models.

It is important to note that most soil erosion models do not estimate intense soil degradation processes involving the detachment of blocks such us gully erosion, landslides and soil mass movements. Soil erosion models are usually calibrated for raindrop, overland flow and rill erosion forms. Box 2.1 provides examples of models for each category.

Sediment yield models: Sediment delivery ratio

Erosion modelling quantifies the potential sediment production of a watershed. However, the most significant and economically meaningful parameter is the watershed sediment yield for a given scenario at a given point in time (i.e. the total amount of sediment that is collected at a certain pour point downstream). This sediment yield can be evaluated as an 'off-site' impact of upland land-use systems. For this simulation we need to define and model four main processes:

1 splash, overland, rill and gully erosion production;
2 scale effects on erosion quantification;
3 transport route for detached sediment; and
4 the sediment delivery ratio.

The sediment delivery ratio is defined as the ratio between effectively delivered sediment and total erosion (0–1) (i.e. the proportion of soil that passes a fixed point in the watershed to the total amount of sediment that is eroded upstream).

The calibration of sediment yield models requires stream sedimentological data for the selected pour control points. This sediment data may be obtained in two ways: first, through direct observation and measuring suspended and

Box 2.1 Soil erosion models

Soil erosion models can range from low data demand to high data demand depending on objectives and data availability. The most commonly used soil erosion models are as follows:

- Models based completely or partly on the empirical Universal Soil Loss Equation (USLE) (Wischmeier and Smith, 1978):
  - USLE (Wischmeier and Smith, 1978) for slopes/areas;
  - USLE-CALSITE (Bradbury, 1995) for basins, raster-based GIS (Idrisi) and field calibration of sediments;
  - RUSLE (Revised Universal Soil Loss Equation) (Renard et al, 1997) for slopes/areas;
  - RUSLE-SAATEC (Lim et al, 2005) for watershed-based GIS (ArcView) and field calibration of sediments.
- Physical models based on physical equations:
  - EUROSEM (European Soil Erosion Model) (Morgan et al, 1998);
  - WEPP (Water Erosion Prediction Programme) (Flanagan and Nearing, 1995);
  - LISEM (Limburg Soil Erosion Model) (De Roo et al, 1994).
- Conceptual models, representing physical processes with empirical equations:
  - SWAT (Soil and Water Assessment Tool), available as an extension of ArcView GIS (Arnold et al, 1995);
  - MMMF (Modified Morgan-Morgan-Finney) (Morgan, 2001);
  - ANSWERS (Beasley et al, 1980);
  - CREAMS (Chemicals, Runoff and Erosion from Agricultural Management Systems) (Foster et al, 1980; Kinsel, 1980)

Most soil erosion models consider only sheet and rill water erosion processes, omitting mass movements and gully erosion sedimentation production, which require estimations by other procedures.

bed load in a given fluvial section; second, the data can be obtained through an interpolation of sediment delivery ratios for a catchment using data from a statistically significant network of sediment stations in the same basin. Both methods require the installation of sedimentological stations in specific stable sections that measure and record suspended and bed sediment load. Sediment delivery ratios are interpolated using geo-statistics from a significant number of field stations for statistical calibration of an entire basin. Once interpolated, sediment delivery ratio values can be obtained for any pixel in the basin through GIS raster processing.

Quantification of soil erosion in the Birrís Basin

Among the available models for quantifying soil erosion (see Box 2.1), we selected the Revised Universal Soil Loss Equation (RUSLE) for potential soil erosion estimation and the CALSITE software for sedimentation calibration. This decision is due in large part to data availability, calibration and comparability. The model was calibrated as a pilot case study for tropical regions because there is infrequently access to the hourly and daily weather records required by conceptual and physical models in the tropics. RUSLE can use regionalized climate data, which permitted us to calculate erosivity based on

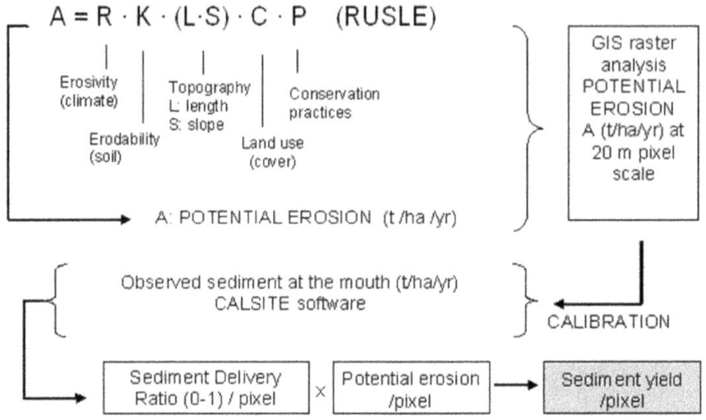

Figure 2.2 *Components of the RUSLE model and CALSITE calibration*

Source: chapter authors

previously established studies (Gómez-Delgado, 2002). The CALSITE software includes a sediment transport routine that also makes it possible to calibrate sediment production for each pixel of the watershed with sediment data measured in the field. Finally, CALSITE has been implemented in Costa Rica, which permitted us to compare our results with those of other studies (ICE, 1999; Saborío, 2000; Gómez-Delgado, 2002; Marchamalo, 2004). Figure 2.2 illustrates the process followed.

In order to use RUSLE, we first needed to develop an accurate map of the land uses in the basin. We started with a baseline map developed by classifying an aerial photograph. In order to properly run RUSLE, we verified this map to ensure that our classification of land uses (cover) and soil factors was correct (Lianes et al, 2009).This provided us with the land-use cover, or the C factor, and the soil factor (K) in the RUSLE equation. Second, we calculated erosivity, the R factor in the RUSLE equation, by using data from a rainfall erosivity analysis of the area (Gómez-Delgado, 2002). The topographic parameters of slope (S) and length of slope (L) were calculated through GIS analysis over a precise digital terrain model. We identified and evaluated conservation practices – the P factor in the equation – that are promoted and adopted in the watershed with potential for greater dissemination to control erosion. Finally, we calculated potential soil erosion production (A) under current landscape soil use using RUSLE (Renard et al, 1997). We calculated and calibrated sediment delivery ratios and sediment yield for actual scenarios using CALSITE software (Bradbury, 1995) with available sedimentological station data. Then, selected conservation practices were modelled in changing land-use scenarios in order to analyse their effects on soil erosion and sedimentation.

During the calibration phase, sediment data is often difficult to acquire. In the Birrís Basin we used the data provided by the Costa Rican Institute of Electricity (ICE) for the Reventazón Basin (Rodríguez, 2001), processed by

Table 2.2 *Reventazón River Basin sedimentological data used in the calibration process*

Station	*Catchment area (km^2)*	*Observed sediment (tonnes ha^{-1} yr^{-1})*
Angostura	1357.4	23.48
Oriente	230.2	17.27
Palomo	373.6	9.39

Note: S. P. Guayabo and La Troya stations were also used to test the calibration, although their records were not long enough to calibrate flow–sediment relationships.
Source: Rodríguez (2001); Gómez-Delgado (2002)

Gómez-Delgado (2002). Table 2.2 lists the results of the calculations of sediment production obtained from the Reventazón River Basin sediment data stations.

This modelling effort included three important processes. First it considered the potential laminar erosion (A in the RUSLE equation; see Figure 2.2) corresponding to the local soil loss processes driven by sheet and rill erosion. In this scenario, soil is eventually carried to the river system or deposited in the delivery path due to the presence of natural or man-built barriers (e.g. depressions, conservation works, planting and contour strips, riparian forest, etc.). We modelled each potential local erosion scenario calibrated with the RUSLE model. Second, we estimated erosion by mass movements, landslides and gullies that RUSLE does not estimate. These forms of erosion are roughly estimated with technical coefficients provided by watershed authorities. Finally, we included sedimentation (yield, S) at the mouth of the watershed. This value corresponds to the amount of sediment washed and exported annually from the basin per unit area. The model must be calibrated with observed sediment yields in several sub-watershed study areas. The results of sediment yield at pixel scale are analysed in raster GIS format (see Figure 2.3). As can be appreciated, over 36 per cent of the basin presents high potential erosion rates that are severe (over 50 tonnes ha^{-1} y^{-1}) in 27 per cent of its surface (1300ha).

For the Reventazón Basin, Gómez-Delgado (2002) suggests that mass movements and landslides account for at least 20 per cent of the total erosion, while ICE reports that mass movements contribute up to 30 per cent of the erosion that reaches the reservoirs in the Reventazón River Basin. Mass movements are triggered by saturation of the soil profiles on slopes by excessive rainfall and are strongly tied to the geology of the area. Changes in land use are thought to have little impact upon these mass movements. The estimated total sediment production for the Birrís Basin was 61 tonnes ha^{-1} y^{-1}. On the other hand, modelled values of potential erosion accounted for 52 tonnes ha^{-1} y^{-1} corresponding to overland, sheet and rill erosion processes. The CALSITE model calibration with observed sediment data confirmed that erosion from other sources, including mass movements and landslides, could cause 15 to 20 per cent of observed sediment.

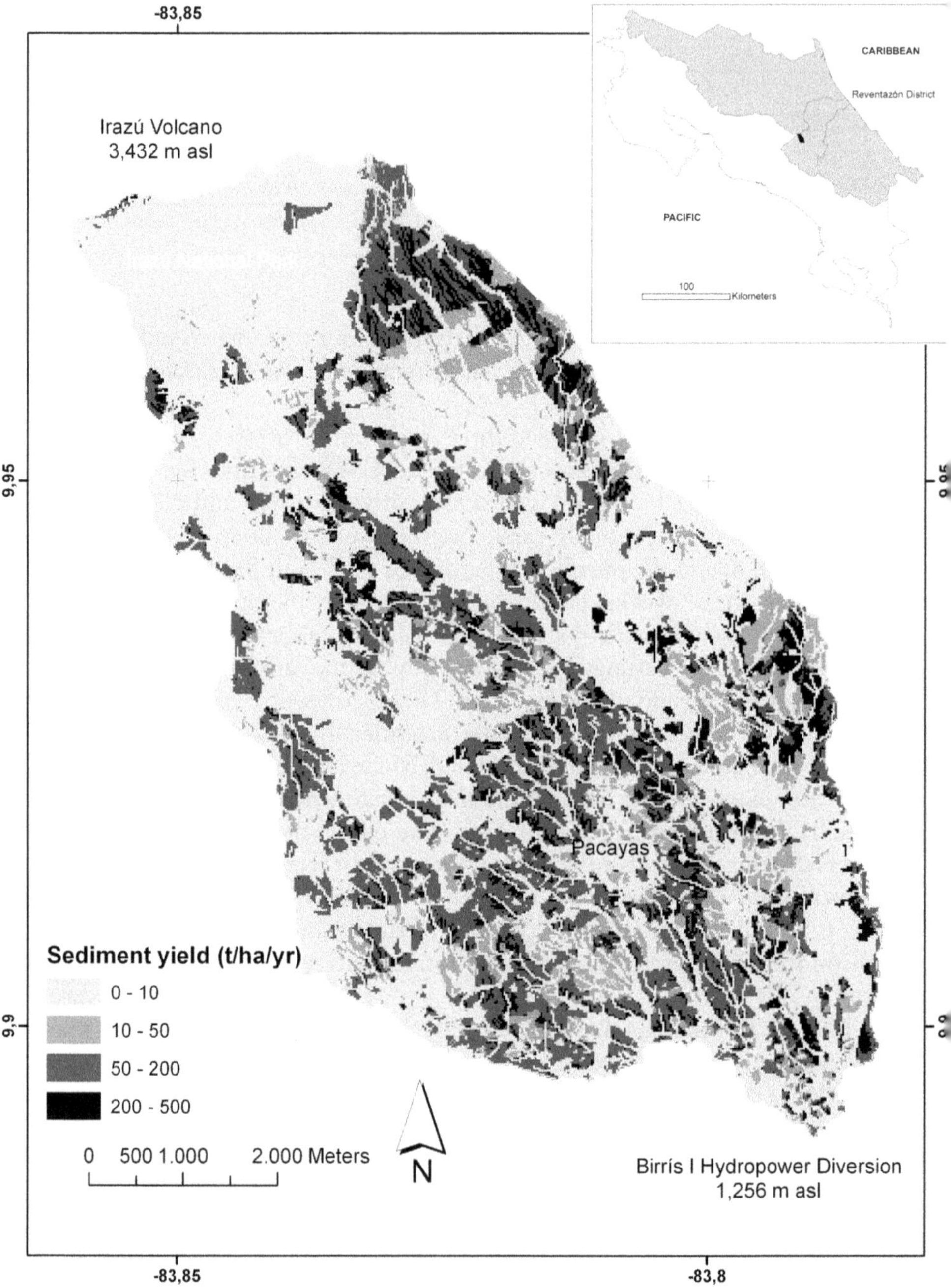

Figure 2.3 *Birrís River Basin sediment yield at pixel scale after RUSLE and CALSITE application*

Source: chapter authors

Land-use conflict methods

We now present an alternative approach to target areas: land-use conflict analysis. This is a qualitative approach implemented in GIS that is simpler and cost effective, and may also be combined with sediment yield modelling.

Sound watershed management requires a means of determining the optimum land use for a specific location, and a classification of land-use capability (Goujon et al, 1968). Land-use suitability assessment is based on a comparison of current use and land capability. Tosi (1972, cited by Lücke, 1986) defined land-use capability as 'the most intensive use that a unit of land can support without deterioration of its productive capability and without excluding lower intensity uses'. This definition, according to Lücke (1999), corresponds to a 'potential use of land'.

During recent decades various schemes have been proposed to classify land-use capability. Some of these are based on analysis of intrinsic land characteristics (e.g. slope, effective soil depth, weather restrictions and soil characteristics), while others classify land conditions in accordance with special use requirements, such as the optimum conditions for a certain crop or agricultural technology. High-value cash crops of Costa Rica, such as bananas, coffee and palm oil, have been classified in this manner.

Land capability classification systems may be qualitative or quantitative (Zinck, 1996). In Latin America, most classification systems are qualitative, derived from the US Department of Agriculture system (USDA, 1965) and based on work by Klingebiel and Montgomery (1961). However, there are alternative classifications, based on factors relevant to different geographical areas, such as Plath (1967) for Central America, Tosi (1972) for Colombia, Sheng (1971) for Jamaica, Van Melle (1984) for Puriscal and Caraigres in Costa Rica, and Delgado (1997) for tropical mountain areas, among others.

The *Evaluation of Current Status of the Territory* (Lücke, 1999) map classifies land use into three categories:

1 appropriate use;
2 underutilized (less intense use than productive capability); and
3 overused (more intensive use).

From the point of view of soil conservation, the underutilized category does not pose a problem because underutilization does not degrade productive land capability (Maldonado and Rodríguez, 1997). However, when taking into account the growing needs of society and the availability of productive land, underutilized land can represent a planning problem in some situations (MAG and FAO, 1996). On the other hand, overused land causes important environmental, social and economic problems involving often irreversible degradation processes. According to Maldonado and Rodríguez (1997), land-use capability should not be the sole criterion for land-use planning. It should also consider economic, social and environmental issues such as ecosystem conservation,

Box 2.2 Land-use capability and conflict in Costa Rica

Land-use capability is officially defined in Costa Rica by the Ministry of Environment and Energy and the Ministry of Agriculture and Livestock (Act No 232,114, dated 13 April 1994; MAG and MIRENEM, 1995). The official methodology is adapted from USDA (1965). Land is classified into eight classes, which have steadily increasing physical limitations, from class I to VIII. Classes I, II, III and IV are for agricultural use; class V is for pasture or woodland; class VI is for permanent vegetation (grass, perennial crops, production forests); class VII is for protection and forest management; and class VIII is preserved. Limitations can occur alone or in combination, and may be due to factors such as slope, soil texture or depth, the presence of stoniness and/or rockiness, drainage, and risks of flooding or climate. Several authors note that this formal system may have limitations for assessing certain land uses that occur in Costa Rica (Van Melle, 1984; Tosi, 1985), being initially designed under US conditions assuming:

- planting annual crops;
- high capital availability;
- labour shortages; and
- intense mechanization of land (USDA, 1965).

Due to these conditions, the US Department of Agriculture (USDA) system is very strict with allowable slopes for cultivation. Much of the land in Costa Rica under perennial crops such as coffee is classified as overuse, with consequent 'punishments' or limitations.

carbon sequestration, water resources protection and scenic beauty, among others, under which an underutilized land can generate more benefits than an agricultural use.

Case study: Land-use conflict analysis in the Birrís River Basin

Returning to the Birrís Basin example, we classified land-use capability with the official Costa Rican classification (MAG and MIRENEM, 1995) adapted from USDA (1965) (see Box 2.2). Conflict in land use occurs when actual land use diverges from the assigned capability class for a specific site. In this case study, we analysed the impact of each degree of conflict upon hydrological services provision following the matrix presented in Table 2.3. In this matrix, conflict values range from 1 to 7, with lower values corresponding to appropriate land-use situations and higher values to higher conflict situations (Marchamalo, 2004). When land use and capability matched, thereby promoting hydrological services production, we assigned an ordinal ranking of 'protective' (1) and 'appropriate' (2). Protective areas are those where any human activity different from natural vegetation affects hydrologic services, and include protected areas (PAs) and very steep and humid lands (class VIII). Adequate land use was assigned the category of 'non-conflict' (3). On the other hand, when land use and capability did not match and negatively affect the production of soil and water services, we assigned the land uses ascending conflict values of: 'low' (4), 'high' (5), 'very high' (6) and 'maximum' (7). Thus,

Table 2.3 *Land-use conflict matrix for land use and land capability for hydrological services provision*

Actual land use	*Land capability class*						
	I/II	*III*	*IV*	*VI*	*VII*	*VIII*	*PA*
Natural forest	2	2	2	2	2	1	1
Regenerating forest	2	2	3	4	5	5	5
Pastures	3	3	3	4	5	6	7
Annual crops	3	3	4	5	6	7	7
Perennial crops	2	3	3	3	4	5	7
Urban/constructed	2	3	4	5	6	7	7
Volcanic ash deposits	4	4	5	5	6	6	6

Note: PA = protected area.

land uses are assigned values ranging between 1 and 7 in order of increasing conflict values.

We processed the data in GIS software by combining land-use maps with land capability maps (ICE, 1999) in order to obtain a land-use conflict map for the basin. For each management unit, we calculated the mean and median land-use conflict, calculated from Table 2.3.

The connectivity approach

The third and final approach we present here for quantifying soil and water services is the connectivity approach. Hydrological connectivity is defined as 'the physical linkage of water or sediment flux within the landscape and the potential for a soil particle to move through the system' (Hooke, 2003). Within the context of soil conservation and water quality, connectivity means the extent to which parts of the landscape are connected through pathways carrying water and sediments. Water will naturally flow downhill and through channels. As it does so, it transports sediment (eroded soil) if there is sufficient energy to carry it. Pathways followed by water continue until the water runs out or it reaches the catchment end point. Sediment will continue to be carried until the water runs out or until all the sediment is deposited because of terrain topography. The longer the pathway, the further away water and sediment will be removed from the slopes, causing 'off-site' impacts downstream and decreasing on-site soil fertility (Hooke et al, 2007).

The main aim of conservation practices under this approach is to minimize connectivity, particularly with regard to sediment in order to reduce soil loss and 'off-site' impacts. Many modern land-use practices tend to increase connectivity, causing increased soil erosion problems. Research has shown that different vegetation types can serve as major storage areas, thereby reducing transmission of the sediment load downstream and minimizing connectivity (Hooke et al, 2007).

The connectivity approach concentrates efforts in priority areas, or 'hotspots', responsible for main hydrological activity in the catchment but at the farm scale. This is a geomorphic-based approach with an economic scope, as it allows for optimizing efforts and resources in priority areas. Hotspots within a farm are locations of concentrated erosion and can be identified as points of high soil loss (Hooke et al, 2007). The location of hotspot areas and pathways may be easily recognized by the landowner as they will often coincide with areas where continuous maintenance is required after any significant rainfall. Areas in the landscape where erosion hotspots and pathways tend to develop include:

- natural depressions/drainage areas and where there is a marked increase in gradient;
- terraces where bank failures frequently occur or terraces that generate significant runoff leading to the formation of rills/gullies;
- tracks and hardened areas where significant runoff takes place during a rainfall event;
- abandoned lands where terrace/bank structures are in a state of decay due to a lack of maintenance.

Generally, soil and water service hotspots can be managed on farm, in some cases by protecting them with vegetation. This can decrease the amount of water that will flow downhill and can even stop sediment transport. Other protection measures can be based on reducing the amount of runoff and sediment from upslope regions – for example, intercepting runoff and sediment in several places where local flux concentration can still be managed effectively and possibly at a lower cost. Most of the sediment is lost from these hotspots in the landscape; therefore, if they can be identified they can be targeted for action (Hooke et al, 2007).

A good reference for this experience was achieved by the RECONDES project (Conditions for Restoration and Mitigation of Desertified Areas Using Vegetation, VI Framework Programme, European Union). The connectivity approach was applied to the Carcavo catchment (Murcia in southeast Spain), located in a semi-arid climate. Documentation and guidelines are available at www.port.ac.uk/research/recondes/practicalguidelines.

Integration of Methodologies: Identification of Soil and Water Conservation Priority Areas in the Birrís River Basin, Costa Rica

In this final section we apply the presented tools to the Birrís River Basin case study. As was previously discussed, due to high sediment yield, a conflict exists between thriving agricultural and dairy production on the steep slopes of the Irazú Volcano and hydropower generation in the lower reaches of the watershed. The erosion caused by agricultural activity is one of the sources of the sediment that negatively affects hydropower generation.

The Birrís River Basin Land-Use Plan commissioned by JASEC (CATIE, 2003) strongly suggested that an initial focus on high-priority areas should be made to ensure cost effectiveness of investments in soil conservation practices. For this purpose, several methodologies were tested and their outputs evaluated by relevant stakeholders, such as land-use planners and those in charge of hydropower generation. Tested methodologies include land-use conflict analysis and sediment yield modelling efforts, both of which were described above.

The study was carried out at two spatial scales: 20m pixel and land management units. Land management units are plots of land whose use and management can be identified and individualized in aerial photo analyses. These units correspond to landowner decision units, and are probably more representative than the 20m pixel used in hydrologic modelling. The conflict value and the average distance of each management unit to the hydrological network (as defined in the official 1:25,000 scale map) was weighted and included as a variable in prioritization. It was assumed that pixels within 50m of the drainage network maintained the total conflict value as calculated. However, for points located greater than 50m from the network, the conflict value was weighted by a coefficient proportional to the distance to the network divided by 50. The average distance from management units to the hydrologic network was calculated as 51m in GIS. Therefore, we selected the threshold value of 50m for weighting land-use conflict.

Land-use capability analysis for the Birrís River Basin (see Figure 2.4) shows that 79 per cent of the basin is not suitable for agriculture or pasture and that 30 per cent of the land in the basin should be devoted to forest management or protection. A total of 12 per cent of the land is included in protected areas (Irazú Volcano National Park and Central Volcanic Range Forest Reserve) and 18 per cent of the land comprises areas classified as class VIII. Actual land use in the basin presents a contrasting situation. The most widespread use is pasture (35 per cent). Crops occupy 32 per cent and forest 28 per cent of the area. Using the land-use conflict matrix outlined in Table 2.3, we see that 28 per cent of the area has a high protective value, coinciding with the forest areas located primarily in protected areas or along rivers. On the

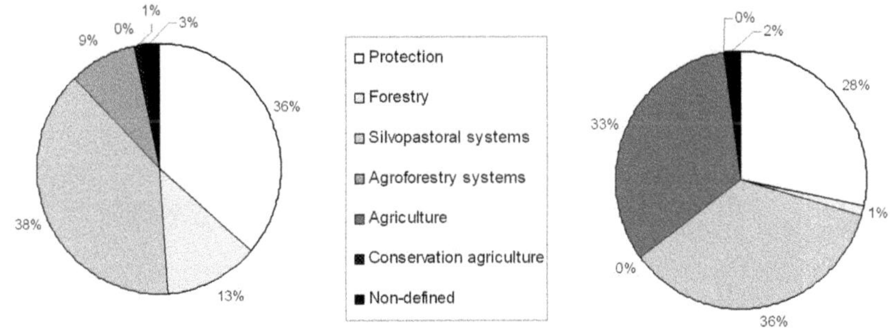

Figure 2.4 *Land-use suitability (*left*) and actual use (*right*) in the Birrís River Basin*

Source: chapter authors

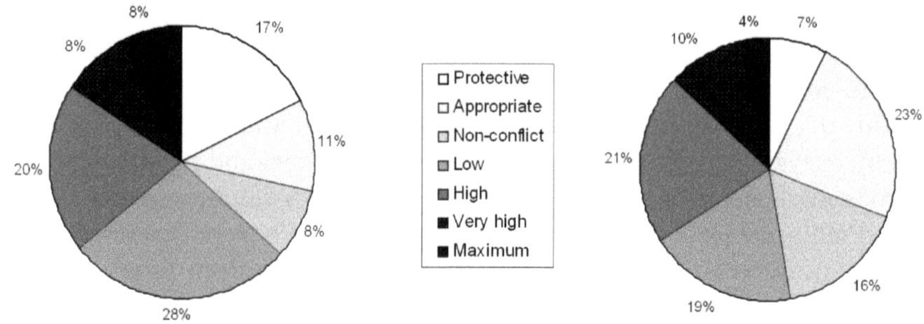

Figure 2.5 *Land-use conflict (*left*) and distance-weighted conflict (*right*) in the Birrís River Basin*

Source: chapter authors

other hand, 37 per cent of the basin presents high to maximum degrees of conflict (see Figure 2.5), coinciding with the 36 per cent of the basin presenting high potential erosion rates (see Figure 2.3).

Property boundaries and landownership data, if available, can be used to facilitate the identification of individual land management decision-making units rather than photo interpretation. Figure 2.6 shows land-use conflicts in a 50m hydrologic network buffer superimposed upon an ownership map. This permits property owners with lands in the most sensitive areas to be identified.

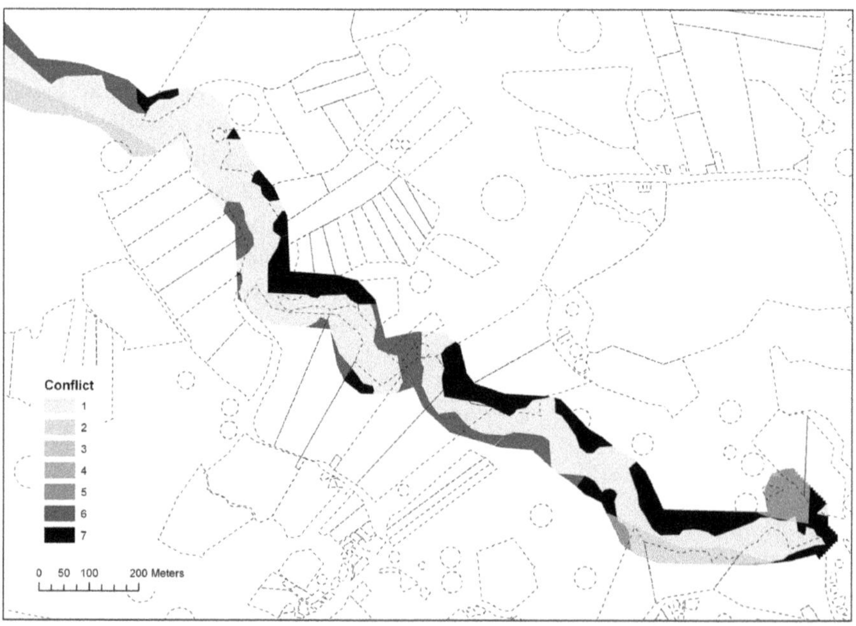

Figure 2.6 *Land-use conflict superimposed upon cadastral information in a riparian buffer*

Source: chapter authors

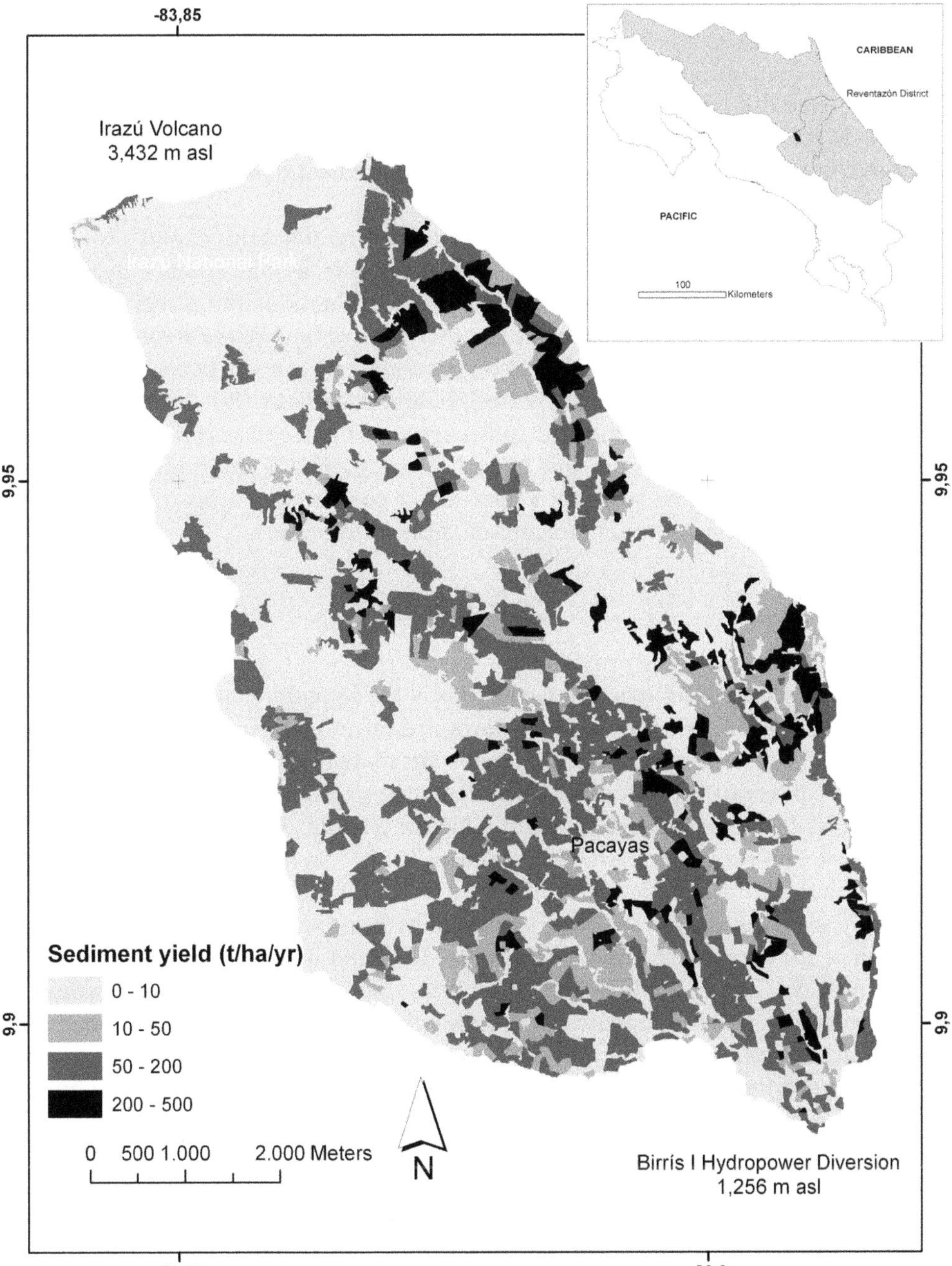

Figure 2.7 *Critical areas identified using sediment yield modelling at land unit scale*

Source: chapter authors

The RUSLE-CALSITE simulation discussed in the previous section was used to determine an estimate of the effective sediment yield in tonnes ha^{-1} y^{-1} for each management unit. This methodology includes climate, soil, topography cover and land-use data, as well as the feasibility of eroded sediment to reach the fluvial network (sediment transport simulation). Data can be analysed at pixel (see Figure 2.3) or land management unit scale, as shown in Figure 2.7.

The combined analysis of results enables targeting critical areas for the catchment. It should be noted that methods coincide, as high conflict areas are also high sediment-producing zones. The usefulness of the integration of methodologies is highlighted by the fact that we now have relevant information about target areas to guide restoration processes, such as the degree and kind of land-use conflict, potential erosion, sediment delivery ratio and sediment yield. The estimated or modelled values serve as a guide in sensitivity analysis and scenarios comparison, but must not be taken as a quantitative truth. Estimated erosion and sediment values can be compared to assess the effect of soil and water conservation measures in different scenarios.

Conclusions

Soil and water services provide important direct benefits to strategic sectors such as on-site conservation of fertile topsoil for agriculture and off-site provision of clean fresh water for hydropower dams. Quantification of soil and water services can be strategically important to increase the cost effectiveness of soil conservation programmes. There are many methods of identifying priority areas for intervention and quantifying the amount of service produced for downstream users (i.e. freshwater sediment loads). An important conclusion is that the appropriate methodology depends on the stakeholder's perspective. More specifically, from the point of view of hydropower producers, the estimation of the amount of sediments produced by land uses is a priority. For the farmer, the loss of fertility and its impacts upon production are important aspects that measure on-site benefits; however, additional models might be required to estimate soil fertility–productivity relationships.

Two different scales can explicitly be considered in modelling soil and water services provision for watershed management plans. The first one is a pixel-based raster map of sediment delivery as an output of the erosion models. This allows for easy recalculation in GIS raster format of sediment produced by different areas of the watershed, allowing for more precise targeting and, if data series are available over time, monitoring programme effectiveness. The second assessment scale utilizes information at the land management unit, which can support the implementation phase of a management plan by targeting specific parcels and owners that have the highest return on investment (i.e. highest 'soil water service/PES investment' ratio).

As discussed earlier, quantification and monitoring are important needs for generalization of reliable hydrological environmental services markets. A stronger effort must be made to fund medium- and long-term monitoring programmes of soil and water services. Each PES project can be considered as an experiment due to the complexity of hydro-social systems, which implies high uncertainty. Each project is an opportunity to enlighten future actions. We need effective project evaluations and adaptive management schemes. Adaptive management is the process whereby management is initiated, evaluated and refined (Holling, 1978; Walters, 1986). Adaptive management is incremental and uses information from monitoring to continually evaluate and modify management practices (Kreshner, 1997).

Soil erosion modelling is evolving, and it is necessary to combine available technology and knowledge, to integrate soil erosion–sedimentation research, and to connect approaches and economic evaluations of on-site and off-site impacts. This will lead to an integrated biophysical and economic evaluation of hydrologic environmental services. An increasing challenge for soil conservation programmes is represented by the uncertainty in precipitation erosivity trends, especially considering climate change (Aguilar et al, 2005). This calls for more attention to the role and needs for monitoring systems that are required for adaptive management.

Acknowledgements

This chapter was prepared with support from the Tropical Forests and Climate Change Adaptation (TroFCCA) research programme, executed by the Tropical Agricultural Research and Higher Education Center in Costa Rica (CATIE) and the Center for International Forestry Research (CIFOR) and funded by the European Commission under contract EuropeAid/ENV/2004-81719. The contents of this chapter are the sole responsibility of the authors and can under no circumstances be regarded as reflecting the position of the European Union. The Birrís Basin study was also supported by the CATIE project: Contribution to the Payment for Environmental Services Policy for the Hydropower sector in Costa Rica, supported by the Danish International Development Agency–International Union for Conservation of Nature (DANIDA–IUCN) Fund for Incidence in Environmental Policies in Mesoamerica.

References

Aguilar, E., Peterson, T. C., Ramírez Obando, P., Frutos, R., Retana, J. A., Solera, M., Soley, J., González García, I., Araujo, R. M., Rosa Santos, A., Valle, V. E., Brunet, M., Aguilar, L., Álvarez, L., Bautista, M., Castañón, C., Herrera, L., Ruano, E., Sinay, J. J., Sanchez, E., Hernandez Oviedo, G. I., Obed, F., Salgado, J. E., Vazquez, J. L., Baca, M., Gutierrez, M., Centella, C., Espinosa, J., Martínez, D., Olmedo, B., Ojeda Espinoza, C. E., Nuñez, R., Haylock, M., Benavides, H. and Mayorga, R. (2005) 'Changes in precipitation and temperature extremes in Central America and northern South America, 1961–2003', *Journal of Geophysical Research*, vol 110, pD23107

Allen, R., Pereira, L. S., Raes, D. and Smith, M. (1998) *Crop Evapotranspiration: Guidelines for Computing Crop Water Requirements*, FAO Irrigation and drainage paper 56, FAO, Rome, Italy

Arnold, J. G., Williams, J. R., Srinivasan, R., King, K. W. and Griggs, R. H. (1995) *SWAT: Soil and Water Assessment Tool, Draft Users' Manual*, USDA–ARS, Temple, TX

Beasley, D. B., Huggins, L. F. and Monke, E. J. (1980) 'ANSWERS: A model for watershed planning', *Transactions of the American Society of Agricultural Engineers*, vol 23–24, pp938–944

Bradbury, P. (1995) *CALSITE Version 3.1:User Manual*, HR Wallingford Limited, UK

CATIE (2003) *Plan de acción 2004–2013 para el manejo de las subcuencas tributarias del sistema hidroeléctrico Birrís*, Turrialba, Costa Rica

DeClerck, F. A., Chazdon, R., Holl, K. D., Milder, J. C., Finegan, B., Martinez-Salinas, A., Imbach, P., Canet, L. and Ramos, Z. (2010) 'Biodiversity conservation in human-modified landscapes of Mesoamerica: Past, present and future', *Biological Conservation*, vol 143, pp2301–2313

De Graaff, J. (2000) 'Downstream effects of land degradation and soil and water conservation', Background paper no 5, in FAO (ed) *Land–Water Linkages in Rural Watersheds*, Electronic workshop, Rome, Italy

De Roo, A. P. J., Wesseling, C. G., Cremers, N. H. D. T., Offermans, R. J. E., Ritsema, C. J. and Oostindie, K. (1994) 'LISEM: A new physically-based hydrological and soil erosion model in a GIS-environment – theory and implementation', in *Variability in Stream Erosion and Sediment Transport* (translated by L. J. Olive, R. J. Loughran and J. A. Kesby), Proceedings of the Canberra Symposium, December 1994, IAHS Publication no 224, pp439–448

Delgado, F. (1997) *Sistema para la evaluación y clasificación de tierras agrícolas y prioridades de conservación de suelos en áreas montañosas tropicales: Un enfoque metodológico*, Serie Suelos y Clima SC-73, CIDIAT, Mérida, Venezuela

Falkenmark, M. and Folke, C. (2002) 'The ethics of socio-ecohydrological catchment management: Towards hydrosolidarity', *Hydrology and Earth System Sciences*, vol 6, no 1, pp1–9

Fisher, B., Turner, R. K. and Morlina, P. (2009) 'Defining and classifying ecosystem services for decision making', *Ecological Economics*, vol 68, pp643–653

Flanagan, D. C. and Nearing, M. A. (1995) *USDA Water Erosion Prediction Project: Hillslope Profile and Watershed Model Documentation*, NSERL Report 10, USDA–ARS National Soil Erosion Research Laboratory, West Lafayette, IN

Foster, G. R., Lane, L. J., Nowlin, J. D., Laflen, J. M. and Young, R. A. (1980) 'A model to estimate sediment yield from field-sized areas: Development of model', in W. G. Knisel (ed) *CREAMS: A Field Scale Model for Chemicals, Runoff, and Erosion from Agricultural Management Systems*, US Department of Agriculture, Science and Education Administration, Conservation Report no 26, pp36–64

Goldman, R. L., Thompson, B. H. and Daily, G. C. (2007) 'Institutional incentives for managing the landscape: Including cooperation for the production of ecosystem services', *Ecological Economics*, vol 65, pp333–343

Gómez-Delgado, F. (2002) *Evaluación de la erosión potencial y producción de sedimentos en tres cuencas de Costa Rica Trabajo de Graduación para obtener el grado de Licenciado en Ingeniería Civil*, UCR, San José, Costa Rica

Goujon, P., Bailly, C., de Vergnette, J., Benoit de Cognac, G., Roche, P., Velly, J. and Celton, J. (1968) 'Conservation des sols en Afrique et a Madagascar', *Bois et Forêts des Tropiques*, no 118–121, pp1–54

Holling, C. S. (1978) *Adaptive Environmental Assessment and Management*, Wiley, New York, NY

Hooke, J. (2003) 'Coarse sediment connectivity in river channel systems: A conceptual framework and methodology', *Geomorphology*, vol 56, no 1–2, pp79–94

Hooke, J., Sandercock, P., Marchamalo, M., van Wesemael, B., Meerkerk, A., Torri, D., Borselli, L., Salvador, M. P., Yáñez, M., Castillo, V., Gónzález, G., Navarro, J. A., Querejeta, J. I., Boix-Fayos, C., Cammeraat, E., Lesschen, J. P., Poesen, J. and de Baets, S. (2007) *Combating Land Degradation by Minimal Intervention: The Connectivity Reduction Approach*, University of Portsmouth, UK

ICE (Instituto Costarricense de Electricidad) (1999) *Plan de Manejo de la Cuenca del Río Reventazón*, San José, Costa Rica

INEC (Instituto Nacional de Estadística y Censos) (2001) *IX Censo Nacional de Población y V de Vivienda del 2000: Resultados Generales*, Instituto Nacional de Estadística y Censos, San José, Costa Rica

Jaubert, M. (2001) *Manejo de la cuenca del río Reventazón*, Foro Cuencas Hidrográficas de Centroamérica, 27–28 September 2001, Colegio de Ingenieros Tecnólogos, San José, Costa Rica

Kinsel, W. G. (ed) (1980) *CREAMS: A Field Scale Model for Chemicals, Runoff, and Erosion From Agricultural Management Systems*, US Department of Agriculture, Conservation Report no 26

Klingebiel, A. A. and Montgomery, P. H. (1961) *Land Capability Classification*, Agriculture Handbook 210, USDA, Soil Conservation Service, Washington, DC

Kreshner, J. L. (1997) 'Monitoring and adaptive management', in J. E. Williams, C. A. Wood and M. P. Dombeck (eds) *Watershed Restoration: Principles and Practices*, American Fisheries Society, Bethesda, MD

Lal, R. et al (1994) *Soil Erosion Research Methods*, Soil and Water Conservation Society and St Lucie Press, FL

Lianes, E., Marchamalo, M. and Roldán, M. (2009) 'Evaluación del factor C de la RUSLE para el manejo de coberturas vegetales en el control de la erosión en la cuenca del Río Birrís, Costa Rica', *Agronomía Costarricense*, vol 33, no 2, pp217–235

Lim, K. J., Sagong, M., Engel, B. A., Tang, Z., Choi, J. and Kim, K.-S. (2005) 'GIS-based sediment assessment tool', *Catena*, vol 64, pp61–80

Lücke, O. (1986) *Consideraciones básicas sobre la aplicación de metodologías de análisis en la planificación del uso de la tierra y la toma de decisiones*, Programa de Manejo de Cuencas Hidrográficas, CATIE, Turrialba, Costa Rica

Lücke, O. (1999) 'Base conceptual y metodológica para los escenarios de ordenamiento territorial', in A. Rodríguez (ed) *Escenarios de uso del territorio para Costa Rica en el año 2025*, Ministerio de Planificación Nacional y Política Económica, Costa Rica

MAG and FAO (Ministerio de Agricultura y Ganadería and Food and Agriculture Organization) (1996) *Agricultura conservacionista: Un enfoque para producir y conservar*, Ministerio de Agricultura y Ganadería, San José, Costa Rica

MAG and MIRENEM (Ministerio de Agricultura y Ganadería and Ministerio del Ambiente y Energía) (1995) *Metodología para la determinación de la capacidad de uso de las tierras en Costa Rica*, MAG and MIRENEM, San José, Costa Rica

Magrin, G., Gay García, C., Cruz Choque, D., Giménez, J. C., Moreno, A. R., Nagy, G. J., Nobre, C. and Villamizar, A. (2007) 'Latin America', in M. L. Parry, O. F. Canziani, J. P. Palutikof, P. J. van der Linden and C. E. Hanson (eds) *Climate Change 2007: Impacts, Adaptation and Vulnerability. Contribution of Working*

Group II to the Fourth Assessment Report of the Intergovernmental Panel on Climate Change, Cambridge University Press, Cambridge, UK, pp581–615

Maldonado, T. and Rodríguez, C. (1997) *Estudio, análisis y cartografía de la capacidad de uso de la tierra en las clases forestales, Costa Rica*, Fundación Neotrópica, San José, Costa Rica

Marchamalo, M. (2004) *Ordenación del territorio para la producción de servicios ambientales hídricos: Aplicación a la cuenca del río Birrís (Costa Rica)*, PhD thesis, Universidad Politécnica de Madrid, Spain

Marchamalo, M. and Romero, C. (2007) 'Participatory decision-making in land use planning: An application in Costa Rica', *Ecological Economics*, vol 63, pp740–748

Morgan, R. P. C. (2001) 'A simple approach to soil loss prediction: A revised Morgan-Morgan-Finney model', *Catena*, vol 44, pp305–322

Morgan, R. P. C., Quinton, J. N., Smith, R. E., Govers, G., Poesen, J. W. A., Auerswald, K., Chisci, G., Torri, D. and Styczen, M. E. (1998) 'The European Soil Erosion Model (EUROSEM): A dynamic approach for predicting sediment transport from fields and small catchments', *Earth Surface Processes and Landforms*, vol 23, pp527–544

Müller, S. (1997) *Evaluating the Sustainability of Agriculture: The Case of the Reventado River Watershed in Costa Rica*, European University Studies: Series 5: Economics and Management, vol 2194, Peter Lang, Frankfurt, Germany

Plath, C. V. (1967) *Productive Capacity of Agricultural Land in Central America*, FAO, Rome, Italy

Renard, K. G., Foster, G. R., Weesies, G. A., McCool, D. K. and Yoder, D. C. (1997) *Predicting Soil Erosion by Water: A Guide to Conservation Planning with the Revised Soil Loss Equation (RUSLE)*, Agricultural Handbook 703, US Government Printing Office, Washington, DC

Rodríguez, C. (2001) *Informe sobre sedimentos: Proyecto Hidroeléctrico Reventazón*, Área de Hidrología, Instituto Costarricense de Electricidad (ICE), San José, Costa Rica

Saavedra, C. (2005) *Estimating Spatial Patterns of Soil Erosion and Deposition in the Andean Region Using Geo-Information Techniques: A Case Study in Cochabamba, Bolivia*, PhD thesis, Universidad de Wageningen, Holanda, The Netherlands

Saborío, J. (2000) *Estudio de erosión para la República de Guatemala*, Centro Agronómico Tropical de Investigación y Enseñanza, Turrialba, Costa Rica

Sanchez-Azofeifa, G. A., Harriss, R. C., Storrier, A. L. and DeCamino-Beck, T. (2002) 'Water resources and regional land cover change in Costa Rica: Impacts and economics', *Water Resources Development*, vol 18, no 3, pp409–424

Sheng, T. C. (1971) *Proyecto de clasificación de la tierra orientada a su tratamiento para tierras marginales montañosas del trópico húmedo*, FAO, Kingston, Jamaica

Swisher, M. E. and Todd-Bockarie, A. (eds) (1996) *Calidad del agua*, Florida Cooperative Extension Service, University of Florida, FL, pp1–7

Tosi, J. A. (1972) *Una clasificación y metodología para la determinación y levantamiento de mapas de la capacidad de uso mayor de la tierra en Colombia*, Serie en Facsímil no 7, Centro Científico Tropical, San José, Costa Rica

Tosi, J. A. (1985) *Sistema para la determinación de la capacidad de uso de las tierras de Costa Rica*, Centro Científico Tropical, San José, Costa Rica

UNCBD (United Nations Convention on Biological Diversity) (2009) *Connecting Biodiversity and Climate Change Mitigation and Adaptation: Report of the Second*

Ad Hoc Technical Expert Group on Biodiversity and Climate Change, Secretariat of the Convention on Biological Diversity, Montreal, Technical Series No 41

USDA (US Department of Agriculture) (1965) *Clasificación por capacidad de uso de las tierras* (translated by F. J. Valencia), fourth edition, Centro Regional de Ayuda Técnica, México

Van Melle, G. (1984) *Estudio sobre la capacidad de uso de la tierra en dos areas de las subregiones Puriscal y Caraigres, Costa Rica*, ASCONA, Cooperación Técnica Holandesa, San José, Costa Rica

Vignola R., Locatelli B., Martinez, C. and Imbach, P. (2009) 'Ecosystem-based adaptation to climate change: What role for policy-makers, society and scientists?', *Mitigation and Adaptation of Strategies for Global Change*, vol 14, pp691–696

Vignola, R., Otarola, M., Marchamalo, M. and Echeverria, J. (2010) 'Land use scenarios for provision of ecosystem services to hydropower production: On-site and off-site benefits of different soil conservation measures in the Birrís watershed, Costa Rica', in UNDP (ed) *Regional Report on Biodiversity and Ecosystems: Why Are They Important for Sustainable Development and Equity in Latin America and the Caribbean*, 2010 UNCBD-COP 10, Japan

Walters, C. J. (1986) Adaptive Management of Renewable Resources, MacMillan, New York, NY

Wischmeier, W. H. and Smith, D. D. (1978) *Predicting Rainfall Erosion Losses: A Guide to Conservation Planning*, US Department of Agriculture Handbook no 537, Washington, DC

Zinck, J. A. (1996) 'La información edáfica en la planificación del uso de las tierras y el ordenamiento territorial', in J. Aguilar, A. Martínez and A. Roca (eds) *Evaluación y manejo de suelos, Junta de Andalucía*, Sociedad Española de Ciencia del Suelo y Universidad de Granada, Spain

3

Measuring Biodiversity

Fabrice A. J. DeClerck and Alejandra Martínez Salinas

Introduction

Recent history has seen the rampant loss of biodiversity that continues largely unchecked to this day. Indeed, even in the geological history of the Earth, such rates of extinction have only been seen on six previous occasions, in each case in relation to tremendous natural disasters such as volcanic eruptions or catastrophic collisions with asteroids. Harvard ecologist E. O. Wilson was once famously asked, of all the calamities which could befall the Earth, which is the greatest? After a brief reflection, Dr Wilson replied (1994):

> *The worst thing that can happen,* will *happen, is not energy depletion, economic collapse, limited nuclear war, or conquest by a totalitarian government. As terrible as these catastrophes would be for us, they can be repaired within a few generations. The one process on-going in the 1980s that will take millions of years to correct is the loss of genetic and species diversity by the destruction of natural habitats. This is the folly our descendents are least likely to forgive us.*

Wilson's answer is important in at least two ways. First, Wilson recognizes the dynamic nature of biodiversity – that is, the loss of a biodiversity is much greater than the loss of any single species, but also represents an evolutionary loss. In other words, the loss of a species includes the loss of the evolutionary history that shaped individual species – it is this history which is the value of biodiversity, and the collective library of information upon which ecosystem services are fundamentally based. As a basic example, let us use aspirin, a common medicine that is familiar to most and used for a variety of ailments. While it is tempting to think that aspirin was formulated by pharmaceutical companies, the truth of the matter is that salicylic acid was formulated by common riparian shrubs of the genus *Salix*, more commonly known as willows. It is also important to recognize that willows did not develop this important compound overnight, nor did they develop the compound as a means of demonstrating good will towards humanity. Rather, salicylic acid was developed on evolutionary timescales as a sort of arms race between plants and plant predators. Anyone who has bitten into an aspirin immediately recognizes the bitter taste imparted by the drug which is primarily meant as a defence mechanism of willows against herbivory.

The main point here is that biodiversity is more than the accumulated richness and diversity of species that inhabit the Earth; rather, biodiversity includes the current interactions between communities of species, as well as the evolutionary history of these interactions, as is demonstrated with the example of willows, their predators and aspirin. The loss of biodiversity represents more than the loss of charismatic species: it represents the loss of this evolutionary history.

In the history of the conservation movement, two main schools of thought have emerged regarding the conservation of biodiversity. The first school argues that biodiversity should be conserved for its own sake – that is, the species that inhabit and share the planet with us have an inherent right to exist, and we as humans have an inherent obligation to watch over and protect the Earth's biodiversity. The origins of the movement are as old as human culture itself; indeed, the biblical references in Genesis first give Adam (the father of humanity in the Judaeo-Christian and Islamic faiths) the right to name each species and to watch over them. Even more importantly, during the Great Flood, God requires Noah to build an ark sufficiently large to protect and conserve a pair of each species that inhabits the Earth (see Levin, 1999, for an interesting discussion of this idea). Some might argue that Noah was the world's first conservation biologist!

However, the philosophy of conservation for conservation's sake has had difficulties during recent years in terms of securing sufficient support to protect endangered species. More recently, the conservation movement has shifted from the single species approach (where conservation efforts target a specific species for conservation – for example, the giant panda) or even an approach that targets protected areas, to one that is much more utilitarian in its approach: ecosystem services. The notion of ecosystem services originated during the 1990s (Daily, 1997) with the explicit aim of demonstrating society's dependence on the functioning ecosystems. The notion came from two fronts: first a series of studies that demonstrated the often positive relationship between biodiversity and ecosystem function, particularly primary productivity, pest control, stability, resistance and resilience (Loreau et al, 2002; Naeem,

2002; Naeem et al, 2009); and, second, a need to communicate the value of biodiversity to the general public in terms distinct from conservation for conservation's sake (Kareiva and Marvier, 2007).

The loss of biodiversity at global levels, but particularly in the tropics where the loss has been the greatest, has initiated renewed efforts to conserve biodiversity both in and out of reserve systems (Harvey et al, 2006, 2008; DeClerck et al, 2010). The most well known of these conservation efforts tied to payment for ecosystem services (PES) include the certification of coffee farms (and an increasingly large array of other land uses) for their conservation values – for example, Starbucks CAFE (Coffee and Farmer Equity) practices, Rainforest Alliance certification and the Smithsonian's Bird Friendly certification. However, the biggest challenge facing these certification schemes lies in understanding how interventions affect the conservation value of these agroecosystems. Biodiversity is a particularly slippery concept to quantify – biological communities abhor vacuums; when species are lost, new ones are quick to replace them, making species richness a particularly difficult and inadequate measure for quantifying biodiversity despite pervasive use of the measure. On the other hand, species-specific measures are also fraught with difficulties. Interventions that may positively impact upon one species may have a negative impact upon another. Species-specific monitoring, particularly of rare species with high conservation value, is particularly expensive, demands a great amount of effort to acquire enough data that can be considered relevant, and also calls for site-specific interventions and measures that cannot be universally applied. In this chapter, we aim to tackle these issues by exploring diverse biodiversity measures and the benefits and pitfalls of each. We make recommendations on combined measures for a more holistic quantification of the conservation value of agricultural landscapes in Mesoamerica, while considering that it is always important to keep in mind what the goal of the project is and that biodiversity can be valued in any number of ways. As a result, we will focus specifically on payment targeting the conservation of biodiversity – that is, payments that aim to alter land uses to protect species of conservation concern or species that are threatened with loss of habitat and extinction. In the Mesoamerican context, this mostly refers to forest-dependent species from multiple taxonomic groups. In this context, one of the most useful tools for quantifying biodiversity is based on comparing the composition of the land use receiving a payment and a reference ecosystem, typically an adjacent protected area or forest area.

What Is Biodiversity?

What is biodiversity and how does one quantify it? The first thing that comes to people's minds when asked this question is that biodiversity is the conglomerate of species that inhabit a specific unit area. The measure that is most commonly used is the number of biological species found in a specific area. In terms of payments for biodiversity conservation, this is frequently translated as how many species are found in a coffee farm (Komar, 2006; Philpott et al, 2008), in a cacao plantation (Harvey and Villalobos, 2007) or in different components of silvopastoral systems, such as live fences, trees scattered in a pasture or riparian buffers (Harvey et al, 2006; Milder et al, 2010).

From a strictly biological point of view, a species is popularly defined as a 'group of organisms capable of interbreeding and producing fertile offspring' and biodiversity is defined as 'the variation of life forms within a given ecosystem, biome, or on the entire Earth' (Wikipedia). In reality, this definition is overly simplistic and excludes the multiple dimensions of biodiversity such as genetic, vertical, horizontal, functional and temporal diversity, all of which have a significant effect on understanding and quantifying the conservation value of agroecosystems, as well as the contribution of biodiversity to the provisioning of ecosystem services within a particular system (see Chapter 10 for a discussion on biodiversity and ecosystem function). In this chapter, however, we focus primarily on the conservation value of biodiversity and how this conservation value might be quantified. This assumes that biodiversity has an inherent value, and that conservation of biodiversity is important for conservation's sake. There is increasing evidence that biodiversity has tremendous value in terms of its capacity to provide ecosystem services, which we also recognize, but which is covered in Chapter 10.

In order to demonstrate the difficulties in measuring biodiversity as an ecosystem service, we use data from the Bird Monitoring Project (BMP) based on the Tropical Agricultural Research and Higher Education Center (CATIE) campus in Turrialba, Costa Rica. For nearly three years, ornithologists and volunteers working with the BMP have captured, measured, banded and released birds in six different land-use areas. These land uses include:

- sugar cane (cane);
- a live fence in a pasture (pasture);
- coffee with poró (*Erythrina poeppigiana*), a leguminous nitrogen-fixing tree (coffee);
- a multi-strata coffee system with poró and laurel (*Cordia alliodora*) which resembles the structure that the Smithsonian would certify (MS coffee);
- a multi-strata cacao (*Theobroma cacao*) system with laurel (*C. alliodora*) and banana (Musaceae) plants (cacao); and
- a well-preserved mature secondary forest which serves as our reference ecosystem (forest).

According to the reference ecosystem, in theory we should be able to evaluate the conservation value of the agricultural land uses by comparing them to this forest – the more our agricultural system approaches the forest reference, the greater its conservation value.

Three days per week, volunteers place ten mist nets (12m long × 2.5m tall, 30mm mesh) in each of the land uses, opening the nets around 5.00 am and closing them around 9.00 am when the sun gets too hot and bird activity

Table 3.1 *Species richness, Simpson and Shannon values for CATIE's bird monitoring sites from 2008 to 2009*

Habitat	*Species richness*	*Shannon (H')*	$e^{H'}$	*Simpson (λ)*	*1/λ*
Forest	32	2.57	13.06	0.14	7.14
MS coffee	82	3.48	32.45	0.06	16.67
Cacao	65	3.07	21.54	0.12	8.33
Coffee	64	3.24	25.53	0.07	14.28
Pasture	61	3.25	25.79	0.06	16.67
Cane	40	2.85	17.28	0.10	10

Notes: Forest = preserved mature secondary forest; MS coffee = multi-strata coffee system with poró (*Erythrina poeppigiana*) and laurel (*Cordia alliodora*); cacao = multi-strata cacao (*Theobroma cacao*) system with laurel (*C. alliodora*) and banana (Musaceae) plants; coffee = coffee with poró (*E. poeppigiana*), a leguminous nitrogen-fixing tree; pasture = live fences in pastures; cane = sugar cane.

declines. The nets are checked every 40 minutes, and all birds that are captured are measured, weighed and given a small leg band with a unique number so that we can identify them if we recapture them at a later date. This data permits us to compare numerous indicators of biodiversity that we will discuss below.

We begin with a brief description of the land uses. In essence, the purpose of the monitoring is to evaluate the conservation value of each land use, which in turn can be used to determine how much a farmer should be paid for conserving biodiversity through interventions, such as agroforestry systems. The simplest system we monitor is sugar cane, comprised of a perennial monoculture that is harvested annually. During the sugar cane's production cycle, the plant can reach heights of 2.5m to 3m, with approximately 1m between rows. Once a year, usually in May, the cane is burned and new plants sprout from the roots (not unlike mowing one's lawn, without the fire, of course). The live fence site is the second land use monitored and consists of a 300m long linear row of trees dominated by a species of the genus *Inga*. These trees form the boundary between two pastures where the trees serve as living posts to which barbed wire is affixed. The pasture is dominated by an improved grass and it was irregularly grazed until a year ago when it was abandoned. The third is coffee with poró (*Erythrina poeppigiana*), which represents the traditional coffee management system in Costa Rica. Coffee plants are interspersed by poró trees whose main trunk is pruned to 2.5m twice a year. Poró is very fast growing, and canopy closure in this plot can vary between 0 to 80 per cent depending on the time since the last pruning. The multi-strata coffee site is a coffee plot that has been abandoned for the past seven years or so. It is dominated by coffee in the understorey, formerly pruned poró (*E. poeppigiana*) trees that now have statures of 12m, and a canopy dominated by laurel (*Cordia alliodora*) with heights of 20m. The cacao (*Theobroma cacao*) site is similar to the abandoned coffee sites with a multi-strata system with cacao plants (2m to 3m) in the understorey, interspersed with banana, and a tree canopy of variable height (10m to 25m). Finally, the forest system serves as our reference system and consists of mature forest with a well-developed understorey and both a complete mid- and upper-level canopy.

So, which of these land uses has the greatest biodiversity value? In the context of payment for ecosystem services (PES) schemes, the question becomes which of these six land uses merits the greatest payment? PES for biodiversity pays for changes in management practices that protect the greatest number of species, or which have the greatest conservation value. In the case of Costa Rican forestry laws, payments are made to landowners as an incentive to maintain existing forest cover, recognizing the lost opportunity that comes with protection. In the case of coffee and agricultural certification, payments are made to encourage farmers to adopt land-use practices that benefit biodiversity, although they may negatively impact upon the economic productivity of the farm – for example, increasing the tree density in a coffee system.

Species Richness

Returning to CATIE's bird monitoring stations, which of the six land uses should receive the greatest payment? The challenge is to determine which of these land uses is making the greatest contribution to the conservation of avian biodiversity. The simplest measures of biodiversity are species richness and species abundance. Species richness, also known as species density, is simply the number of species found in a specific land use or area. Species abundance, in contrast, is total number of individual birds, irrespective of species, found in a specific land use or area. Figure 3.1 presents the bird richness and abundance values from the BMP. There are a couple of points to note here. First, most readers will be surprised to find that the forest has the lowest species richness and abundance of any of the land uses monitored, whereas the pasture has the greatest species richness and is second in abundance values when we consider the mean number of species and individuals captured in 100 hours of mist-netting (one year of data during 2009). However, if we keep collecting data and we use species accumulation curves (years 2008 and 2009), which account for differential sampling efforts, the greatest species richness and abundance is found in our multi-strata coffee agroforestry system ($r = 82$; $n = 665$), which is second to the forest in terms of structural complexity (vertical structure) (see Figure 3.2). Note, however, that there is no significant difference ($p < 0.05$) between the pasture site and our multi-strata cacao agroforestry systems and coffee in association with poró. Finally, the sites with lowest values considering species richness and abundance were sugar cane and our forest plot, which did not show any significant difference ($p < 0.05$) between each other or with the rest of our sites. If we were to use these richness and abundance values as indicators of conservation value, live fences in pasture would have greater value than our forest plot and would be the target of conservation efforts.

What causes these counter-intuitive trends? Certainly, we would assume that conservation efforts should target forest conservation rather than pasture systems. First, it is important to understand the method by which biodiversity is quantified. In this case, the bird community was quantified using mist nets. Mist nets are quite good at capturing and quantifying birds that inhabit the lower strata of different land uses (< 3m), which works fine for most agricultural systems where the majority of the plant biomass is located below 5m in height. In the forest, however, where most of the biomass is located above the nets, this presents a significant bias. A different method of evaluating species richness and/or abundance, such as point counts, might have given a less biased indication. The point count method allows you to account for presence/absence of species in a given habitat. During a point count, the observer stands in a fixed point and registers all birds that can be identified by sight or sound during a set time period (typically ten minutes). The observer can survey as many points as possible during a day as long as they are at least 150m to 200m apart to prevent overlapping of individuals and/or overestimating. The disadvantage of point counts, however, is that they are heavily dependent on the availability of a qualified ornithologist capable of recognizing species by sight or by sound (Ralph et al, 1996). The same bias against forests persists since visibility is often more limited in a forest than in agricultural land uses. Nonetheless, several studies demonstrate that agricultural land uses can have greater richness and abundance than forest systems, in part because of their disturbed nature that frees resources for avian biodiversity and because the combination of different land uses found in agricultural

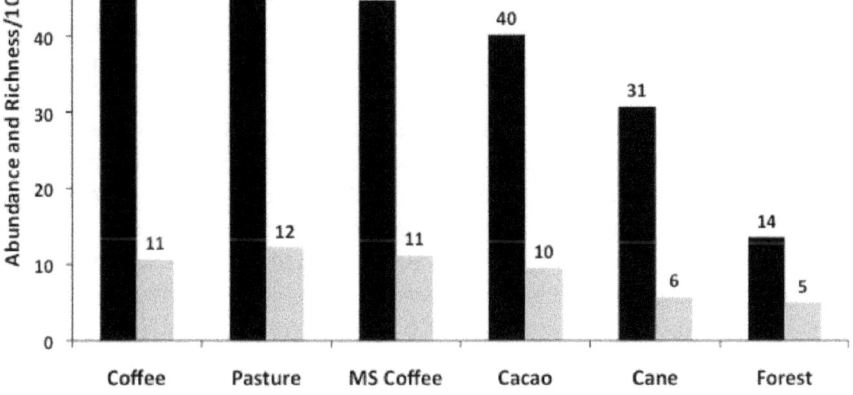

Figure 3.1 *Species richness and abundance by land use in the CATIE Bird Monitoring Project (BMP)*

Note: The values have been normalized by calculating the richness and abundance per 100 hours of netting.
Source: chapter author

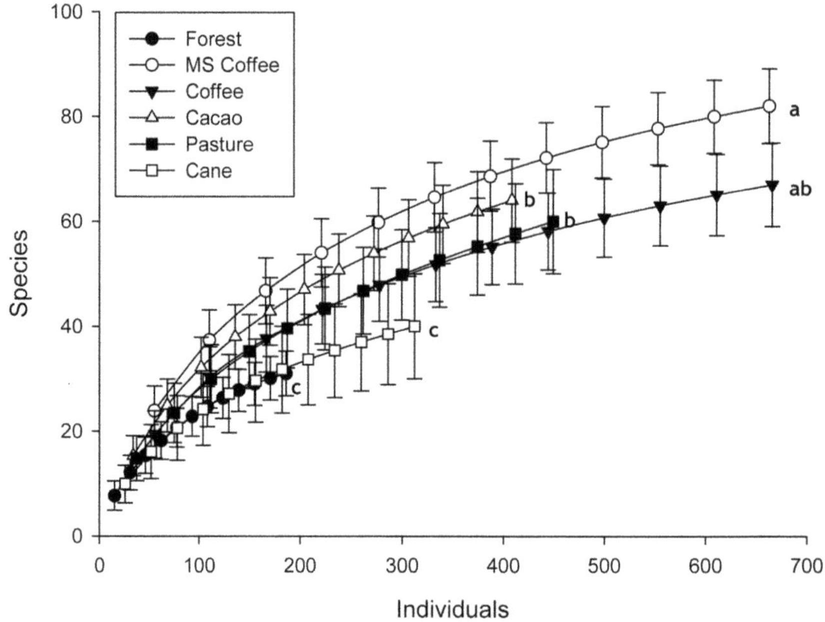

Figure 3.2 *Accumulation curves showing species richness and abundance of birds for BMP sites for 2008 to 2009 data*

Notes: Land-use codes are the same as those used in Table 3.1. Different letters mean significant differences between sites.
Source: chapter author

systems provides different habitats for different species, also known as edge effects. If we used species richness and abundance as a means of quantifying the conservation value of agricultural landscapes, then the agroforestry systems would all merit equal payments. Of the agricultural sites, only the sugar cane would potentially be excluded since it is the one that shows the lowest number of species and individuals when compared to the other agricultural habitats.

Despite the biases we have just discussed, much information is held in Figures 3.1 and 3.2. For example, we can still ask ourselves why the live fences in the pasture have such high richness and abundance. A closer look at the project data shows that few of these species are regularly captured in this land use, and that there are many rare species. Of particular interest in the live fences is that on occasion we find a forest-dependent species indicating that the linear nature of the live fence may contribute local connectivity. This brings us to three other important indicators of biodiversity conservation: species diversity, species composition and habitat quality.

Species Diversity

Take a look at Figure 3.3. Which of the two communities has the greatest biodiversity? If we use species richness and abundance as our indicators of biodiversity, the two communities are equal. Each has 5 species and 100 individuals. Yet, intuitively most people would agree that these two communities are distinct and that community A has the greatest species richness. This is because in comparison to community B, community A has the most even distribution of species (20 of each species). Several indicators of species diversity, in contrast to species richness, take into account the evenness of the distribution of species in a community. The two most common indicators of diversity are the Simpson and Shannon indices of diversity. To understand these indices, imagine that the species from Figure 3.3 are all placed in a brown paper bag and that we randomly take an individual out of this bag one by one. The Simpson index is simply the probability that two individuals pulled sequentially from the bag are of the same species. For example, in community A, where there are 20 individuals from each of these 5 species, this probability is 0.20, or a 20 per cent chance of drawing the same species. In contrast, in community B, which is dominated by woodpeckers, the Simpson value is 0.95, indicating a 95 per cent chance that we pull a woodpecker from the bag. Greater Simpson values indicate a more uneven distribution of species in a community.

The Shannon index is similar, but measures the average degree of uncertainty in predicting what species an individual chosen at random from a collection of S species and N individuals will belong to. The uncertainty increases with increasing species richness and evenness. For example, in community A, where there is an even distribution of individuals per species, we really have very little clue as to what the next species drawn from the bag will be. In contrast, in community B, we can be pretty confident that we will draw a woodpecker. These two indices are helpful, but must be used with caution when used alone, as it is possible that a community with very different degrees of species richness and abundances has the same diversity values.

Shannon and Simpson values can be relatively difficult to interpret (e.g. what is a species diversity of 0.14?). However, a rather simple transformation of each index can make interpretation simpler. The exponential (e) of the Shannon index, or one divided by the Simpson index, gives us an estimation of the *effective* species diversity: a measure of the degree to which proportional abundances are distributed among the species, or a measure of the number of species in the sample where each species is weighted by its abundance. The transformed Shannon value gives us the number of *abundant* species, whereas the transformed Simpson value gives us an estimate of the number of very *abundant* species (Ludwig and Reynolds, 1988).

Again, let us have a closer look at bird monitoring data (see Table 3.1). If we look at the effective species diversity with either the Shannon or the Simpson index, we find that the greatest species diversity with the Shannon

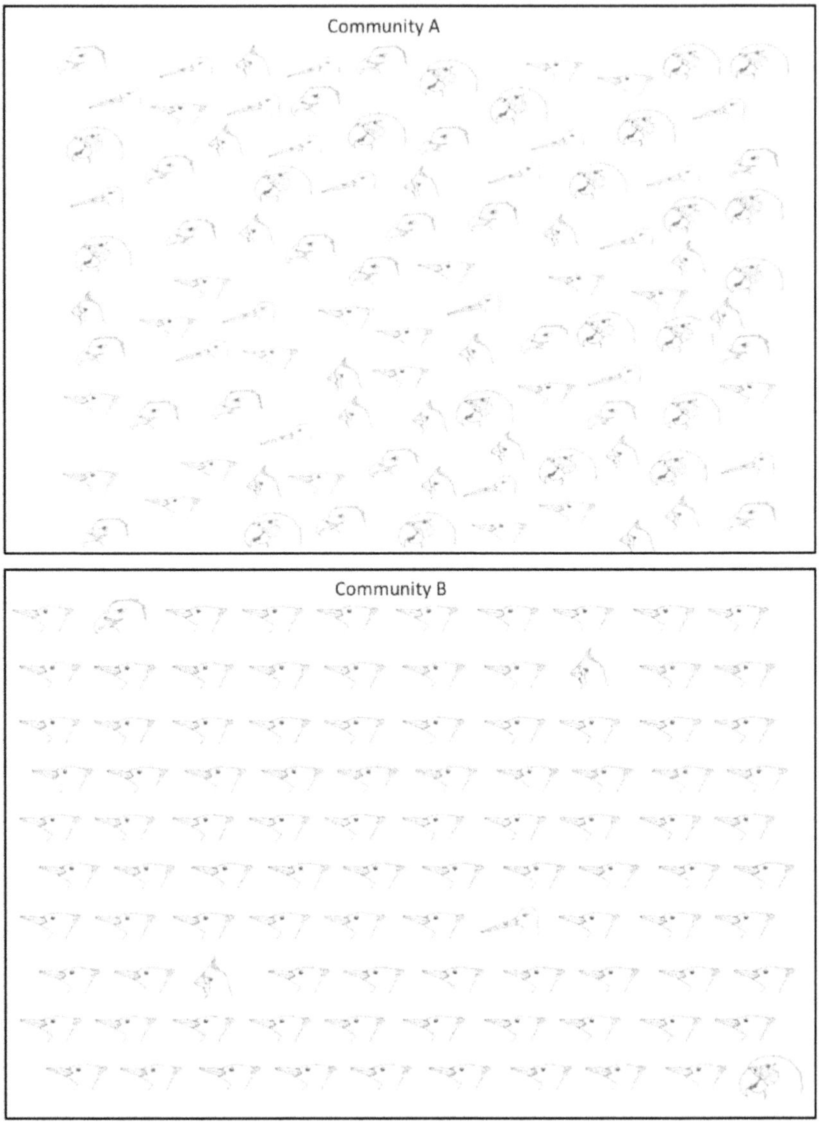

Figure 3.3 *Communities can be quantified with diversity measures*

Note: Look at both bird communities. Each has 5 species and 100 individuals and therefore is equal from the point of view of species richness and abundance. However, it would be difficult to argue that they are biologically equal. Using diversity measures such as the Shannon and Simpson indices, we find that community A has a higher Shannon value as there is an equal chance of pulling each of the species at random (20 per cent). In community B, the Shannon value is low because we are most likely to draw a woodpecker at random. The Simpson value is the probability that two individuals drawn at random are the same species, the probability of which is greater in community B than in A. In general, we would say that community A has greater species diversity based on both of these indices.
Source: chapter author

index is in the multi-strata coffee (H' = 3.48), indicating that it has 32 abundant species, and 17 very abundant species. If we compare this to its absolute species richness of 82 species, we can infer that a little more than half of the species are not very abundant. On the other hand, of the 61 species found in pasture, 26 are abundant and 17 are very abundant, indicating that in comparison to all the other sites, species found in the live fence are common and abundant. We can compare this with the forest site, which has an absolute species richness of 32 species. As with the coffee site, about half of these are abundant and one third are very abundant.

Species Composition

As we see from the examples given above, even with measures of species richness, abundance and diversity, we only get a very partial picture of what is going on and which land use has the greatest conservation value. In truth, a combination of these measures must be used, as well as a strong understanding of the methods used (and the biases) to quantify this biodiversity.

Let us go back once more to CATIE's bird monitoring data to consider community composition as a means of quantifying biodiversity. Remember that the forest plot serves as our reference ecosystem and that we may have been surprised by the very low species richness and abundance in the forest when compared to the other ecosystems. Some of this is due to the bias of mist nets; however, this does not explain all of these differences. Figure 3.4 is a principal components analysis (PCA) using metric multidimensional scaling (MMS) that arranges information on species composition by land use along two principal axes (note, however, that the analysis has *n* dimensions; we are limited to displaying a maximum of three axes on a flat sheet of paper and in this particular case we are showing only two). Land uses that are more similar in their avian composition will be closer to one another and the lines connecting the different habitats in Figure 3.4 serve to indicate the minimum difference in composition between land uses. This is also known as the 'minimum spanning tree'. Land uses can overlap in terms of species composition if there is overlap in the species found in each land use. In contrast, land uses that have few species in common will be located at greater distances from one another in Figure 3.4 (e.g. the forest and the sugar cane) and will have little or no overlap.

What we see in Figure 3.4 is that all of the agricultural land uses are located along the x, or horizontal, axis, with very little space between each land use (especially between the coffee systems). The cacao plot is located on the far left, followed by the multi-strata coffee. The coffee with poró plot is mixed in with all the agricultural land uses in the middle, indicating that it shares many species with these land uses. Our pasture and sugar cane sites are located on the far right, and are within close proximity to each other in the figure, indicating that many species are shared between these two land uses. Figure 3.4 provides us with an easy way to visualize the differences in composition between the sites. It also shows that the habitat that is closest in species

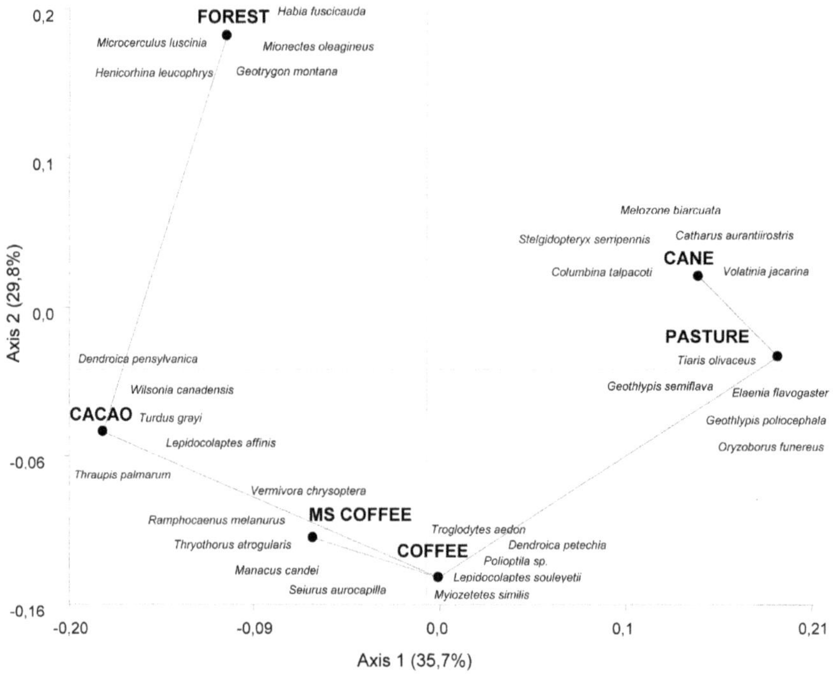

Figure 3.4 *Metric multidimensional scaling analysis (MMS) showing spatial distribution of sampling sites according to species composition*

Note: Land-use codes are the same as those used in Table 3.1. The species names in this graphic are only included to show what species was considered important for each one of the habitats after an importance value index analysis; it does not show any additional information about species overlapping in the different systems or spatial distribution of the species on this graphic.
Source: chapter author

composition to our reference forest system is the cacao agroforest. More importantly, it shows that our reference forest system is unique in species composition when compared to all of the agricultural land uses (physically separated from the rest in Figure 3.4). The greatest distance between the agricultural land uses is found between the cacao and the sugar cane, which makes sense when we consider that these are also very structurally distinct land uses (one is tree and shrub based, and the other a grass monoculture). If we use this type of compositional analysis to quantify the conservation value of an agricultural land use, the cacao site would merit the greatest payment, and the sugar cane, the smallest, or none at all. Note, however, than no single agricultural land use is able to replicate the unique composition found in the forest system. Data from the CATIE monitoring programme indicates that 6 per cent of the bird species found in the forest are not found in any other land use; this latter is really important when discussing the value of remnant forest fragments within an agricultural matrix, keeping in mind that the importance of these fragments will be also tied to their size and shape, which is going to be relevant for bird species richness and composition (Daily et al, 2001).

Using these comparative measures, and making payments for biodiversity conservation, it is clear that the cacao and coffee agroforestry systems make the greatest contribution to conserving bird biodiversity even though the live fence had greater species richness.

Habitat Quality as a Surrogate Measure of Biodiversity

Monitoring and measuring biodiversity is an expensive and time-consuming process. For example, CATIE's bird monitoring effort which we have been using as an example throughout included literally thousands of hours of work in the field, and tens of thousands of dollars dedicated to a single farm. Other measures, such as point counts, are certainly more cost effective, but provide less information and still require a competent ornithologist to correctly identify species. Biodiversity is also quite sensitive to sampling effort; a few minutes in the field quickly identifies the most common species – but a much greater effort is needed to find the rare species that are the focus of most conservation efforts. While continued work on the conservation value of different land uses is necessary to improve their contribution to species of conservation concern, surrogate indicators do exist and are often used. In fact, most certification standards used these surrogates to quantify the conservation value of coffee farms.

Recall that with Figure 3.4 we considered the similarity in composition between different land uses, and we rationalized the similarity in bird species composition between the forest and the cacao plot by the similarity in the form and structure of their vegetation. Both the cacao and the forest contain relatively high tree species richness; however, there are significant and glaring differences. Nevertheless, compared to the other land uses, the cacao plantation comes closest to sharing the bird community composition of the forest. Numerous studies have confirmed that increasing tree species richness, vertical strata and canopy closure tends to increase species richness and abundance of birds. For example, the work conducted by Perfecto et al (1996), Greenberg et al (1997a, 1997b) and Cruz-Angón and Greenberg (2005) was key to understanding the role of vertical structure on bird composition and conservation, one of the most important conclusions being that having trees within coffee systems was important; but even more important was the type of trees (diversity of trees) and number of vertical strata available (shrubs, medium-sized trees and canopy). On the other hand, the work of Gillespie et al (2000), Cárdenas et al (2003) and Harvey et al (2006) showed that greater richness and diversity of bird species were associated with greater arboreal cover, and that there is a relationship between presence/absence of species in occupied wooded habitats and overall arboreal cover.

Coffee certification standards have adopted these vegetation surrogates of conservation value in order to facilitate quantifying the conservation value of agricultural land uses. Three particular certification standards include biodiversity conservation as conservation criteria: Smithsonian Bird Friendly, Rainforest Alliance, and Starbucks CAFE (Coffee and Farmer Equity) practices. Although these examples focus on coffee certification as a means of paying for conservation value, the lessons learned from these examples can also be applied to other payment schemes.

Summary of Coffee Standards for Biodiversity

Payment for biodiversity conservation in the agricultural systems of Mesoamerica largely revolves around the certification of specific agricultural crops. The most important of these are Rainforest Alliance, Starbucks CAFE and, to a much lesser extent, Smithsonian Bird Friendly. We include Smithsonian's certification because it is the only standard which focuses solely on the conservation role of coffee farms, in contrast to other standards which are much more broadly adopted but which include indicators other than biodiversity conservation. Other standards also exist; but most of these do not explicitly address how to quantify the biodiversity value of agricultural systems.

Rainforest Alliance

The Rainforest Alliance certification is operated through the Sustainable Agriculture Network (SAN), and the standards are freely available to anyone interested (see www.rainforest-alliance.com). Rainforest Alliance (RA) began in 1989 with the Smart Wood seal; however, they started working with agricultural crops in 1993 (almost exclusively with banana). The first RA-certified farm was declared in 1998. According to RA, between 1992 and 2009, almost 800 certificates for more than 31,000 farms (600,000ha) in 24 countries have met the SAN standards for 22 crops, including coffee, cocoa, banana, tea, pineapple, flowers and foliage, and citrus. SAN is currently in the final phase of outlining similar standards for cattle production systems. The goal of the certification is to promote efficient agriculture, biodiversity conservation and sustainable community development by creating social and environmental standards. These standards are based on ten principles, two of which are ecosystem conservation and wildlife protection.

For ecosystem conservation, the certification includes eight criteria, two of which must be met at all times:

1 identification, protection and restoration of existing natural ecosystems on the farm; and
2 prohibition of the destruction of natural ecosystems due to farm management.

The remaining six criteria can be summarizes as follows:

3 Production areas cannot negatively impact upon adjacent conservation areas.
4 Harvesting of threatened or endangered species is not permitted and timber extraction requires an approved sustainable management plan.
5 Natural areas and production areas must be separated by vegetated protected zones.
6 Aquatic ecosystems must be protected and separated from production areas; water from aquatic systems cannot be diverted to the production systems.
7 The farm must establish and maintain vegetation barriers between crop areas and areas of human activity, as well as between production areas and on the edges of public or frequently travelled roads passing through the farm.
8 Where agroforestry replaces native forest, shade within the agroforestry systems must be homogeneous on the farm and consist of 12 native species per hectare, on average, with at least two canopy strata and a canopy density of at least 40 per cent.

The second principle that Rainforest Alliance includes regarding biodiversity conservation is 'wildlife protection'. This may be more related to biodiversity conservation in the traditional sense since the goal of this criterion is to ensure that the farms serve as refuges for resident and migratory wildlife, particularly species that are threatened or endangered (as we will see, there is little evidence of the latter). This is primarily accomplished by protecting natural areas that contain habitat (food, shelter and reproductive space). The criteria found under this goal include:

- an inventory of wildlife and wildlife habitat found on the farm;
- protection of habitat on the farm;
- prohibition of hunting on the farm;
- an inventory of wildlife held in captivity on the farm, with the aim of reducing their numbers;
- a stipulation that farmers may not breed wild animals in captivity;
- appropriate permits for the farm from the relevant authorities in order to release wildlife to natural habitats.

Regarding biodiversity conservation, the three first criteria are clearly the most important.

Starbucks CAFE Practices

Starbucks CAFE (Coffee and Farmer Equity) practices are probably the most adopted certification standard in Mesoamerica, in large part because of their direct tie to the Starbucks market. Not surprisingly, the first criterion required by Starbucks is ensuring the quality of the product. This is followed by a series of other criteria that include:

- a financial audit ensuring the sustainability of the enterprise;
- criteria regarding social responsibility in the tradition of Fair Trade certification;
- coffee production and environmental leadership.

This last criterion on environmental leadership includes several components, including:

- protection of hydrological resources;
- soil protection;
- biodiversity conservation;
- reduction of agrochemical use, including integrated pest management; and
- a detailed environmental plan rather than *ad hoc* decision-making.

We focus here on the third criterion: biodiversity conservation. The first criterion is more of a recommendation, rather than an obligation, encouraging farmers to maintain an average of 40 per cent native tree cover on the farm with at least three distinguishable strata or layers and a diversity of species. Starbucks, as with many of the other certifiers, recognizes that the effects of shade on coffee quantity, quality and diseases is still heavily debated and site specific (see Chapter 4). The criteria also suggest native tree cover in favour of exotic species unless the exotic species demonstrate their value in conserving native fauna. Farmers are encouraged to maintain a tree cover that is high, diverse and multi-strata; however, they also state: 'In some regions, due to local conditions, natural tree cover may not be appropriate or possible.' The second criterion is species specific and could be termed wildlife management. This requires identifying species found in and around the farm, particularly those that are of particular conservation concern, and protecting these by limiting hunting and identifying portions of the farm that serve as critical habitat. Vulnerable species are identified by the International Union for Conservation of Nature's (IUCN's) Red List of endangered species. Best practices include setting aside portions of the farm as 'protected' where natural regeneration or the conservation of existing native vegetation are promoted. Farmers need to identify those species that are particularly vulnerable and found on the farm based on the IUCN Red List (see www.redlist.org), and should develop management plans for these species in conjunction with local authorities or conservation organizations.

The third criterion for biodiversity conservation under the Starbucks certification is habitat approach and targets the conservation or restoration of conservation areas on or around the farm. The wording in the summary of this criterion is important: 'in order to conserve or increase biodiversity and ecosystem functions, areas of high ecological value must be identified and managed with an emphasis in conservation'. Starbucks recognizes the relationship between biodiversity conservation and ecosystem functioning, permitting the farmer to stack ecosystem services such as water conservation, soil erosion control and carbon sequestration, for example, on areas of the farm most suited to the provisioning of these services (see Chapter 10). The evaluation criteria for this service includes demonstrating that forest cover has not been reduced on the farm, that critical areas for conservation have been identified, and that if no such areas exist, a plan for restoring such areas is developed and implemented. Interestingly, Starbucks does not exclude coffee production areas from conservation areas – that is, when coffee is cultivated under a high canopy cover comprised of primary forest species, this area can be counted as a conservation area. This would be the equivalent of Moguel and Toledo's (1999) rustic coffee. Conservation, rather than production, however, must remain the primary management strategy in these areas.

Smithsonian's Bird Friendly

The last certification standard that we review here is the Smithsonian's Bird Friendly, which created their norm and officially launched their seal in 1999. Of the three, this is the least utilized certification; however, it was developed primarily by biologists and was instrumental both in promoting the idea of certification as a conservation tool, as well as identifying how biodiversity conservation can be quantified in coffee farms. It is equally important to mention that this seal was initially developed to help provide quality habitat for Neotropical birds, especially Neotropical migrants that spend their winters in southern Mexico and Central and South America. The Smithsonian certification is unique in several ways. First, it is the only certification standard that requires additional organic certification. Second, the Smithsonian's certification is the only certification that includes very specific suggestions as to which trees should be included or not. For example, the Smithsonian prohibits the use of exotic species such as *Pinus*, whose natural distribution extends to northern Nicaragua, and *Grevillea*. Although it permits the use of *Inga* spp., a popular coffee shade tree, it requires multiple species of *Inga* with different flowering phenologies in order to provide flower and fruit resources to wildlife year round. The standard states that no single species of *Inga* should constitute more than 50 per cent of the *Inga* abundance. It also recommends that the genus *Inga* not comprise more than 70 per cent of the total canopy cover. Finally, it disfavours the dominant use of deciduous species such as *Erythrina* or *Gliricidia* because these do not provide foliage during the dry season, which coincides with the presence of migratory species in the Neotropics.

As with the other standards, the Smithsonian recognizes that the acceptable density of shade on a plantation is site specific and is particularly tied to elevation; that lower-elevation plantations are typically more amenable to shade than high-elevation plantations, where cloud cover significantly reduces solar radiation. As with Starbucks and Rainforest Alliance certification, the target shade cover is 40 per cent. Pruning of shade trees, a common practice in coffee systems, is recommended only for the rainy season in order to avoid affecting the primary nesting season in the tropics and the return migration of most Neotropical migrants. Backbone species, or the dominant shade of the plantation, should be permitted to attain heights of 12m to 15m. The certification also recommends that the shade be divided between two primary strata or layers: a dominant strata of trees that reach 12m to 15m in height, which can be comprised of timber species, and a lower strata of trees, which can include useful fruit trees. Canopy cover should be evenly divided between both strata (20 per cent each).

Smithsonian certification is also the only standard that recommends the conservation of secondary vegetation – primary epiphytes (plants that grow in the canopy of other plants) such as bromeliads, orchids, ferns, vines and lianas – to ensure year-round food supplies for many birds. Finally, the certification suggests that living barriers be maintained between coffee parcels, around the perimeter of the coffee plantation, and along waterways. 5m of natural vegetation is recommended on both edges of streams, and 10m of vegetation for rivers.

Indicators of Biodiversity and Certification

Several points merit discussion regarding certification standards and how they quantify biodiversity. The first is that although Rainforest Alliance and Starbucks both require inventories of biodiversity on the farm, this is typically limited to the tree species found on farm rather than an inventory of all taxonomic groups. Second, no standard uses species richness, species diversity or community composition and indicators of biodiversity of conservation criteria. Third, all three certification standards primarily target the conservation of plants with the implicit understanding that conserving plant diversity leads to a concomitant conservation of animal diversity. Although this is not explicit in the Rainforest and Starbucks certification standard, by far the majority of research conducted on the conservation value of coffee has been conducted on birds.

Rainforest Alliance and Starbucks both recognize that not all species have the same conservation value, as indicated earlier, and specifically require management plans for species of conservation concern, most notably species that are listed under the IUCN Red List as vulnerable or endangered. Although in most cases there is little probability of these species being found in coffee plantations since most critically endangered species are nearly, by definition, forest dependent, the inclusion of this variable does leave room for specific management activities targeting the conservation of these species when they are found.

All three standards focus on the management of shade within the productive portions of the farm through agroforestry systems that strive for a minimum of 40 per cent tree cover, three strata and increasing the species richness of trees. Studies have shown that such interventions do, indeed, increase the diversity and abundance birds, specifically of migratory species that seem particularly fond of coffee systems. This is less the case with mammals. In an unpublished study of mammals found in coffee agroforests, we found only 2 small- and medium-sized mammals within a coffee agroforest, compared with up to 12 in forests directly adjacent to the coffee, including ocelots (*Leopardus pardalis*). Likewise, a similar study led by Thomas Husband and David Abedon of the University of Rhode Island found that mammal diversity is highest in forest fragments preserved within the farm. Both Starbucks and Rainforest Alliance make provisions for this by prohibiting reductions of on-farm tree cover up to three years prior to certification. They also require that conservation areas be protected and/or restored on farm. Ecological studies support this criterion as even coffee plantations that meet recommended shade criteria will never equal the conservation value of forests. The Smithsonian has not included this criterion possibly in light of the focus on migratory species, which are particularly well suited to coffee systems, as demonstrated by CATIE's BMP data and previous research (Estrada and Coates-Estrada, 2005; Komar, 2006; Philpott et al, 2008).

All standards recognize that management practices other than shade affect conservation value, and all three recommend either the total elimination of agrochemicals (Smithsonian) or the better use of these chemicals in terms of reducing their application and eliminating their application in sensitive areas, such as around waterways. The Smithsonian is the only standard that recognizes the impact of pruning, more importantly of the timing of the pruning, suggesting that pruning is limited to the rainy season when birds and their young are less vulnerable.

Quantifying Biodiversity, from Science to Practice

At the beginning of this chapter, we discussed different indicators of biodiversity, including species richness, diversity and composition, as well as surrogates of biodiversity, such as land-use type. We followed this with a discussion of certification of agricultural practices and the standards used to quantify how these farms contribute to biodiversity conservation. What should be apparent at this point is that the means of quantifying biodiversity from the ecological and farm management points of view are quite distinct. None of the standards require a certain level of species richness, diversity or shared species composition with adjacent forest. Rather, all rely on surrogate measures that focus on tree cover, both in terms of tree density and structure in the productive portions of the farm, and the conservation of protected areas adjacent to the farm. Indeed, certification serves as the middle man between ecological measures of conservation and measures that can easily be performed in a farm audit. Measuring the vegetation characteristics of a farm is significantly simpler and more cost effective than measuring avian, mammal, reptile or amphibian biodiversity, even though these latter taxonomic groups are significantly more threatened than plant groups in the Neotropics.

For the most part, these surrogates work. Numerous studies have found that increasing tree species richness, density and number of strata are well correlated with increasing animal (particularly avian) biodiversity. Indeed, many of these studies were conducted by the same biologists who developed the standards (Perfecto et al, 1996; Greenberg et al, 1997a, 1997b; Gillespie et al, 2000; Cárdenas et al, 2003; Cruz-Angón and Greenberg, 2005; Harvey et al, 2006). While there is no doubt that certification of agricultural systems makes significant contributions to biodiversity conservation, we are still a long way from understanding how these systems can better contribute to the conservation of species of critical concern. These often require specific information on the distribution of endangered species, as well as on the specific requirements of the species in question. For example, many of the bird species most sensitive to agricultural intensification in the tropics (as well as elsewhere) are cavity nesting birds such as parrots, trogons, toucans and woodpeckers. Of the three standards discussed here, only the Smithsonian mentions the importance of maintaining snags (standing dead trees) within the agroforest. Second, we need more information on how these farms can better contribute to taxonomic groups other than birds. Mammals appear to avoid all but the most rustic coffee systems. Indicators such as the proportion of forest cover on the farm and, more importantly, the degree of connectivity found on a coffee farm or between forested areas adjacent to the coffee farm are absent from any standard focusing on biodiversity despite the fact that this would be a relatively easy indicator to measure, and could most likely be tied to the buffer requirements.

Finally, there is a distinct lack of information on how the timing of management event affects biodiversity. Aside from the Smithsonian requirement that pruning only be conducted during the rainy season, there is no mention of this temporal indicator in any other standard, where the same restrictions might not only apply to pruning, but also to the application of agrochemicals.

Habitat versus Connectivity and Other Biodiversity Needs

One of the points that we made in describing community composition as a measure of conservation value is that 6 per cent of the species found in the forest were not found in any other land use. It is more than likely that conventional agroforestry systems, where production is the primary goal, will never provide suitable habitat for these species. However, conservation interventions in agroecosystems do not need to be restricted to providing habitat. Rather, agroforests can provide connectivity functions in addition to habitat functions and contribute to biodiversity conservation by facilitating the movement of forest-dependent species between forest patches.

One of the reasons that the species richness values for the live fence example used in this study were so high was exactly this reason. During the three years of monitoring in CATIE's live fences, as well as in the coffee agroforests, we occasionally captured forest-dependent species, such as the

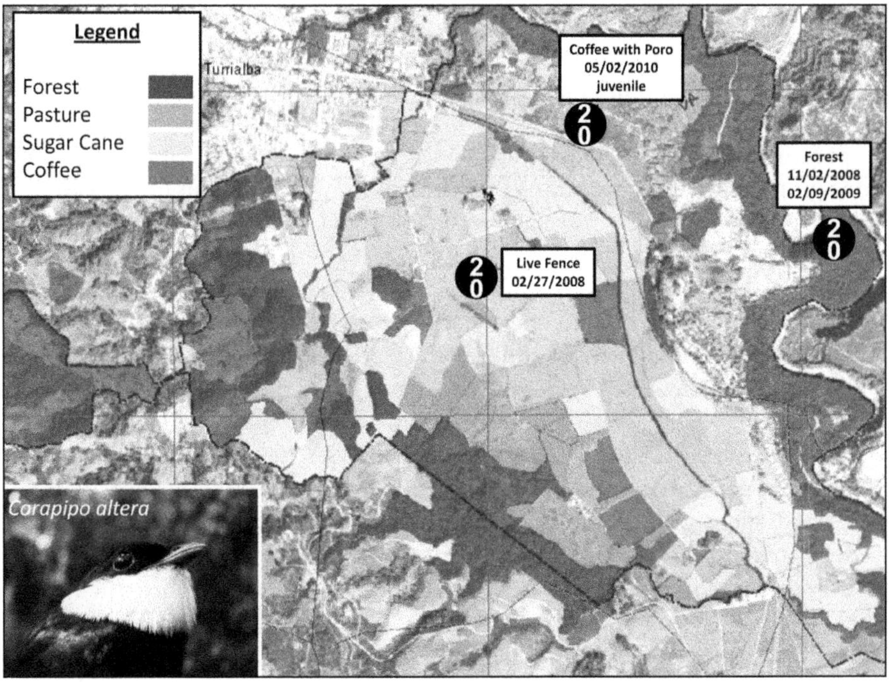

Figure 3.5 *Some species, such as the white-ruffed manakin (*Corapipo altera*), are considered forest dependent, but may also be found to move between forest patches through the use of agroforestry systems*

Note: Capturing these events is rare and demonstrates the importance of Neotropical agroforestry systems for maintaining connectivity. Most individuals captured are young males, supporting the hypothesis that they disperse in search of new territories. Marked points in the figure indicate captures of these species both in the forest as well as when dispersing.
Source: chapter author

white-ruffed manakin (*Corapipo altera*) (see Figure 3.5), captured in pasture, and the chestnut-backed antbird (*Myrmeciza exsul*), captured in our MS coffee plot. Although these captures are rare, and would be nearly impossible to observe without long-term monitoring efforts, they indicate that these agroforests may play an additional role in facilitating the movement of individuals between forest patches. Typically, the forest-dependent species that we capture are juveniles venturing away from their natal forest patch in search of a new territory. The success of these individuals lies in their ability to traverse the agricultural matrix to unoccupied territories. The role that agroforests play in facilitating is bound to be important, though little studied – even more if we considered that research conducted on fragmented landscapes has shown very few observations of forest-dependent species (Daily et al, 2001; Cárdenas et al, 2003; Lang et al, 2003; Robinson et al, 2004; Harvey et al, 2006; Taylor, 2006; Vilchez-Mendoza et al, 2008; Milder et al, 2010).

Similarly, in our MS coffee site, we regularly captured the white-collared manakin (*Manacus candei*), which is likewise dependent on secondary forest vegetation. Curiously, all of the individuals that we have captured in this site are either females or juvenile males, according to their plumage. Only on one occasion have we captured an adult male. Although untested, this observation suggests that this coffee agroforest may provide habitat for a subset of the population. Much information is lacking as to how species of conservation concern use different portions of the landscape during different life phases.

Conclusions

Biodiversity remains threatened by agricultural intensification. The loss of biodiversity is particularly of concern in the tropics, which are renowned for particularly high levels of species richness and high levels of endemism (species that have very small distribution ranges and can only be found in very specific areas of the world). Significant strides have been made to protect biodiversity in national parks and reserves; however, it is evident that these efforts will fail to protect Neotropical biodiversity in the absence of significant conservation efforts in the surrounding agricultural matrix. Increasing research on the conservation value of agroecosystems in the Neotropics supports the notion that biodiversity can be conserved in agricultural landscapes (Gardner et al, 2009; DeClerck et al, 2010), particularly when farmers are able to increase on-farm tree diversity and structural complexity, and when little disturbed forest fragments are conserved within the landscape.

In this chapter we have demonstrated that although quantifying biodiversity in agricultural systems presents its challenges, the use of multiple metrics, including species composition, can provide appropriate means of evaluating the contribution of agricultural systems to biodiversity conservation. When groups of species are the primary conservation target, on-farm interventions that strive to mimic the structure and diversity of local native vegetation are appropriate surrogates of measuring the conservation value of the system. When specific species, particularly forest-dependent species, are the conservation target, specific research and interventions that provide the habitat requirement for the species in question will be needed.

Payment for ecosystem services schemes in the Neotropics have focused primarily on these on-farm interventions as the primary means of protecting Neotropical biodiversity. From the point of view of increasing species richness and abundance, these strategies have largely been successful, though additional efforts are required to increase their implementation and adoption by farmers. Unfortunately, these strategies have been much less successful at demonstrating their capacity to protect forest-dependent species, or species of particular conservation concern (Komar, 2006; Milder et al, 2010).

Part of the problem lies on the focus at the farm scale, which occurs primarily for logistical reasons – it is much easier to work with a farmer or a co-operative in order to target conservation efforts than it is to work with multiple actors within a landscape. It is essential that we build on these widely accepted measures of on-farm conservation value discussed here in order to focus on landscape-scale measures. Of particular interest, and need, are methods for quantifying the landscape-scale conservation value of the agricultural landscape. Some agricultural land uses may have low conservation value, but may facilitate the movement of forest-dependent species between forest patches. Quantifying the conservation value of agricultural landscapes, including the contribution of different land uses to habitat and connectivity, will be essential to maintaining Neotropical diversity for future generations.

Acknowledgements

We are grateful for the financial support provided by the European Union via the CAFNET (Connecting, Enhancing and Sustaining Environmental Services and Market Values of Coffee Agroforestry in Central America, East Africa and India) and PolicyMix (Assessing the Role of Economic Instruments in Policy Mixes for Biodiversity Conservation and Ecosystem Services Provision) collaborative projects. Support was also provided by the Pôle de Compétences en Partenariat (PCP) programme Agroforestry Systems with Perennial Crops, CIRAD–CATIE–INCAE–Bioversity–CABI–Promecafé.

References

Cárdenas, G., Harvey, C. A., Ibrahim, M. and Finegan, B. (2003) 'Diversidad y riqueza de aves en diferentes hábitats en un paisaje fragmentado en Cañas, Costa Rica', *Agroforestería en las Américas*, vol 10, pp39–40

Cruz-Angón, A. and Greenberg, R. (2005) 'Are epiphytes important for birds in coffee plantations? An experimental assessment', *Journal of Applied Ecology*, vol 42, pp150–159

Daily, G. C. (1997) *Nature's Services: Societal Dependence on Natural Ecosystems*, Island Press, Washington, DC

Daily, G. C., Ehrlich, P. R. and Sánchez-Azofeifa, A. (2001) 'Countryside biogeography: Use of human-dominated habitats by the avifauna of southern Costa Rica', *Ecological Applications*, vol 11, no 1, pp1–13

DeClerck, F. A., Chazdon, R., Holl, K. D., Milder, J. C., Finegan, B., Martinez-Salinas, A., Imbach, P., Canet, L. and Ramos, Z. (2010) 'Biodiversity conservation in human-modified landscapes of Mesoamerica: Past, present and future', *Biological Conservation*, vol 143, pp2301–2313

Estrada, A. and Coates-Estrada, R. (2005) 'Diversity of Neotropical migratory landbird species assemblages in forest fragments and man-made vegetation in Los Tuxtlas, Mexico', *Biodiversity and Conservation*, vol 14, no 7, pp1719–1734

Gardner, T. A., Barlow, J., Chazdon, R., Ewers, R. M., Harvey, C. A., Peres, C. A. and Sodhi, N. S. (2009) 'Prospects for tropical forest biodiversity in a human-modified world', *Ecology Letters*, vol 12, no 6, pp561–582

Gillespie, T., Grijalva, A. and Farris, C. (2000) 'Diversity, composition, and structure of tropical dry forests in Central America', *Plant Ecology*, vol 147, pp37–47

Greenberg, R., Bichier, P., Cruz-Angón, A. and Reitsma, R. (1997a) 'Bird populations in shade and sun coffee plantations in Central Guatemala', *Conservation Biology*, vol 11, no 2, pp448–459

Greenberg, R., Bichier, P. and Sterling, J. (1997b) 'Bird populations in rustic and planted shade coffee plantations of Eastern Chiapas, Mexico', *Biotropica*, vol 29, no 4, pp501–514

Harvey, C. A., and Villalobos, J. A. G. (2007) 'Agroforestry systems conserve species-rich but modified assemblages of tropical birds and bats', *Biodiversity and Conservation*, vol 16, no 8, pp2257–2292

Harvey, C. A., Medina, A., Merlo Sánchez, D., Vilchez, S., Hernández, B., Sáenz, J. C., Maes, J. M., Casanoves, F. and Sinclair, F. L. (2006) 'Patterns of animal diversity in different forms of tree cover in agricultural landscapes', *Ecological Applications*, vol 16, no 5, pp1986–1999

Harvey, C. A., Komar, O., Chazdon, R., Ferguson, B. G., Finegan, B., Griffith, D. M., Martinez-Ramos, M., Morales, H., Nigh, R., Soto-Pinto, L., Van Breugel, M. and Wishnie, M. (2008) 'Integrating agricultural landscapes with biodiversity conservation in the Mesoamerican hotspot', *Conservation Biology*, vol 22, no 1, pp8–15

Kareiva, P. and Marvier, M. A. (2007) 'Conservation for the people', *Scientific American*, October, p9

Komar, O. (2006) 'Ecology and conservation of birds in coffee plantations: A critical review', *Bird Conservation International*, vol 16, no 1, pp1–23

Lang, I., Gormley, L., Harvey, C. A. and Sinclair, F. (2003) 'Composición de la comunidad de aves en cercas vivas de Río Frío, Costa Rica', *Agroforestería en las Américas*, vol 10, no 39–40, pp86–92

Levin, S. (1999) *Fragile Dominion*, Perseus Books, Santa Fe, CA

Loreau, M., Naeem, S. and Inchausti, P. (eds) (2002) *Biodiversity and Ecosystem Functioning, Synthesis and Perspectives, Oxford Biology*, Oxford University Press, Oxford.

Ludwig, J. A. and Reynolds, J. F. (1988) *Statistical Ecology: A Primer on Methods and Computing*, Wiley and Sons, New York, NY

Milder, J. C., DeClerck, F., Sanfiorenzo, A., Sanchez, D., Tobar, D. and Zuckerberg, B. (2010) 'Effects of tree cover, land use, and landscape context on biodiversity conservation in an agricultural landscape in western Honduras', *Ecosphere*, vol 1, no 1, pp1–22

Moguel, P. and Toledo, V. M. (1999) 'Biodiversity conservation in traditional coffee systems of Mexico', *Conservation Biology*, vol 13, no 1, pp11–21

Naeem, S. (2002) 'Ecosystem consequences of biodiversity loss: The evolution of a paradigm', *Ecology*, vol 83, no 6, pp1537–1552

Naeem, S., Bunker, D. E., Hector, A., Loreau, M. and Perrings, C. (2009) *Biodiversity, Ecosystem Functioning, and Human Well-Being: An Ecological and Economic Perspective*, Oxford Biology, Oxford

Perfecto, I., Rice, R. A., Greenberg, R. and Van Der Voort, M. E. (1996) 'Shade coffee: A disappearing refuge for biodiversity', *BioScience*, vol 46, no 8, pp598–608

Philpott, S. M., Arendt, W. J., Armbrecht, I., Bichier, P., Diestch, T. V., Gordon, C., Greenberg, R., Perfecto, I., Reynoso-Santos, R., Soto-Pinto, L., Tejeda-Cruz, C., Williams-Linera, W., Valenzuela, J. and Zolotoff, J. M. (2008) 'Biodiversity loss in Latin American coffee landscapes: Review of the evidence on ants, birds, and trees', *Conservation Biology*, vol 22, no 5, pp1093–1105

Ralph, J., Geupel, G., Pyle, P., Martin, T., DeSante, D. and Milá, B. (1996) *Manual de métodos de campo para el monitoreo de aves terrestres*, Forest Service, US Department of Agriculture, Pacific Southwest Research Station, Albany, CA

Robinson, D., Angehr, G., Robinson, T., Petit, L., Petit, D. and Drawn, J. (2004) 'Distribution of bird diversity in a vulnerable Neotropical landscape', *Conservation Biology*, vol 18, no 2, pp510–518

Taylor, R. (2006) *Birds Using a Contemporary Neotropical Landscape: The Effects of Forest Fragmentation and Agricultural Landscape Structure on Neotropical Birds*, PhD thesis, University of Wales–CATIE, Bangor, Wales

Vilchez-Mendoza, S., Harvey, C. A., Sánchez Merlo, D., Medina, A., Hernández, B. and Taylor, R. (2008) 'Diversidad y composición de aves en un agropaisaje de Nicaragua', in C. Harvey and J. Sáenz (eds) *Evaluación y conservación de biodiversidad en paisajes fragmentados de Mesoamérica*, Instituto Nacional de Biodiversidad (INBio), Santo Domingo de Heredia, Costa Rica, pp547–576

Wilson, E. O. (1994) *Naturalist*, Warner Books, New York

4

Ecological Mechanisms for Pest and Disease Control in Coffee and Cacao Agroecosystems of the Neotropics

Jacques Avelino, G. Martijn ten Hoopen and Fabrice A. J. DeClerck

Introduction

Pests and diseases can have impacts upon most ecosystem services (ES), including food production and yield (Cheatham et al, 2009). Food production, according to the Millennium Ecosystem Assessment (MEA, 2005) is one of the most important ES. In severe cases, pest and disease damage results in plantation abandonment, famine and emigration. The Irish potato famine, caused by potato blight (*Phytophthora infestans*) in Ireland between 1846 and 1851, is one of the best-known cases. In the tropics, examples of disasters caused by pests include the coffee leaf rust (*Hemileia vastatrix*) in Ceylon (now Sri Lanka) during the 1870s, the coffee wilt disease (*Gibberella xylarioides*) in Central Africa during the 1940s and 1950s, and *Fusarium oxysporum*, the causal agent of Panama disease on bananas in Latin America during the 1960s. More recently, two closely related cacao pathogens, *Moniliophthora perniciosa* (previously *Crinipellis perniciosa*), the causal agent of witches' broom, and *Moniliophthora roreri*, the causal agent of frosty pod rot, have wreaked havoc among cocoa producers in the Neotropics. After the arrival of witches' broom in Bahia in 1989, Brazil moved from being the second largest cocoa producer in the world (374,000 tonnes in 1988) to the fifth rank in 2000, with a 47 per cent decrease in production despite a 6 per cent increase of the planted area during the same period (Meinhardt et al, 2008; FAO, 2010). Similarly, frosty pod rot led to the almost total disappearance of cacao cultivation in Costa Rica during the 1970s.

After World War II, industrialized countries, while intensifying crop production, have avoided such disasters through increased use of pesticides. However, intensive pesticide use includes major negative externalities, such as human health problems, pollution, reductions in the populations of beneficial organisms, and the emergence of secondary diseases or pathogen resistance to pesticides (Wilson and Tisdell, 2001; Jackson, 2002; Tilman et al, 2002; Leach and Mumford, 2008; Geiger et al, 2010). Therefore, in order to reduce pesticide use and limit these negative externalities, the question of how to stimulate ecological mechanisms of pest and disease control seems very relevant in industrialized countries.

In less-developed tropical countries, the situation is somewhat different. In general, tropical agriculture does not consume the quantities of pesticides of industrialized countries (Abhilash and Singh, 2009). In the Neotropics, we often find low-input and low-yield (partly due to plant pest and disease impacts) smallholder systems with minimal or no use of pesticides. This is the case for almost all cacao producers and many coffee producers in Central America. Therefore, in the Neotropics ecological mechanisms of pest and disease control must be better used not only in order to reduce pesticide use, but also to increase yields of the very common low-input systems by reducing pest and disease impact.

The high vulnerability of agroecosystems to pests and diseases, compared to natural ecosystems, has been related to the loss of biodiversity and simplification of agroecosystems. As a consequence, increasing biodiversity has been proposed to decrease pest and disease risks (Andow, 1991; Altieri, 1999; Tilman et al, 2002; Bianchi et al, 2006; Cheatham et al, 2009; Malezieux et al, 2009). In Neotropical agroecosystems, particularly in agroforestry systems, plants, birds, insects, fungi and bacteria contribute to pest and disease control and consequently supply important environmental services (Cheatham et al, 2009). However, the myriad of interactions between plants, birds, insects, fungi and bacteria, the environment in which they live, and the way in which these interactions influence diseases and pests cycles in natural systems are extremely complicated and poorly understood. Future research should focus on elucidating these interactions to capitalize on biological control services (BCS) (Coll, 2009).

In this chapter we review examples of control effects, emphasizing the different pathways involved. We do not intend to describe all existing ecological mechanisms of pest and disease control, but highlight those present in coffee and cacao agroecosystems: two important crops of the Neotropics. It is important to note that although increasing biodiversity generally leads to better control of pests and diseases, some negative effects also exist and will be discussed. Finally, the question of quantification of pest and disease regulation services provided by these tropical agroecosystems will be examined.

Genetic Diversity, Disease and Pest Resistance: The Need for *In-Situ* Conservation

Plants and co-evolved pathogens in their centre of origin have a high genetic diversity because of continuous mutual adaptation. Therefore, centres of origin for specific crops provide an important source of dynamically evolving resistance genes that are exploitable for plant breeding (Cheatham et al, 2009). In the Neotropics, this is especially true for cacao (*Theobroma cacao*), which is native to central and western Amazonia and has its centre of diversity in the upper Amazon Basin in Peru and Ecuador (Motamayor et al, 2008), although its centre of domestication is Mexico and Central America (Evans et al, 1998). The importance of *in-situ* conservation is particularly apparent when considering the current phytosanitary situation of cacao in Brazil. Since the outbreak of witches' broom in Bahia in 1989, approximately 150,000ha of susceptible varieties have been replaced with resistant descendants of Scavina clones as an emergency measure. Some of these descendants have shown a decreased resistance, probably the result of an increase in frequency of strains of witches' broom capable of overcoming Scavina's resistance (Gramacho and Pires, 2009). An additional problem of increasing importance in Bahia is *Ceratocystis* wilt caused by the fungus *Ceratocystis cacaofunesta* (Baker Engelbrecht and Harrington, 2005), which, similar to witches' broom and frosty pod rot, probably originated in northern South America, after which it spread to other cacao-growing regions (Engelbrecht et al, 2007). Because much of the germplasm that has been selected for resistance to witches' broom and other diseases is susceptible to *Ceratocystis* wilt, the impact of the wilt will increase as these materials are put in place. Since frosty pod rot, witches' broom and probably *Ceratocystis* wilt originated on cacao in the Americas, resistant sources should be present within cacao's centre of origin.

Plant Species Diversity and Pest and Disease Control Services

Modification of host plant densities and physical barriers for pests and pathogens

The first probable effect of introducing plant diversity in agroecosystems is modification of host plant densities. It is well known that agriculture has favoured monocultures where crops are more perceptible to pests and available to diseases (Burdon and Chilvers, 1982; Altieri, 1999; Malezieux et al, 2009). Tropical crops are not exempt. Several relationships between high densities and greater pest and disease attack intensities have been mentioned for coffee. For instance, short distances between rows of coffee trees have been reported to facilitate American leaf spot disease (*Mycena citricolor*) spread, whose dispersal distance is very short (Avelino et al, 2007). This probably explains why this disease is considered to be severe almost only in Costa Rica, a country where coffee plantation densities are among the highest in the world. A similar relationship, again in Costa Rica, has been found in nematodes (*Meloidogyne exigua* and *Pratylenchus coffeae sensu lato*), organisms known for their very limited mobility. Increased contact between roots in high-density systems probably facilitates nematode spread (Avelino et al, 2009). Another disease favoured by high densities is *Rosellinia* root rot in coffee and cacao. *Rosellinia* dispersal is through root contact and is characterized by circular pattern patches caused by the infection process.

Introducing non-host vegetation in agroecosystems at plot and landscape scales can help to intercept pests and diseases when dispersing or spreading. Windbreaks, hedges and woody borders particularly influence insect dispersal (Pasek, 1988; Bhar and Fahrig, 1998; Schroth et al, 2000; Sciarretta and Trematerra, 2006). Barrier effects have also been reported for wind-borne, splash-borne or soil-borne pathogens in intercrop systems (Michel et al, 1997; Bannon and Cooke, 1998; Gomez-Rodriguez et al, 2003; Schoeny et al, 2010). According to Schroth et al (2000), shade trees in tropical agroforestry systems can potentially serve as barriers to pest and pathogen spread. Shade trees reduce wind speed in coffee and cacao plantations (Beer et al, 1998), and probably affect different wind-borne pathogens such as coffee leaf rust, frosty pod rot or witches' broom on cacao. Evans (1981), for example, noted that cacao grown in unshaded blocks without windbreaks demonstrated increased turbulence, especially at plantation edges, and that these currents favoured inoculum movement of *Moniliophthora perniciosa*, with a resulting increase in both pod and flower infections.

Similar effects can be found at landscape scales. Some landscape elements constitute dispersal-limiting barriers or, on the contrary, can serve as dispersal-favouring corridors (Altieri, 1999; Plantegenest et al, 2007). This is the connectivity paradox: connectivity for desirable organisms may also favour the spread of noxious organisms and make their control more difficult (Zadoks, 1999). The effect of connectivity on pest and pathogen dispersal has been suggested in only a few studies, mostly in temperate forests (Perkins and Matlack, 2002; Condeso and Meentemeyer, 2007). In those cases, landscape fragmentation with non-host plots was proposed to create barriers to pest and pathogen movement. There is no evidence of such effects on tropical crop pests and pathogens. However, the potential of fragmenting agricultural landscapes with forest corridors that serve as barriers for agricultural pests seems promising for less mobile pathogens or pests, particularly host-specific species such as the coffee berry borer. For noxious organisms with high dispersal abilities, such as coffee leaf rust, witches' broom and frosty pod rot of cacao, isolating susceptible plots may be insufficient to avoid infection.

Modifications of microclimate

Introduction of plant diversity into agroecosystems can deeply modify the microclimate, particularly that of understorey vegetation. Air and soil temperatures are generally buffered; wind speed and solar radiation are reduced, whereas relative humidity, plant organ wetness and soil humidity are increased (Ong et al, 1991; Olasantan et al, 1996; Beer et al, 1998; Staver et al, 2001; Avelino et al, 2004; DaMatta, 2007; Lott et al, 2009). These changes can directly affect different processes of the pest and pathogen life cycles and create adverse conditions for their development (Schroth et al, 2000; Gomez-Rodriguez et al, 2003; Schoeny et al, 2010). In Cameroon, shade trees have been suggested to reduce coffee berry disease (*Colletotrichum kahawae*) incidence by intercepting rainfall and reducing raindrop impact intensity upon coffee trees, consequently limiting the splash dispersal of propagules (Mouen Bedimo et al, 2008). In the Neotropics, coffee blight (*Phoma costarricencis*) can be reduced by establishing windbreaks and shade trees in the plantation (Muller et al, 2004). Trees buffer coffee plants from winds, which has a dual effect. First, the trees provide a degree of insulation from cold winds. Second, the trees protect the coffee plants from mechanical damage by reducing wind speed. Cold temperatures and infection entry points caused by wounded leaves and stems are two of the conditions that favour infection by the coffee blight. Shade also lessens coffee's brown eye spot disease (*Cercospora coffeicola*) (Echandi, 1969; Staver et al, 2001). According to Echandi (1969), *C. coffeicola* has a high temperature requirement for germination (30°C) and plant tissue colonization is favoured when soil moisture is low, two conditions frequently encountered at full sun exposure that can be corrected by intercropping coffee with shade trees.

Microclimate modifications due to shade may also indirectly affect pests and diseases through host physiology changes. That is the case with coffee leaf rust, whose epidemics are more intense when coffee yield is high (Avelino et al, 2004, 2006), probably because of the translocation of phenolic compounds from leaves to fruits. This condition is often reached at full sun exposure (see Figure 4.1). As a consequence, a high degree of shade may negatively affect the development of this disease through the effects of shade on yield (Avelino et al, 2004, 2006), which largely depends on light availability. Similarly, the overbearing disease (*Colletotrichum* spp.), which affects overproducing plants and causes branch dieback, can be almost completely eliminated by implementing an appropriate shade to regulate yield (Muller et al, 2004). A similar effect has been shown in the case of witches' broom. Since cacao grows more vigorously and produces more flowers and pods when grown under full sun, it becomes particularly vulnerable to witches' broom because this pathogen affects actively growing tissue (Evans, 1998). It has been demonstrated in Costa Rica that Neotropical mirids (*Monalonion* spp.) caused more damage on full sun-grown cacao compared with shaded cacao (Villacorta, 1977). A possible explanation, similar to the case of witches' broom, is that full sun-grown

Figure 4.1 *High yields reached in a full-sun plantation predisposed coffee leaves to severe attacks by rust (*Hemileia vastatrix*) despite copper-based fungicide applications (left); a great proportion of the leaf area was infected (right)*

Source: J. Avelino

cacao is more vigorous and provides more food sources for cacao mirids than shade-grown cacao (Babin et al, 2010). Based on these results, it was suggested that homogeneous shade provided by large forest trees should reduce mirid damage more effectively.

Improvement of soil physical and chemical characteristics

Plant diversity can improve physical and chemical soil characteristics, which can have an indirect impact upon pest and disease control. In coffee-based agroforestry systems, di-nitrogen (N_2)-fixing leguminous trees are frequently used for coffee shading. Litter fall and pruned branches can contain up to 340kg of nitrogen ha^{-1} y^{-1} (Beer et al, 1998). As a consequence, total soil nitrogen (N) for the superficial soil layer was found to be higher in coffee intercropped with leguminous trees, compared to coffee monocultures (Hergoualc'h et al, 2008). In general, plants with good nutritional status and good growth are better able to replace diseased leaves or roots. This effect has been suggested for coffee leaf rust and American leaf spot disease on coffee, where disease intensities and the annual number of fertilizer applications were negatively associated (Avelino et al, 2006, 2007). Plant nutrition can also

Figure 4.2 *The appearance of cocoa pod rot, caused by Phytophthora, is delayed by the presence of a litter layer (left) when compared with a situation where no litter layer is present (right)*

Note: Only the young cocoa pods (cherelles) closest to the soil have been affected. Both pictures were taken on 25 May 2010.
Source: G. M. ten Hoopen

increase resistance of plants to pathogens. Disease severity of facultative parasites, which prefer senescent tissues and kill host plant cells in order to feed, normally decreases with nitrogen applications (Dordas, 2008). This has been found for brown eye spot disease (Pozza et al, 2001) and branch dieback associated with *Colletotrichum* spp. infections on coffee (Muller et al, 2004). Other elements such as potassium (K), phosphorus (P), manganese (Mn), boron (B), chlorine (Cl), sulphur (S) and silicon (Si) have also been cited to increase plant resistance to diseases (Walters and Bingham, 2007; Dordas, 2008). Moreover, the litter layer which forms from natural leaf fall and pruning residues can affect diseases. Experimentations in Cameroon showed that leaf litter acts as a physical barrier for the soil-borne primary inoculum of cacao *Phytophthora* pod rot (*P. megakarya*) and delays the first appearance of diseased pods by three to four weeks compared with the situation of no leaf litter present (see Figure 4.2) (ten Hoopen et al, 2009). Although a delay of three to four weeks may not seem very long, it reduces the number of fungicide applications by one or two compared to the official spray recommendation.

Biological Control Services (BCS)

Bird communities and pest control services

Possibly one of the most interesting advances in community ecology in relation to pest control as an ecosystem function explores the relationship between avian communities and arthropod removal. Although literature on this function is still scarce, there is increasing evidence that avian communities play an important role in pest control. In the Neotropics, the majority of this evidence is limited to coffee agroforests. The data from these studies overwhelmingly support three primary conclusions. First, they show that increasing tree density and diversity has a concomitant increase in bird community richness and abundance (Harvey et al, 2008). Second, the subset of these studies that test the effects of bird communities on arthropod removal find strong results, indicating that birds can contribute upwards of 80 per cent arthropod removal in coffee agroforests (Van Bael et al, 2008) and that this removal is strongly correlated with bird species richness. And, third, studies of arthropod removal in coffee systems all implicate Neotropical migratory species as the main drivers of this function (Komar, 2006). Two meta-analyses of enclosure studies in various agroforests likewise show that the species and functional richness of bird communities is strongly correlated to arthropod removal (Van Bael et al, 2008; Philpott et al, 2009); however, no distinction is made as to whether these are beneficial insects or crop pests. Several studies indirectly test the effects of bird predation by counting the number of damaged leaves inside and outside of enclosures. Greenberg et al (2000) and Johnson et al (2009) reported up to 50 per cent increase in coffee leaf damage inside bird-proof exclosures. Similarly, a study in shade-grown cacao showed that insectivorous birds slightly reduced herbivore damage to cacao foliage (Van Bael et al, 2007). In a study of Jamaican Blue Mountain coffee, Kellermann et al (2008), focusing on the impacts of bird communities upon the coffee berry borer, found infestations up to 14 per cent lower in coffee plants exposed to foraging birds compared to plants excluded from them. More importantly, their analysis of the stomach content of the three most important predators, black-throated blue warbler (*Dendroica caerulescens*), American redstart (*Setophaga ruticilla*) and prairie warbler (*D. discolour*), showed that the coffee borer comprised 53, 56 and 44 per cent of the stomach content, respectively. Although many studies indicate that bird abundance is strongly tied to shade in coffee agroforests, there is some debate as to whether this effect is carried over to pest control. Perfecto et al (2004) found that lepidopteran removal was twice as high in a coffee farm with diverse shade as in a coffee farm with simple shade, although there is no significant difference in bird diversity between the two systems. However, the total abundance of birds was twice as high, with three times as many insectivores in the diverse shade system. DeClerck and Martínez Salinas (see Chapter 3 in this book) likewise found no significant difference in migratory bird richness between simple shade and diverse shade systems, although abundance values are slightly greater in the diverse shade.

Additional foraging resources in multi-strata systems provided in the canopy may explain the difference (Van Bael et al, 2008). Many other studies, however, found no effect of farm tree cover on arthropod removal (Greenberg et al, 2000; Johnson et al, 2009; Philpott et al, 2009), with equal proportions of arthropods removed in sun coffee systems as well as shaded coffee systems. This may be partially explained by the high dispersal ability of birds, permitting them to seek forest cover for roosting, yet move into sun coffee systems for foraging.

Arthropods and pest control services

Pests are known to be especially severe in crops that do not originate in the region in which they are cultivated because they lack predators and parasitoids. Contrary to Africa and Asia where cacao mirids (*Sahlbergella singularis* and *Distantiella theobroma*) and cacao pod borer (*Conopomorpha cramerella*) are very serious new encounter pests of cacao, in the Neotropics cacao pests are of lesser importance than diseases (Vargas et al, 2005; Delabie et al, 2007). A reason could be that in cacao plantations in Africa and Asia, natural control mechanisms do not suffice to reduce the impact of these pests to economically supportable levels due to the exogenous nature of the cacao tree. According to Delabie et al (2007), the scarcity of insect pests in Bahian cacao systems (Brazil) is probably due to a predatory function exercised by ants and to the limited use of insecticides that most likely contributes to an insect equilibrium between pests and predators. Similarly, coffee berry borer is especially severe in Latin America and almost negligible in Africa (Wegbe et al, 2003), where coffee and the coffee berry borer originate and where numerous parasitoids of the borer can be found (Vega et al, 1999). Three of them, *Prorops nasuta*, *Cephalonomia stephanoderis* and *Phymastichus coffea*, were introduced from Africa to Latin America to help regulate coffee berry borer populations (Barrera et al, 1990; Espinoza et al, 2009).

In the Neotropics the most well-known cacao insect pests are mirids of the genus *Monalonion* and thrips. Mirids of the genus *Termatophylidae* are known predators of thrips (McCallan, 1975) and ants control thrips as well as mirids (Philpott and Armbrecht, 2006; Delabie et al, 2007). Ants are an important component of tropical biodiversity. They comprise a large fraction of the animal biomass and are among the major pest predators in tropical agroforestry systems (Philpott and Armbrecht, 2006). In Bahia, Brazil, two ant species (*Azteca chartifex spiriti* and *Ectatomma tuberculatum*) are recognized for their role as bio-control agents of cacao thrips and mirids (Delabie et al, 2007). In Mesoamerican coffee agroecosystems, ant species *A. instabilis* (Perfecto and Vandermeer, 2006) and *Solenopsis* cf. *picea* (Armbrecht and Gallego, 2007) have been reported to prey on coffee berry borer. Perfecto and Vandermeer (2006) showed that the presence of coccids (*Coccus viridis*) was important for coffee berry borer control through a complex mutual relationship between ants and the coccids. Predation of the Mediterranean fruit fly

(*Ceratitis capitata*) by the ants *Solenopsis* sp. and *Pheidole geminata* has also been reported in coffee plantations (Armbrecht and Perfecto, 2003). In addition, coffee leaf damage by leaf miners (*Leucoptera coffeella*) was significantly lower where abundance of twig-nesting ants was higher (De la Mora et al, 2008). Shaded coffee and cacao agroforests provide a refuge for biodiversity and enhance natural pest control. For instance, in Brazil, a high diversity of shade trees in cacao agroforests has been reported to favour parasitoid populations, which are potential natural enemies of cacao pests, due to increased resources (Sperber et al, 2004). On the contrary, reduction of plant diversity in coffee and cacao agroecosystems has led to losses of ant diversity due to microclimatic changes and nest site limitation (Philpott and Armbrecht, 2006). Ants appear to be important predators mainly in shaded plantations, as reported by Armbrecht and Gallego (2007) in the case of coffee berry borer predation. Ant species richness of ground-foraging ants in coffee plantations has also been shown to decrease with distance from forest fragments (Perfecto and Vandermeer, 2002). Similarly, Armbrecht and Perfecto (2003) found a reduction of twig-nesting ant species richness as the distance from forest fragments increased in coffee plantations shaded by *Inga*. However, in both cases, the reduction in ant species richness was limited or even reversed in coffee plantations with diversified and dense shade. These results provide evidence of the need to consider a broader scale (beyond farm boundaries) in order to foster biological control in agroecosystems (Coll, 2009).

Microorganisms and pest and disease control services

Many fungi and bacteria contribute to natural bio-control of pests and diseases. Four key mechanisms are sometimes involved together: parasitism, antibiosis, competition for resources and induced resistance. Beneficial microorganisms often cited are *Beauveria bassiana*, *Metarhizium anisopliae*, *Lecanicillium* spp., *Trichoderma* spp., *Bacillus* spp. and *Pseudomonas* spp. (Jacques et al, 1993; Verma et al, 2007; Lugtenberg and Kamilova, 2009; Vega et al, 2009). These organisms are either found as epiphytes, endophytes or sometimes both. Endophytes are especially interesting as they colonize host tissues without causing harm and can even establish mutualistic associations with plants. Mycorrhizal fungi, which form symbiotic associations with plant roots, may help plants to better tolerate stresses, especially biotic stresses, mostly by providing the plant with more nutrients, which leaves the trees in better general health (as discussed earlier) (Harrier and Watson, 2004).

A large variety of fungal and bacterial epiphytes are present on the aerial parts of cacao and coffee. Some of these epiphytes are pathogen antagonists. For instance, epiphytic fungal parasites of the genera *Clonostachys* and *Fusarium*, potential agents for bio-control of black pod (*Phytophthora palmivora*) and frosty pod rot, were detected on cacao flowers and pods in Costa Rica (ten Hoopen et al, 2003). Additionally, a study by Melnick et al (2008) showed that bacteria of the genus *Bacillus* were capable of colonizing cacao leaves, primarily as epiphytes but also as endophytes. Their presence led to a significant decrease in disease severity when the leaves were challenged with *P. capsici*. Moreover, it seemed that one *Bacillus* isolate brought about a disease suppression mechanism of induced systemic resistance in cacao. Natural bio-control of pathogens by beneficial microorganisms can be established through complex ecological webs. Bio-control of *Hemileia vastatrix* appears to be enhanced by ant-coccid mutualism favouring the presence of abundant scale insects in coffee trees infected by *Lecanicillium lecanii*, which is, additionally, a hyper-parasite of the coffee rust fungus (Vandermeer et al, 2009).

Similar to the leaf surface (phyllosphere), the root surface (rhizosphere) contains many soil-borne disease antagonists. Castro (1995) and Mendoza García et al (2003) showed that fungal antagonists could help to control *Rosellinia* spp., causal agents of soil-borne diseases of coffee and cacao. In addition, Castro-Toro and Rivillas-Osorio (2002) showed the usefulness of mycorrhiza in controlling *Rosellinia* root rot. Similarly, Vaast et al (1998) demonstrated that mycorrhizal coffee plants exhibited enhanced tolerance to the root lesion nematode *Pratylenchus coffeae*. Lesions were fewer and more localized in mycorrhizal coffee plants than in non-mycorrhizal controls. Soil is also an important reservoir for entomopathogenic fungi. For instance, coffee berry borer is susceptible to *Beauveria bassiana*, which has been found to infect the pest naturally throughout America, with levels of infection reaching 44 per cent in some cases (Carrion and Bonet, 2004; Vera-Montoya et al, 2007; Monzon et al, 2008). This fungus has been used as an inundative biological control agent of coffee berry borer, particularly in Colombia (Benavides-Machado et al, 2003; Posada-Florez, 2008). The main challenge of biological control with *B. bassiana* is to find strains with a high ability to infect insects in diverse habitats. First, the fungus is considered a facultative insect pathogen and some strains have a limited effect on coffee berry borer. Moreover, *B. bassiana* genotypes are highly associated with their habitat, which indicates that habitat is an important selection factor (Bidochka et al, 2002). The preceding may explain some failures in control when using elite clonal strains in habitats very different from those in which they originated. A new strategy for improving the biological control of coffee berry borer is the use of strain mixtures adapted to different environmental conditions (Cruz et al, 2006).

Increasing evidence suggest that fungal endophytes of cacao, particularly the genera *Trichoderma* and *Clonostachys*, may be of interest in protecting cacao against pests and diseases (Arnold et al, 2003; Bailey et al, 2008; Mejia et al, 2008). Several new species of endophytic *Trichoderma*, cacao pathogen antagonists, have been described recently (Samuels et al, 2006; Hanada et al, 2008). One such endophytic *Trichoderma* species, *T. stromaticum* (Samuels et al, 2000), probably a co-evolved antagonist of *Moniliophthora perniciosa*, is now available as a commercial product with promising results for the management of witches' broom in Brazil (Loguercio et al, 2009a). This interesting application also shows why conservation of the biodiversity of host plants and their associated fungi in their natural habitat (*in-situ* conservation, discussed above) is of such great importance.

Bio-control by fungi of pests and diseases affecting plant aerial organs seems to be improved by shade, which intercepts solar radiation and promotes wetness in the plantation, which in turn favours propagule viability and infection. In particular, Staver et al (2001) suggested that *Beauveria bassiana* and *Lecanicillium lecanii* found better microclimatic conditions for their development and bio-control in coffee plantations under managed shade compared to full sun exposure. Similarly, although mixed bio-control inocula (*Clonostachys rosea* and *Trichoderma* spp.) were able to reduce incidences of frosty pod rot, witches' broom and black pod rot (*Phytophthora palmivora*) in shaded and non-shaded cacao in Peru, the reduced disease loss attributed to the control only resulted in a net economic return in shaded plantations (Krauss and Soberanis, 2001). Different isolates of *Trichoderma stromaticum*, antagonist of *Moniliophthora perniciosa*, have also been reported to respond differently to microclimate variation in terms of sporulation and antagonism at different cacao canopy levels (Loguercio et al, 2009b). Litter and pruning residues of shade trees may also play an important role in soil pest and disease control through the improvement of soil microbial activity levels, particularly that of antagonists (Wardle et al, 1995; Altieri, 1999). This effect has been suggested in Costa Rica to explain the negative relationship between soil organic matter content and population densities of *Meloidogyne exigua* in coffee roots (Avelino et al, 2009). In Papua New Guinea, leaf litter mulch has been reported to reduce the survival of *P. palmivora* under cacao trees by accelerating substrate decomposition and by stimulating the activity of antagonistic and hyper-parasitic microbes (Konam and Guest, 2002).

Undesirable effects of biodiversity

Even though the overall balance of biodiversity at plot and landscape scales seems to favour noxious organism control, especially pest control (Andow, 1991; Altieri, 1999; Bianchi et al, 2006), some undesirable effects exist. Most negative effects of biodiversity are related to support plants serving as alternative hosts for pathogens or altering microclimate conditions in favour of pests and diseases (see Figure 4.3).

Some plant species may constitute alternative hosts or reservoirs for pests and pathogens at plot and landscapes scales (Schroth et al, 2000; Plantegenest et al, 2007). Actually, several pests and diseases of coffee and cacao are not specific to these two crops. For instance, *Phytophthora palmivora*, *P. citrophthora* and *P. capsici*, which cause cacao black pod rot in the Neotropics, can affect numerous hosts that are often found in or near cacao plantations, such as citrus species, squash crops, chillies or papaya. Similarly, *Moniliophthora roreri* is able to infect different plants of the *Theobroma* and *Herrania* genera, including *T. bicolor* (Evans et al, 2003), which is often intercropped with cacao trees,

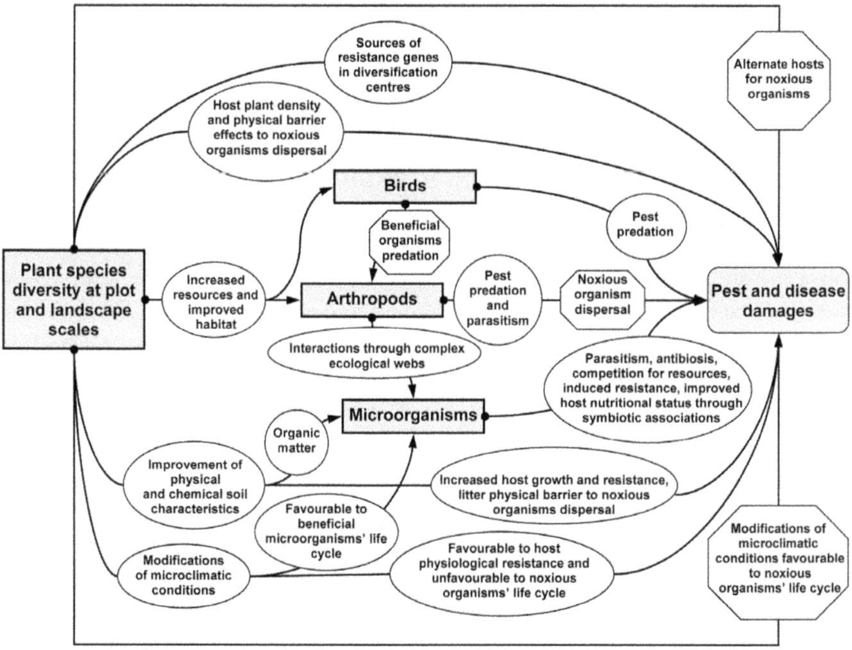

Figure 4.3 *A schematic overview of pathways and effects of biodiversity (rectangles) on pests and diseases of coffee and cacao agroforests: Despite some undesirable effects of biodiversity (octagons), the overall balance of biodiversity at plot and landscape scales seems to favour control of noxious organisms (ovals)*

Source: chapter authors

especially in Mexico. Other examples are the pathogens *Corticium salmonicolor* (now *Erythricium salmonicolor*), causal agent of pink disease, *Corticium koleroga*, responsible for the white thread blight, and *Rosellinia* spp. that affect a great number of plant species. Some of them are forestry species that could be used as shade trees in rustic systems (Benchimol et al, 2001; Roux and Coetzee, 2005; ten Hoopen and Krauss, 2006). Similarly, *Mycena citricolor* is able to attack at least 150 plant species belonging to 45 families, including legume trees of the genus *Inga* (Sequeira, 1958), which is commonly used as shade tree in Mesoamerica. However, Sequeira (1958) did not observe formation of pathogen propagules on most of these susceptible plants. Coffee berry borer, despite being considered specific to coffee, is able to find refuge and reproduce in other fruits (Damon, 2000; Gumier-Costa, 2009). *Xylella fastidiosa*, causal agent of the leaf scorch disease on coffee, also has a wide range of hosts and insect vectors. On citrus, *X. fastidiosa* causes the citrus variegated chlorosis (CVC). Li et al (2001) demonstrated that a strain of *X. fastidiosa* isolated from citrus plants was pathogenic for coffee plants, indicating that citrus, which is frequently planted adjacent to or intercropped with coffee, can be a source of inoculum for coffee.

Some pests and pathogens may be favoured by the microclimatic conditions of shaded plantations. In coffee plantations, shade is known to favour *Mycena citricolor* (Avelino et al, 2007), *Corticium koleroga* and *Corticium salmonicolor* (Schroth et al, 2000), probably because shading promotes higher wetness. Black pod rot of cacao is also favoured by increased wetness in shaded plantations (Beer et al, 1998; Schroth et al, 2000). Likewise, coffee berry borer seems to be favoured by shade (Bosselmann et al, 2009), particularly by dense shade (Feliz Matos et al, 2004), possibly in relation to the higher relative humidity which increases insect longevity and fecundity (Baker et al, 1994). The effects of shade on pests and diseases are usually not clear, as shade may favour a given process of the life cycle of a noxious organism and hamper another process at the same time. The balance of these antagonistic effects is variable and often controversial, as is the case of shade effects on coffee rust. As mentioned before, a high shade percentage may reduce coffee rust attacks by moderating yields, which could partly explain results obtained by Soto-Pinto et al (2002) in Mexico. However, shade also buffers temperatures, intercepts light and increases moisture in the plantation, all factors favourable to the infection process (Avelino et al, 2004), which can probably explain the opposite results observed in Central America (Staver et al, 2001; Avelino et al, 2006). Antagonistic effects of shade on black pod rot also exist. Work by Monteith and Butler (1979) and Butler (1980) showed that wind speed plays an important role in the duration of condensed water on cocoa pods, and pod wetness duration affects black pod rot. Increasing wind speed in the cocoa canopy by reducing shade tree density and even by making occasional openings in the cacao canopy will reduce pod wetness duration, which in turn reduces the occurrence of black pod rot. Yet, balanced against this will be the changing pattern of rainfall. Without a shade canopy, direct rain hitting the pod surface and the soil may increase intensity and velocity of raindrop splash, which, when combined with increased wind velocity, could result in increased long-range spore dispersal (Evans, 1998).

Natural enemies of pests and pathogens may also affect other beneficial organisms. For instance, it is difficult to know whether bird predation on arthropod communities has a significant net positive, negative or neutral effect since arthropods fall within multiple trophic levels (herbivores, predators and parasites). Greenberg et al (2000) noted that bird predation was mostly affecting leaf chewers (Orthoptera). However, predatory arthropods such as spiders were also significantly affected. Moreover, beneficial organisms can contribute to pathogen dispersal. The ant *Ectatomma tuberculatum*, which protects cacao from mirids and thrips, also facilitates the dissemination of the economically important black pod rot (Delabie et al, 2007).

Quantifying and Valuing Pest and Disease Control Services

Managing plots, farms and landscapes for the provision of pest and disease control services is far from straightforward and simple. Traditional control mechanisms target oversimplification: identify the problem and apply a targeted solution (literally the case when using agrochemicals and even when implementing biological control). Tropical countries are well known globally for their hyper-diversity, with numerous and complex inter-specific interactions that are the crux of natural pest and disease control (see Figure 4.3). Gliessman (1998) suggests that, ultimately, understanding the ecological basis for how diversity operates and taking advantage of complexity, rather than striving to eliminate it, is the only strategy leading to sustainability. Pest control as an ecosystem service is based on embracing this complexity. The question remains as to how to quantify these services, and to what degree of understanding the specifics of this complexity is needed. One of the advantages of ecosystem service-based management strategies is that they are based on complex adaptive systems that target the overall health of the plantation, rather than targeting any single disease. Managing for biodiversity, in theory, should increase the resistance of the parcel to the arrival of new pest and disease outbreaks, and should maintain these below epidemic levels (Andow, 1991; Altieri, 1999; Tilman et al, 2002; Bianchi et al, 2006; Cheatham et al, 2009; Malezieux et al, 2009).

Pest and disease control services can be quantified in terms of avoided crop losses. When assessing crop losses, the absolute reference is the attainable yield – that is, the yield and quality obtained using the fully available technology without any losses due to pests and diseases (see Figure 4.4) (Chiarappa, 1971; Zadoks and Schein, 1979; Savary et al, 2006). The attainable yield is theoretically independent of economic factors. Each pest and pathogen considered separately causes specific crop losses according to its particular effect on plant organs. Crop losses due to certain pests and diseases are relatively easy to quantify at different scales (plantation, regional and national). This is especially true for those cases where the disease or pest affects the harvested part of the plant. It becomes more difficult in cases where diseases and pests have an indirect effect on production, as with coffee rust or mirids of cacao that cause a reduction in leaf area with concomitant reductions in photosynthetic ability. This difficulty is even greater when considering the impacts upon product quality, especially sensory quality, whose development at the field level is still incompletely understood. In addition, different pests and pathogens are normally found in combination. Crop losses caused by different injury profiles, such as combinations of different injury levels caused by pests and diseases (see Figure 4.4), are generally less than the sum of the losses caused by each pest and disease as interactions among injuries occur (Savary et al, 1997, 2006). Despite its importance, reliable information on crop losses caused by pests and

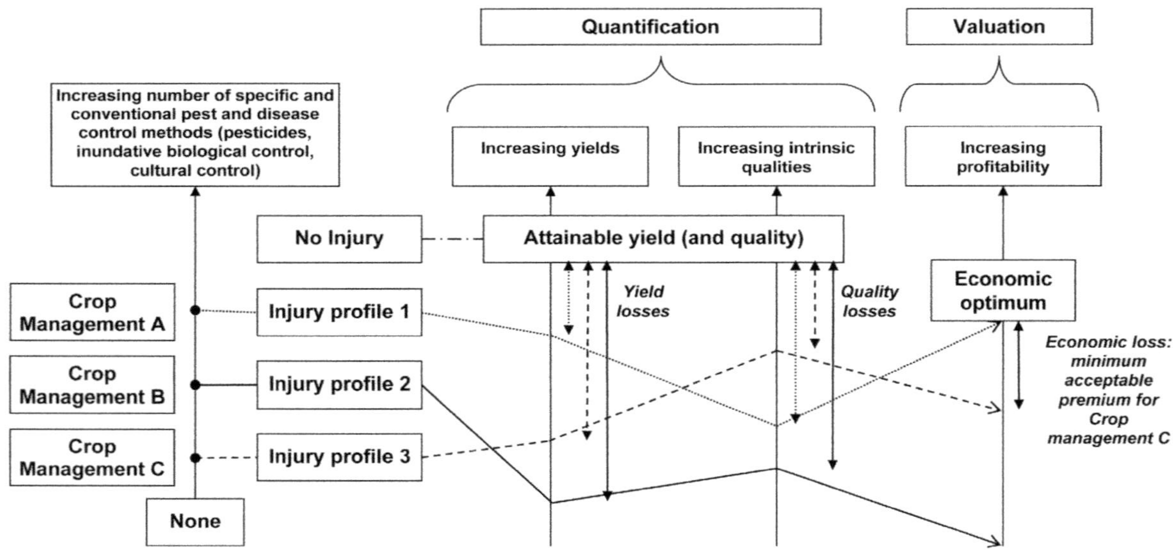

Figure 4.4 *Quantifying and valuing pest and disease control services as a function of crop losses and economic losses*

Notes: The figure represents the sequence of events leading to economic losses in a given physical context under different management scenarios. Crop management A is an intensified system whose yield is close to the economic yield (i.e. the yield obtained using only cost-effective practices). Most of the pest and disease attacks reach low to medium levels. Crop management B is an imperfectly managed plot. Its yield is low due to pest and disease attacks and low fertility management. Most of the pest and disease attacks reach medium to high levels. Pests and diseases are the same as in case A. In crop management C, an ecological strategy of pest and disease control is employed. Pests and diseases presented in cases A and B reach low levels and product quality is improved. However, other pests and diseases have emerged and yield is affected. Although pest and disease control services primarily affect injury levels caused by pests and diseases, services must be quantified in terms of avoided crop losses (yield and quality). In these hypothetical scenarios, the ecological crop management C enables avoidance of quality losses only. The minimum acceptable premium to ensure sustainability of crop management C is calculated utilizing the economic optimum as a reference.
Injury profile: a combination of injury levels caused by a range of pests and diseases (Savary et al, 2006).
Attainable yield: the yield and quality obtained using available technology (eventually non-economic production methods), which means without any losses, particularly without losses due to pests and diseases (Chiarappa, 1971; Zadoks and Schein, 1979; Savary et al, 2006). Quality may include sanitary attributes, as well as physical, chemical and sensory attributes of the product. Normally, high yields are not associated with high physical, chemical and sensory qualities.
Source: chapter authors

diseases is scarce (Savary et al, 2006). This lack of information is a handicap when quantifying and valuing pest and disease control services.

Managing trade-offs remains one of the biggest limitations to improving management of pest and disease control services. In this review, we highlighted several cases where management interventions that promote the control of one pest lead to the rise of a second pest. For example, increasing tree cover in coffee plantations provides habitat for beneficial insects and insectivorous birds, yet can alter the plantations' microclimatic conditions to favour fungal pathogens. Here, environmental management to reduce one impact, pest damage, could result in a different impact: disease losses. We might consider that there is a gain in pest and disease control services only if the new pathogen is less destructive than the one now under control.

Although yield differences between different management options are important, valuing pest and disease control services should also include the direct and indirect costs of the management intervention. Crop losses must be translated into economic terms, which should also include the financial cost of management, including labour and materials as well as the cost of maintaining the service through time. In that case, the reference for the calculation of economic losses is the economic optimum yield, such as the yield obtained using only cost-effective practices (see Figure 4.4) (Chiarappa 1971; Zadoks and Schein, 1979; Savary et al, 2006). In an interesting study, Bisseleua et al (2009) tried to establish a link between biodiversity and net income for cacao plantations in Cameroon. Their study showed that there was no simple trade-off between biodiversity and net income. Although simplification, reducing or eliminating shade is often recommended for increasing cacao yield, their study showed that the highest yield and income corresponded to 40 to 50 per cent shade cover. In part this is because reducing shade led to an increase in insect herbivory and a reduction in ant diversity.

As an illustration of how pest management as an ecosystem service can be quantified based on crop losses, consider three crop management (CM) patterns ranging from A to C according to the number of specific and conventional pest and disease control methods employed (see Figure 4.4). Crop management A (CM-A) is an intensified system whose yield is close to the economic optimum yield. It is characterized by a high number of pest and disease control methods, including pesticides, inundative biological control and/or cultural control. Most of the pest and disease attacks reach low to medium levels. Crop management B (CM-B) is an imperfectly managed plot. Its yield is low due to pests and diseases attacks and low fertility management. Most of the pest and disease attacks reach medium to high levels. Pests and diseases are the same as in case A. In contrast, in crop management C (CM-C), an ecological strategy of pest and disease control is employed. Pests and diseases present in cases A and B reach low levels. However, other pests and diseases may emerge and affect yield negatively.

The injury profiles of each of these management scenarios are distinct and are not equivalent in terms of crop losses. Each crop management, with its corresponding injury profile, leads to a distinct proportion of the attainable yield. In these hypothetical scenarios, CM-A most closely reaches this yield, whereas CM-C offers a lower yield, but a greater yield than the imperfectly managed plot (CM-B). As a consequence, despite controlling some pests and diseases, the hypothetical CM-C case does not provide a sufficient pest and disease control service in terms of avoided yield losses. The service becomes obvious only when considering product quality, which is improved in CM-C with respect to CM-A and CM-B.

Unlike many other ecosystem services discussed in this book (e.g. carbon and hydrological services), farmers are the direct recipients in terms of reducing losses to pests and pathogens. However, maximizing pest and disease control services can have important indirect benefits received off farm or in neighbouring communities through improved product quality, environmental health or increased water quality due to a reduced reliance on agrochemicals (Wilson and Tisdell, 2001; Abhilash and Singh, 2009). Other possible impacts are the improved health of farm labourers, the increased pollinator populations, or even the additional carbon sequestered when agroforestry systems add to the qualities of the system. These indirect services can be considered as intrinsic qualities of the system. By intrinsic value we mean the additional ecosystem services and benefits that are provided by the ecologically based pest and disease control services. This intrinsic quality does not necessarily contribute to the attainable yield. However, this quality has value that must be taken into account in potential decreased economic value associated with ecologically based pest and disease control services. When the intrinsic quality includes services received off farm, then additional payment schemes such as certification or direct payments should be considered.

In order to ensure the sustainability of CM-C and its benefits for the farm and society, a premium has to be paid. The minimum acceptable premium can be calculated based on the differential between the economic optimum, the highest profitability (CM-A) and the profitability of CM-C. This differential is the economic loss accepted by CM-C to provide services to the farm and society and that have to be at least compensated by the premium. However, this minimum premium can be reduced substantially when the highest profitability is low – for instance, in case of low international prices of the product, indicating that additional payments for other services provided might be necessary.

Conclusions

There is still insufficient knowledge of the ecological mechanisms of pest and disease control. Understanding the complexity of these mechanisms and of their interrelationships is a necessary step still missing in order to apply them to the management of agroecosystems. This is certainly one of the greatest challenges for sustainable agriculture in the coming decades. As long as we fail to understand how this complexity works in its entirety, it is likely that agroecosystems will remain highly simplified compared to natural ecosystems. These simplified systems will continue to yield ecological imbalances or pest and disease outbreaks that will lead to yield losses if not corrected by specific interventions such as pesticide applications, with all its negative externalities.

Agriculture must increase its production if it is to feed, clothe and fuel the growing global population while simultaneously reducing its environmental impact. We think this goal cannot be reached if we only consider intensified agriculture, which despite its high yields, is overly reliant on pesticides. We suggest that a two-pronged approach is needed. First, pesticide use in intensified systems must obviously be reduced, while retaining as high a yield as possible. This can be achieved by optimizing chemical control, but also by accompanying this optimization with ecologically based pest and disease control interventions. We must, however, be prepared for possible and/or temporary yield reductions. Second, yields of rustic or low technology systems must be increased while maintaining ecological functions of pest and disease control at high levels. Better management of biodiversity associated with the agroecosystem could probably help to attain these levels. Payments for services have an important role to play in promoting the development of systems that are less dependent on pesticides, while maintaining or even improving yield and quality.

Acknowledgements

We are grateful for the financial support provided by the European Union via the CAFNET (Connecting, Enhancing and Sustaining Environmental Services and Market Values of Coffee Agroforestry in Central America, East Africa and India) collaborative project and by CIRAD through the Omega3 Ecological Mechanisms of Pest and Disease Management Optimized to Sustainably Improve Agrosystem Productivity project. Support was also provided by the Pôle de Compétences en Partenariat (PCP) Agroforestry Systems with Perennial Crops project, CIRAD–CATIE–INCAE–Bioversity–CABI–Promecafé.

References

Abhilash, P. C. and Singh, N. (2009) 'Pesticide use and application: An Indian scenario', *Journal of Hazardous Materials*, vol 165, pp1–12

Altieri, M. A. (1999) 'The ecological role of biodiversity in agroecosystems', *Agriculture, Ecosystems & Environment*, vol 74, pp19–31

Andow, D. A. (1991) 'Vegetational diversity and arthropod population response', *Annual Review of Entomology*, vol 36, pp561–586

Armbrecht, I. and Gallego, M. C. (2007) 'Testing ant predation on the coffee berry borer in shaded and sun coffee plantations in Colombia', *Entomologia Experimentalis et Applicata*, vol 124, pp261–267

Armbrecht, I. and Perfecto, I. (2003) 'Litter-twig dwelling ant species richness and predation potential within a forest fragment and neighboring coffee plantations of contrasting habitat quality in Mexico', *Agriculture, Ecosystems & Environment*, vol 97, pp107–115

Arnold, A. E., Mejia, L. C., Kyllo, D., Rojas, E. I., Maynard, Z., Robbins, N. and Herre, E. A. (2003) 'Fungal endophytes limit pathogen damage in a tropical tree', *Proceedings of the National Academy of Sciences of the United States of America*, vol 100, pp15649–15654

Avelino, J., Willocquet, L. and Savary, S. (2004) 'Effects of crop management patterns on coffee rust epidemics', *Plant Pathology*, vol 53, pp541–547

Avelino, J., Zelaya, H., Merlo, A., Pineda, A., Ordonez, M. and Savary, S. (2006) 'The intensity of a coffee rust epidemic is dependent on production situations', *Ecological Modelling*, vol 197, pp431–447

Avelino, J., Cabut, S., Barboza, B., Barquero, M., Alfaro, R., Esquivel, C., Durand, J. F. and Cilas, C. (2007) 'Topography and crop management are key factors for the development of American leaf spot epidemics on coffee in Costa Rica', *Phytopathology*, vol 97, pp1532–1542

Avelino, J., Bouvret, M.-E., Salazar, L. and Cilas, C. (2009) 'Relationships between agro-ecological factors and population densities of *Meloidogyne exigua* and *Pratylenchus coffeae sensu lato* in coffee roots, in Costa Rica', *Applied Soil Ecology*, vol 43, pp95–105

Babin, R., ten Hoopen, G. M., Cilas, C., Enjalric, F., Yede, Gendre, P. and Lumaret, J. P. (2010) 'Impact of shade on the spatial distribution of *Sahlbergella singularis* in traditional cocoa agroforests', *Agricultural and Forest Entomology*, vol 12, pp69–79

Bailey, B. A., Bae, H., Strem, M. D., Crozier, J., Thomas, S. E., Samuels, G. J., Vinyard, B. T. and Holmes, K. A. (2008) 'Antibiosis, mycoparasitism, and colonization success for endophytic *Trichoderma* isolates with biological control potential in *Theobroma cacao*', *Biological Control*, vol 46, pp24–35

Baker, P. S., Rivas, A., Balbuena, R., Ley, C. and Barrera, J. F. (1994) 'Abiotic mortality factors of the coffee berry borer (*Hypothenemus hampei*)', *Entomologia Experimentalis et Applicata*, vol 71, pp201–209

Baker Engelbrecht, C. J. and Harrington, T. C. (2005) 'Intersterility, morphology and taxonomy of *Ceratocystis fimbriata* on sweet potato, cacao and sycamore', *Mycologia*, vol 97, pp57–69

Bannon, F. J. and Cooke, B. M. (1998) 'Studies on dispersal of *Septoria tritici pycnidiospores* in wheat-clover intercrops', *Plant Pathology*, vol 47, pp49–56

Barrera, J. F., Baker, P. S., Valenzuela, J. E. and Schwarz, A. (1990) 'Introduction of two African parasitoid species to Mexico for biological control of the coffee borer *Hypothenemus hampei* (Ferrari) (Coleoptera: Scolytidae)', *Folia Entomologica Mexicana*, vol 79, pp245–247

Beer, J., Muschler, R., Kass, D. and Somarriba, E. (1998) 'Shade management in coffee and cacao plantations', *Agroforestry Systems*, vol 38, pp139–164

Benavides-Machado, P., Bustillo-Pardey, A. E., Cárdenas-Murillo, R. and Montoya-Restrepo, E. C. (2003) 'Análisis biológico y económico del manejo integrado de la broca del café en Colombia', *Cenicafé*, vol 54, pp5–23

Benchimol, R. L., Poltronieri, L. S., Trindade, D. R. and Albuquerque, F. C. (2001) 'White-thread blight: Five new hosts in the state of Para, Brazil', *Fitopatologia Brasileira*, vol 26, p778

Bhar, R. and Fahrig, L. (1998) 'Local vs. landscape effects of woody field borders as barriers to crop pest movement', *Conservation Ecology (online)*, vol 2, www.consecol.org/vol2/iss2/art3

Bianchi, F., Booij, C. J. H. and Tscharntke, T. (2006) 'Sustainable pest regulation in agricultural landscapes: A review on landscape composition, biodiversity and natural pest control', *Proceedings of the Royal Society B: Biological Sciences*, vol 273, pp1715–1727

Bidochka, M. J., Menzies, F. V. and Kamp, A. M. (2002) 'Genetic groups of the insect-pathogenic fungus *Beauveria bassiana* are associated with habitat and thermal growth preferences', *Archives of Microbiology*, vol 178, pp531–537

Bisseleua, D. H. B., Missoup, A. D. and Vidal, S. (2009) 'Biodiversity conservation, ecosystem functioning, and economic incentives under cocoa agroforestry intensification', *Conservation Biology*, vol 23, pp1176–1184

Bosselmann, A. S., Dons, K., Oberthur, T., Olsen, C. S., Raebild, A. and Usma, H. (2009) 'The influence of shade trees on coffee quality in small holder coffee agroforestry systems in Southern Colombia', *Agriculture, Ecosystems & Environment*, vol 129, pp253–260

Burdon, J. J. and Chilvers, G. A. (1982) 'Host density as a factor in plant disease ecology', *Annual Review of Phytopathology*, vol 20, pp143–166

Butler, D. R. (1980) 'Dew and thermal lag: Measurements and an estimate of wetness duration on cocoa pods', *Quarterly Journal of the Royal Meteorological Society*, vol 106, pp539–550

Carrion, G. and Bonet, A. (2004) 'Mycobiota associated with the coffee berry borer (Coleoptera: Scolytidae) and its galleries in fruit', *Annals of the Entomological Society of America*, vol 97, pp492–499

Castro, B. L. (1995) 'Antagonismo de algunos aislamientos de *Trichoderma koningii* originados en suelo Colombiano contra *Rosellinia bunodes*, *Sclerotinia sclerotiorum* y *Pythium ultimum*', *Fitopatologia Colombiana*, vol 19, pp7–18

Castro-Toro, A. M. and Rivillas-Osorio, C. A. (2002) '*Entrophospora colombiana*, *Glomus manihotis* y *Burkholderia cepacia* en el control de *Rosellinia bunodes*, agente causante de la llaga negra del cafeto', *Cenicafé*, vol 53, pp193–218

Cheatham, M. R., Rouse, M. N., Esker, P. D., Ignacio, S., Pradel, W., Raymundo, R., Sparks, A. H., Forbes, G. A., Gordon, T. R. and Garrett, K. A. (2009) 'Beyond yield: Plant disease in the context of ecosystem services', *Phytopathology*, vol 99, pp1228–1236

Chiarappa, L. (1971) *Crop Loss Assessment Methods: FAO Manual on the Evaluation and Prevention of Losses by Pests, Diseases and Weeds*, FAO/CAB, Slough, UK

Coll, M. (2009) 'Conservation biological control and the management of biological control services: Are they the same?', *Phytoparasitica*, vol 37, pp205–208

Condeso, T. E. and Meentemeyer, R. K. (2007) 'Effects of landscape heterogeneity on the emerging forest disease sudden oak death', *Journal of Ecology*, vol 95, pp364–375

Cruz, L. P., Gaitan, A. L. and Gongora, C. E. (2006) 'Exploiting the genetic diversity of *Beauveria bassiana* for improving the biological control of the coffee berry borer through the use of strain mixtures', *Applied Microbiology and Biotechnology*, vol 71, pp918–926

DaMatta, F. M. (2007) 'Ecophysiology of tropical tree crops: An introduction', *Brazilian Journal of Plant Physiology*, vol 19, pp239–244

Damon, A. (2000) 'A review of the biology and control of the coffee berry borer, *Hypothenemus hampei* (Coleoptera: Scolytidae)', *Bulletin of Entomological Research*, vol 90, pp453–465

De la Mora, A., Livingston, G. and Philpott, S. M. (2008) 'Arboreal ant abundance and leaf miner damage in coffee agroecosystems in Mexico', *Biotropica*, vol 40, pp742–746

Delabie, J. H. C., Jahyny, B., do Nascimento, I. C., Mariano, C. S. F., Lacau, S., Campiolo, S., Philpott, S. M. and Leponce, M. (2007) 'Contribution of cocoa plantations to the conservation of native ants (Insecta: Hymenoptera: Formicidae) with a special emphasis on the Atlantic Forest fauna of southern Bahia, Brazil', *Biodiversity and Conservation*, vol 16, pp2359–2384

Dordas, C. (2008) 'Role of nutrients in controlling plant diseases in sustainable agriculture: A review', *Agronomy for Sustainable Development*, vol 28, pp33–46

Echandi, E. (1969) 'La chasparria de los cafetos causada por el hongo *Cercospora coffeicola* Berk. and Cooke', *Turrialba*, vol 9, pp54–67

Engelbrecht, C. J., Harrington, T. C. and Alfenas, A. (2007) 'Ceratocystis wilt of cacao: A disease of increasing importance', *Phytopathology*, vol 97, pp1648–1649

Espinoza, J. C., Infante, F., Castillo, A., Perez, J., Nieto, G., Pinson, E. P. and Vega, F. E. (2009) 'The biology of *Phymastichus coffea* LaSalle (Hymenoptera: Eulophidae) under field conditions', *Biological Control*, vol 49, pp227–233

Evans, H. C. (1981) 'Witches' broom disease: A case study', *Cocoa Growers' Bulletin*, pp5–19

Evans, H. C. (1998) 'Disease and sustainability in the cocoa agroecosystem', in *First International Workshop on Sustainable Cocoa Growing*, Smithsonian Tropical Research Institute, Smithsonian Migratory Bird Center, Panama City

Evans, H. C., Krauss, U., Rios Rutz, R., Zecevich Acosta, T. and Arevalo-Gardini, E. (1998) 'Cocoa in Peru', *Cocoa Growers' Bulletin*, pp7–22

Evans, H. C., Holmes, K. A. and Reid, A. P. (2003) 'Phylogeny of the frosty pod rot pathogen of cocoa', *Plant Pathology*, vol 52, pp476–485

FAO (United Nations Food and Agriculture Organization) (2010) *FAOSTAT Statistical Database*, http://faostat.fao.org

Feliz Matos, D., Guharay, F. and Beer, J. (2004) 'Incidence of the coffee berry borer (*Hypothenemus hampei*) in coffee plants under different shade types in San Marcos, Nicaragua', *Agroforesteria en las Americas*, vol 41/42, pp56–61

Geiger, F., Bengtsson, J., Berendse, F., Weisser, W. W., Emmerson, M., Morales, M. B., Ceryngier, P., Liira, J., Tscharntke, T., Winqvist, C., Eggers, S., Bommarco, R., Pärt, T., Bretagnolle, V., Plantegenest, M., Clement, L. W., Dennis, C., Palmer, C., Oñate, J. J., Guerrero, I., Hawro, V., Aavik, T., Thies, C., Flohre, A., Hänke, S., Fischer, C., Goedhart, P. W. and Inchausti, P. (2010) 'Persistent negative effects of pesticides on biodiversity and biological control potential on European farmland', *Basic and Applied Ecology*, vol 11, pp97–105

Gliessman, S. R. (1998) 'Agroecology: Researching the ecological processes in sustainable agriculture', in C. H. Chou and K. T. Shan (eds) *Frontiers in Biology: The Challenge of Biodiversity, Biotechnology, and Sustainable Agriculture*, Academia Sinica Taipei, Taiwan

Gomez-Rodriguez, O., Zavaleta-Mejia, E., Gonzalez-Hernandez, V. A., Livera-Munoz, M. and Cardenas-Soriano, E. (2003) 'Allelopathy and microclimatic modification of intercropping with marigold on tomato early blight disease development', *Field Crops Research*, vol 83, pp27–34

Gramacho, K. P. and Pires, J. L. (2009) 'Evolution of *Moniliophthora perniciosa* in southeast of Bahia, Brazil', in *6th INCOPED Seminar Sharing Crop Protection Technologies for Sustainable Cocoa*, INCOPED, Bali, Indonesia

Greenberg, R., Bichier, P., Angon, A. C., MacVean, C., Perez, R. and Cano, E. (2000) 'The impact of avian insectivory on arthropods and leaf damage in some Guatemalan coffee plantations', *Ecology*, vol 81, pp1750–1755

Gumier-Costa, F. (2009) 'First record of the coffee berry borer, *Hypothenemus hampei* (Ferrari) (Coleoptera: Scolytidae), in para nut, *Bertholletia excelsa* (Lecythidaceae)', *Neotropical Entomology*, vol 38, pp430–431

Hanada, R. E., Souza, T. D., Pomella, A. W. V., Hebbar, K. P., Pereira, J. O., Ismaiel, A. and Samuels, G. J. (2008) '*Trichoderma martiale* sp. nov., a new endophyte from sapwood of *Theobroma cacao* with a potential for biological control', *Mycological Research*, vol 112, pp1335–1343

Harrier, L. A. and Watson, C. A. (2004) 'The potential role of arbuscular mycorrhizal (AM) fungi in the bioprotection of plants against soil-borne pathogens in organic and/or other sustainable farming systems', *Pest Management Science*, vol 60, pp149–157

Harvey, C. A., Komar, O., Chazdon, R., Ferguson, B. G., Finegan, B., Griffith, D. M., Martinez-Ramos, M., Morales, H., Nigh, R., Soto-Pinto, L., Van Breugel, M. and Wishnie, M. (2008) 'Integrating agricultural landscapes with biodiversity conservation in the Mesoamerican hotspot', *Conservation Biology*, vol 22, pp8–15

Hergoualc'h, K., Skiba, U., Harmand, J. M. and Henault, C. (2008) 'Fluxes of greenhouse gases from Andosols under coffee in monoculture or shaded by *Inga densiflora* in Costa Rica', *Biogeochemistry*, vol 89, pp329–345

Jackson, W. (2002) 'Natural systems agriculture: A truly radical alternative', *Agriculture, Ecosystems & Environment*, vol 88, pp111–117

Jacques, P., Delfosse, P., Ongena, M., Lepoivre, P., Cornélis, P., Koedam, N., Neirinckx, L. and Thonart, P. (1993) 'Les mécanismes biochimiques développés par les Pseudomonas fluorescents dans la lutte biologique contre les maladies des plantes transmises par le sol', *Cahiers Agricultures*, vol 2, pp301–307

Johnson, M. D., Levy, N. J., Kellermann, J. L. and Robinson, D. E. (2009) 'Effects of shade and bird exclusion on arthropods and leaf damage on coffee farms in Jamaica's Blue Mountains', *Agroforestry Systems*, vol 76, pp139–148

Kellermann, J. L., Johnson, M. D., Stercho, A. M. and Hackett, S. C. (2008) 'Ecological and economic services provided by birds on Jamaican Blue Mountain coffee farms', *Conservation Biology*, vol 22, pp1177–1185

Komar, O. (2006) 'Ecology and conservation of birds in coffee plantations: A critical review', *Bird Conservation International*, vol 16, pp1–23

Konam, J. K. and Guest, D. I. (2002) 'Leaf litter mulch reduces the survival of *Phytophthora palmivora* under cocoa trees in Papua New Guinea', *Australasian Plant Pathology*, vol 31, pp381–383

Krauss, U. and Soberanis, W. (2001) 'Biocontrol of cocoa pod diseases with mycoparasite mixtures', *Biological Control*, vol 22, pp149–158

Leach, A. W. and Mumford, J. D. (2008) 'Pesticide environmental accounting: A method for assessing the external costs of individual pesticide applications', *Environmental Pollution*, vol 151, pp139–147

Li, W. B., Pria, W. D., Teixeira, C., Miranda, V. S., Ayres, A. J., Franco, C. F., Costa, M. G., He, C. X., Costa, P. I. and Hartung, J. S. (2001) 'Coffee leaf scorch caused by a strain of *Xylella fastidiosa* from citrus', *Plant Disease*, vol 85, pp501–505

Loguercio, L. L., de Carvalho, A. C., Niella, G. R., De Souza, J. T. and Pomella, A. W. V. (2009a) 'Selection of *Trichoderma stromaticum* isolates for efficient biological control of witches' broom disease in cacao', *Biological Control*, vol 51, pp130–139

Loguercio, L. L., Santos, L. S., Niella, G. R., Miranda, R. A. C., Souza, J. T., Collins, R. T. and Pomella, A. W. V. (2009b) 'Canopy-microclimate effects on the antagonism between *Trichoderma stromaticum* and *Moniliophthora perniciosa* in shaded cacao', *Plant Pathology*, vol 58, pp1104–1115

Lott, J. E., Ong, C. K. and Black, C. R. (2009) 'Understorey microclimate and crop performance in a *Grevillea robusta*-based agroforestry system in semi-arid Kenya', *Agricultural and Forest Meteorology*, vol 149, pp1140–1151

Lugtenberg, B. and Kamilova, F. (2009) 'Plant-growth-promoting Rhizobacteria', *Annual Review of Microbiology*, vol 63, pp541–556

Malezieux, E., Crozat, Y., Dupraz, C., Laurans, M., Makowski, D., Ozier-Lafontaine, H., Rapidel, B., de Tourdonnet, S. and Valantin-Morison, M. (2009) 'Mixing plant species in cropping systems: Concepts, tools and models. A review', *Agronomy for Sustainable Development*, vol 29, pp43–62

McCallan, E. (1975) 'Miridae of the genus Termatophylidea (Hemiptera) as predators of cacao thrips', *Entomophaga*, vol 20, pp389–391

MEA (Millenium Ecosystem Assessment) (2005) *Ecosystems and Human Wellbeing: Synthesis*, Island Press, Washington, DC

Meinhardt, L. W., Rincones, J., Bailey, B. A., Aime, M. C., Griffith, G. W., Zhang, D. P. and Pereira, G. A. G. (2008) '*Moniliophthora perniciosa*, the causal agent of witches' broom disease of cacao: What's new from this old foe?', *Molecular Plant Pathology*, vol 9, pp577–588

Mejia, L. C., Rojas, E. I., Maynard, Z., Van Bael, S., Arnold, A. E., Hebbar, P., Samuels, G. J., Robbins, N. and Herre, E. A. (2008) 'Endophytic fungi as biocontrol agents of *Theobroma cacao* pathogens', *Biological Control*, vol 46, pp4–14

Melnick, R. L., Zidack, N. K., Bailey, B. A., Maximova, S. N., Guiltinan, M. and Backman, P. A. (2008) 'Bacterial endophytes: *Bacillus* spp. from annual crops as potential biological control agents of black pod rot of cacao', *Biological Control*, vol 46, pp46–56

Mendoza García, R. A., ten Hoopen, G. M., Kass, D. C. J., Sánchez Garita, V. A. and Krauss, U. (2003) 'Evaluation of mycoparasites as biocontrol agents of *Rosellinia* root rot in cocoa', *Biological Control*, vol 27, pp210–227

Michel, V. V., Wang, J. F., Midmore, D. J. and Hartman, G. L. (1997) 'Effects of intercropping and soil amendment with urea and calcium oxide on the incidence of bacterial wilt of tomato and survival of soil-borne *Pseudomonas solanacearum* in Taiwan', *Plant Pathology*, vol 46, pp600–610

Monteith, J. L. and Butler, D. R. (1979) 'Dew and thermal lag: Model for cocoa pods', *Quarterly Journal of the Royal Meteorological Society*, vol 105, pp207–215

Monzon, A. J., Guharay, F. and Klingen, I. (2008) 'Natural occurrence of *Beauveria bassiana* in *Hypothenemus hampei* (Coleoptera: Curculionidae) populations in unsprayed coffee fields', *Journal of Invertebrate Pathology*, vol 97, pp134–141

Motamayor, J. C., Lachenaud, P., Mota, J., Loor, R., Kuhn, D. N., Brown, J. S. and Schnell, R. J. (2008) 'Geographic and genetic population differentiation of the Amazonian chocolate tree (*Theobroma cacao* L)', *PLoS One*, vol 3, p8

Mouen Bedimo, J. A., Njiayouom, I., Bieysse, D., Nkeng, M. N., Cilas, C. and Notteghem, J. L. (2008) 'Effect of shade on Arabica coffee berry disease development: Toward an agroforestry system to reduce disease impact', *Phytopathology*, vol 98, pp1320–1325

Muller, R. A., Berry, D., Avelino, J. and Bieysse, D. (2004) 'Coffee diseases', in J. N. Wintgens (ed) *Coffee: Growing, Processing, Sustainable Production: A Guidebook for Growers, Processors, Traders, and Researchers*, Wiley–VCH, Weinheim, Germany, pp491–545

Olasantan, F. O., Ezumah, H. C. and Lucas, E. O. (1996) 'Effects of intercropping with maize on the micro-environment, growth and yield of cassava', *Agriculture, Ecosystems & Environment*, vol 57, pp149–158

Ong, C. K., Subrahmanyam, P. and Khan, A. A. H. (1991) 'The microclimate and productivity of a groundnut millet intercrop during the rainy season', *Agricultural and Forest Meteorology*, vol 56, pp49–66

Pasek, J. E. (1988) 'Influence of wind and windbreaks on local dispersal of insects', *Agriculture, Ecosystems & Environment*, vol 22–23, pp539–554

Perfecto, I. and Vandermeer, J. (2002) 'Quality of agroecological matrix in a tropical montane landscape: Ants in coffee plantations in southern Mexico', *Conservation Biology*, vol 16, pp174–182

Perfecto, I. and Vandermeer, J. (2006) 'The effect of an ant–hemipteran mutualism on the coffee berry borer (*Hypothenemus hampei*) in southern Mexico', *Agriculture, Ecosystems & Environment*, vol 117, pp218–221

Perfecto, I., Vandermeer, J. H., Bautista, G. L., Nunez, G. I., Greenberg, R., Bichier, P. and Langridge, S. (2004) 'Greater predation in shaded coffee farms: The role of resident neotropical birds', *Ecology*, vol 85, pp2677–2681

Perkins, T. E. and Matlack, G. R. (2002) 'Human-generated pattern in commercial forests of southern Mississippi and consequences for the spread of pests and pathogens', *Forest Ecology and Management*, vol 157, pp143–154

Philpott, S. M. and Armbrecht, I. (2006) 'Biodiversity in tropical agroforests and the ecological role of ants and ant diversity in predatory function', *Ecological Entomology*, vol 31, pp369–377

Philpott, S. M., Soong, O., Lowenstein, J. H., Pulido, A. L., Lopez, D. T., Flynn, D. F. B. and DeClerck, F. (2009) 'Functional richness and ecosystem services: Bird predation on arthropods in tropical agroecosystems', *Ecological Applications*, vol 19, pp1858–1867

Plantegenest, M., Le May, C. and Fabre, F. (2007) 'Landscape epidemiology of plant diseases', *Journal of the Royal Society Interface*, vol 4, pp963–972

Posada-Florez, F. J. (2008) 'Production of *Beauveria bassiana* fungal spores on rice to control the coffee berry borer, *Hypothenemus hampei*, in Colombia', *Journal of Insect Science*, vol 8, insectscience.org/8.41

Pozza, A. A. A., Martinez, H. E. P., Caixeta, S. L., Cardoso, A. A., Zambolim, L. and Pozza, E. A. (2001) 'Influence of the mineral nutrition on intensity of brown-eye spot in young coffee plants', *Pesqui. Agropecu. Bras.*, vol 36, pp53–60

Roux, J. and Coetzee, M. P. A. (2005) 'First report of pink disease on native trees in South Africa and phylogenetic placement of *Erythricium salmonicolor* in the homobasidiomycetes', *Plant Disease*, vol 89, pp1158–1163

Samuels, G. J., Pardo-Schultheiss, R., Hebbar, K. P., Lumsden, R. D., Bastos, C. N., Costa, J. C. and Bezerra, J. L. (2000) '*Trichoderma stromaticum* sp. nov., a parasite of the cacao witches broom pathogen', *Mycological Research*, vol 104, pp760–764

Samuels, G. J., Suarez, C., Solis, K., Holmes, K. A., Thomas, S. E., Ismaiel, A. and Evans, H. C. (2006) '*Trichoderma theobromicola* and *T. paucisporum*: Two new species isolated from cacao in South America', *Mycological Research*, vol 110, pp381–392

Savary, S., Srivastava, R. K., Singh, H. M. and Elazegui, F. A. (1997) 'A characterisation of rice pests and quantification of yield losses in the rice-wheat system of India', *Crop Protection*, vol 16, pp387–398

Savary, S., Teng, P. S., Willocquet, L. and Nutter, F. W. (2006) 'Quantification and modeling of crop losses: A review of purposes', *Annual Review of Phytopathology*, vol 44, pp89–112

Schoeny, A., Jumel, S., Rouault, F., Lemarchand, E. and Tivoli, B. (2010) 'Effect and underlying mechanisms of pea-cereal intercropping on the epidemic development of ascochyta blight', *European Journal of Plant Pathology*, vol 126, pp317–331

Schroth, G., Krauss, U., Gasparotto, L., Aguilar, J. A. D. and Vohland, K. (2000) 'Pests and diseases in agroforestry systems of the humid tropics', *Agroforestry Systems*, vol 50, pp199–241

Sciarretta, A. and Trematerra, P. (2006) 'Geostatistical characterization of the spatial distribution of *Grapholita molesta* and *Anarsia lineatella* males in an agricultural landscape', *Journal of Applied Entomology*, vol 130, pp73–83

Sequeira, L. (1958) 'The host range of *Mycena citricolor* (Berk C Curt) Sacc', *Turrialba*, vol 8, pp136–147

Soto-Pinto, L., Perfecto, I. and Caballero-Nieto, J. (2002) 'Shade over coffee: Its effects on berry borer, leaf rust and spontaneous herbs in Chiapas, Mexico', *Agroforestry Systems*, vol 55, pp37–45

Sperber, C. F., Nakayama, K., Valverde, M. J. and Neves, F. D. (2004) 'Tree species richness and density affect parasitoid diversity in cacao agroforestry', *Basic and Applied Ecology*, vol 5, pp241–251

Staver, C., Guharay, F., Monterroso, D. and Muschler, R. G. (2001) 'Designing pest-suppressive multistrata perennial crop systems: Shade-grown coffee in Central America', *Agroforestry Systems*, vol 53, pp151–170

ten Hoopen, G. M. and Krauss, U. (2006) 'Biology and control of *Rosellinia bunodes*, *Rosellinia necatrix* and *Rosellinia pepo*: A review', *Crop Protection*, vol 25, pp89–107

ten Hoopen, G. M., Rees, R., Aisa, P., Stirrup, T. and Krauss, U. (2003) 'Population dynamics of epiphytic mycoparasites of the genera *Clonostachys* and *Fusarium* for the biocontrol of black pod (*Phytophthora palmivora*) and moniliasis (*Moniliophthora roreri*) on cocoa (*Theobroma cacao*)', *Mycological Research*, vol 107, pp587–596

ten Hoopen, G. M., Techou, Z., Mbarga, J. B., Benel Ngue, S., Tompe Kamtcha, L. and Cilas, C. (2009) 'Impact of barriers on the onset of a *Phytophthora megakarya* epidemic in cocoa', in *6th INCOPED Seminar Sharing Crop Protection Technologies for Sustainable Cocoa*, INCOPED, Bali, Indonesia

Tilman, D., Cassman, K. G., Matson, P. A., Naylor, R. and Polasky, S. (2002) 'Agricultural sustainability and intensive production practices', *Nature*, vol 418, pp671–677

Vaast, P., Caswell-Chen, E. P. and Zasoski, R. J. (1998) 'Influences of a root-lesion nematode, *Pratylenchus coffeae*, and two arbuscular mycorrhizal fungi, *Acaulospora mellea* and *Glomus clarum*, on coffee (*Coffea arabica* L.)', *Biology and Fertility of Soils*, vol 26, pp130–135

Van Bael, S. A., Bichier, P. and Greenberg, R. (2007) 'Bird predation on insects reduces damage to the foliage of cocoa trees (*Theobroma cacao*) in western Panama', *Journal of Tropical Ecology*, vol 23, pp715–719

Van Bael, S. A., Philpott, S. M., Greenberg, R., Bichier, P., Barber, N. A., Mooney, K. A. and Gruner, D. S. (2008) 'Birds as predators in tropical agroforestry systems', *Ecology*, vol 89, pp928–934

Vandermeer, J., Perfecto, I. and Liere, H. (2009) 'Evidence for hyperparasitism of coffee rust (*Hemileia vastatrix*) by the entomogenous fungus, *Lecanicillium lecanii*, through a complex ecological web', *Plant Pathology*, vol 58, pp636–641

Vargas, A., Somarriba, E. and Carballo, M. (2005) 'Population dynamics of the thrip (*Monalonion dissimulatum* Dist.) and pod damage in organic cacao plantations in Alto Beni, Bolivia', *Agroforesteria en las Americas*, pp72–76

Vega, F. E., Mercadier, G., Damon, A. and Kirk, A. (1999) 'Natural enemies of the coffee berry borer, *Hypothenemus hampei* (Ferrari) (Coleoptera: Scolytidae) in Togo and Cote d'Ivoire, and other insects associated with coffee beans', *African Entomology*, vol 7, pp243–248

Vega, F. E., Goettel, M. S., Blackwell, M., Chandler, D., Jackson, M. A., Keller, S., Koike, M., Maniania, N. K., Monzon, A., Ownley, B. H., Pell, J. K., Rangel, D. E. N. and Roy, H. E. (2009) 'Fungal entomopathogens: New insights on their ecology', *Fungal Ecology*, vol 2, pp149–159

Vera-Montoya, L. Y., Gil-Palacio, Z. N. and Benavides-Machado, P. (2007) 'Identificación de enemigos naturales de *Hypothenemus hampei* en la zona cafetera central colombiana', *Cenicafé*, vol 58, pp185–195

Verma, M., Brar, S. K., Tyagi, R. D., Surampalli, R. Y. and Valero, J. R. (2007) 'Antagonistic fungi, *Trichoderma* spp.: Panoply of biological control', *Biochemical Engineering Journal*, vol 37, pp1–20

Villacorta, A. (1977) 'Annual fluctuation of populations of *Monalonion annulipes* Sig. and its relation to dieback of *Theobroma cacao* in Costa Rica', *Anais da Sociedade Entomologica do Brasil*, vol 6, pp215–223

Walters, D. R. and Bingham, I. J. (2007) 'Influence of nutrition on disease development caused by fungal pathogens: Implications for plant disease control', *Annals of Applied Biology*, vol 151, pp307–324

Wardle, D. A., Yeates, G. W., Watson, R. N. and Nicholson, K. S. (1995) 'The detritus food-web and the diversity of soil fauna as indicators of disturbance regimes in agro-ecosystems', *International Symposium on Soil Biodiversity*, E Lansing, MI, 3–6 May, vol 170, pp35–43

Wegbe, K., Cilas, C., Decazy, B., Alauzet, C. and Dufour, B. (2003) 'Estimation of production losses caused by the coffee berry borer (Coleoptera: Scolytidae) and calculation of an economic damage threshold in Togolese coffee plots', *Journal of Economic Entomology*, vol 96, pp1473–1478

Wilson, C. and Tisdell, C. (2001) 'Why farmers continue to use pesticides despite environmental, health and sustainability costs', *Ecological Economics*, vol 39, pp449–462

Zadoks, J. C. (1999) 'Reflections on space, time, and diversity', *Annual Review of Phytopathology*, vol 37, pp1–17

Zadoks, J. C. and Schein, R. D. (1979) *Epidemiology and Plant Disease Management*, Oxford University Press, New York, NY

5

Services from Plant–Pollinator Interactions in the Neotropics

Lucas Alejandro Garibaldi, Nathan Muchhala, Iris Motzke, Liliana Bravo-Monroy, Roland Olschewski and Alexandra-Maria Klein

Introduction

The Neotropics, with its large expanses of rainforests, forests and woodland savannas, includes some of the most diverse places on Earth (Kricher, 1999; Myers et al, 2000). A large proportion of plant and animal species in Neotropical communities are unique, including several pollinator species, which provide essential services to human welfare. In general, pollinators are known to enhance the sexual reproduction of the majority of angiosperms (Kearns et al, 1998) and can be important for the production of many crop species (McGregor, 1976; Klein et al, 2007; Aizen et al, 2009a). There is a wide array of arthropod and vertebrate pollinator species in the Neotropics, although we know little about their natural history and contribution to pollination (Kevan and Imperatriz-Fonseca, 2002; Freitas et al, 2009).

This chapter reviews studies on pollination services in the Neotropics, with an emphasis on crop pollination. We briefly describe the main taxa involved in pollination, followed by a list of the main crops grown in the Neotropics and a description of how many they rely on biotic pollination. Because methods vary across studies, key methodologies to determine pollination services are summarized. Finally, we discuss management options to improve pollination services at the farm and landscape scale, and socio-economic drivers affecting pollination.

Major Pollinator Taxa

Pollination by animals plays a vital role for plant reproduction in the tropics, where it is estimated that more than 98 per cent of plants are animal pollinated (Bawa, 1990). However, in general, information on pollinator communities and the diversity of taxonomic guilds in the Neotropics is incomplete (Freitas et al, 2009). In this section we give examples of the major pollinator taxa in comparison to other regions.

Similar to the Old World, bees play a major role in pollination of Neotropical plants (Roubik, 1995). Around 5000 bee species are thought to occur in the Neotropics, including 391 eusocial stingless bee species (Meliponini), an important pollinating bee taxa (Slaa et al, 2006). The invasive Africanized honey bee, *Apis mellifera scutellata* Lepeletier, is widespread throughout the Neotropics. Although presumed to compete with native bees, evidence is still controversial (Roubik, 2009). Other important invertebrate pollinators are wasps (Hymenoptera), beetles (Coleoptera), moths and butterflies (Lepidoptera) and flies (Diptera).

Pollinators in the Neotropics seem to be as diverse as in other tropical areas (Roubik, 1995); but species composition and identity are highly distinct. For example, in South America, coffee (*Coffea arabica* L.) is predominately visited by the non-native Africanized honey bee, but also by a high diversity of stingless bees (Klein et al, 2008a). In contrast, coffee-visiting bee species in Southeast Asia include the native eastern honey bee (*Apis cerana* Fabricius), the giant honey bee (*A. dorsata* Fabricius), the honey bee (*A. nigrocincta* Smith), a close relative of the eastern honey bee, few stingless bee species, and a high diversity of solitary species (Klein et al, 2008a; Klein, 2009).

Among vertebrate pollinators, birds, especially hummingbirds, followed by bats play the most important role for many wild flowers in the Neotropics. There are more than 300 hummingbird species confined to the Neotropics (Bawa, 1990). In agricultural systems, hummingbirds visit papaya (*Carica papaya* L.) and banana (*Musa* sp.) flowers (Free, 1993); but their role in crop pollination is not well documented. In other areas of the world, sunbirds (Palaeotropical and Pacific), sugarbirds (South Africa) and honeyeaters (Australasia) fill the ecological niche of hummingbirds in the Neotropics (Roubik, 1995; Ortega-Olivencia et al, 2005). Nectar-feeding bats are the second most widespread vertebrate pollinators in Neotropical rainforests, especially for many wild trees and epiphytes, but also for locally important crops (see Box 5.1).

Biotic Pollination and Crop Production

Biotic pollination is important for many crop species in the Neotropics. Altogether 44 crops and 4 commodities (method as in Klein et al, 2007) represent 99 per cent (98 and 1 per cent, respectively) of the total crop production in the Neotropics in 2007 (FAOSTAT, 2009). Of these, 29 (70 per cent) crops

Box 5.1 Bat pollination in the Neotropics

Bat pollination is restricted to the tropics and subtropics; plant-visiting bat species do not occur in temperate regions (Koopman, 1981; Fleming and Muchhala, 2008). Bats adapted to a nectarivorous diet occur in two distantly related families: the Phyllostomidae in the Neotropics and Pteropodidae in the Palaeotropics. Of these, bat species of the sub-family Glossophaginae are the most morphologically and ecologically specialized; they possess elongated snouts, highly extensible tongues and the ability to hover in front of flowers like hummingbirds (Helversen, 1993; Winter and Helversen, 2003).

In the Neotropics, nectar bats are known to pollinate flowers from 360 species of plants in 159 genera from 44 families (Geiselman et al, 2002; Fleming et al, 2009). The majority of these are trees and epiphytes, including many conspicuous members of local ecosystems, such as canopy-emergent Bombacaeae trees in rainforests and large columnar cacti (e.g. saguaro, organ pipe cacti) in arid regions. Although numerically a relatively small proportion of total angiosperm diversity, bat-pollinated plant species cannot be serviced as effectively by other pollinator taxa because specialized floral adaptations are required to attract, fit and reward bats: chiropterophilous flowers typically are physically robust and well exposed beyond the foliage, have wide bell-shaped flowers or a 'brush' morphology, open nocturnally, and produce a strong odour and copious nectar (Helversen, 1993; Muchhala, 2007; Fleming et al, 2009). Although such adaptations require large investments in floral structures compared to other pollination systems, bats provide two important advantages as pollinators. First, they can carry large amounts of pollen in their hairs (Law and Lean, 1999; Muchhala and Thomson, 2010). Second, they can disperse this pollen over extremely long distances. For instance, paternity analyses reveal that pollen was transferred up to 18km between individuals of the bat-pollinated kapok tree (*Ceiba pentandra*) (Dick et al, 2008). Such long-distance pollen dispersal improves gene flow, as evidenced in low genetic subdivision for bat-pollinated plant species (Roesel et al, 1996; Hamrick et al, 2002).

A number of bat-pollinated plants in the Neotropics provide economically important products. The kapok tree, which is pantropical and bat pollinated throughout its range (Elmqvist et al, 1992; Gribel et al, 1999; Nathan et al, 2005), produces silky fibres which are used in bedding and cushion materials. Many bat-pollinated cacti throughout the Americas produce edible fruits that are sold in local and international markets, often as jellies or jams (Anderson, 2001). Bat-pollinated dragon-fruit and other fruits of the cactus genus *Hylocereus* are now cultivated worldwide, both as food and as ornamental plants (Valiente-Banuet et al, 2007). Fruits of *Stenocereus griseus* (Haw.) Buxb. are harvested by indigenous communities, which also use the cacti for construction materials and as living fences (Nassar et al, 1997; Villalobos et al, 2007). The seed set of agaves, from which the well-known liquor tequila is derived, drops to less than 5 per cent in the absence of bat pollinators (Howell and Roth, 1981; Molina-Freaner and Eguiarte, 2003). Finally, many ornamental plants rely on bat pollination, such as *Cobaea scandens* Cav. and *C. trianae* Hemsl. (Polemoniaceae) (Vogel, 1969).

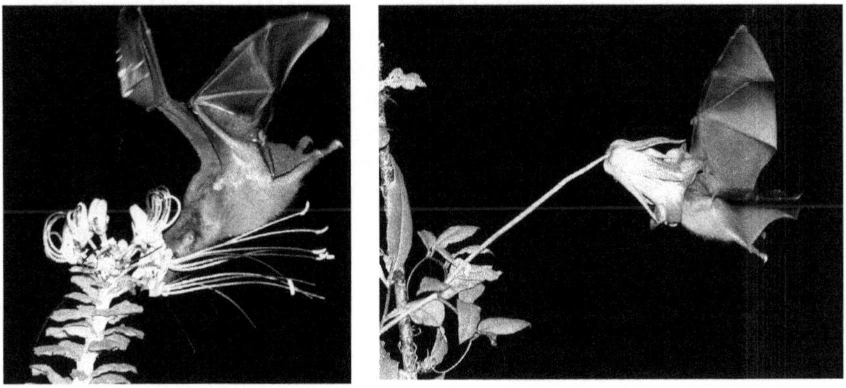

Figure 5.1 Anoura geoffroyi *Gray, 1838 pollinating* Cleome anamola *Kunth* (left) *and the ornamental* Cobaea trianae *Hemsl.* (right)

Source: N. Muchhala

increase their seed or fruit production in the presence of animal pollination. In the following discussion we highlight the leading animal-pollinated crops in terms of cultivation area, and give further examples of highly pollinator-dependent crops.

The most important pollinator-dependent crops exotic to the Neotropics are coffee, coconut, citrus, mango, and soybean (Table 5.1; see Box 5.2 for details on coffee pollination; FAOSTAT, 2009). For example, soybean is the second most cultivated crop in the Neotropics. Primarily self-compatible, flower-visiting insects, such as honey bees, have been shown to increase soybean production, measured in kilograms per hectare (kg ha^{-1}), between 38 and 58 per cent for some varieties in Brazil (Chiari et al, 2005, 2008). Given the importance of this crop, more research on its pollination system across countries and varieties is urgently needed.

The most important native Neotropical crops dependent totally or to certain degrees on insect pollination are cocoa, common bean, guava and cashew (see Table 5.1). Cocoa, for example, is generally highly self-incompatible and depends heavily on insect pollination, although a few self-compatible varieties exist (Falque et al, 1996). Tiny midges of the Ceratopogonidae and Cecidomyiidae families are predominantly responsible for pollination of the cocoa varieties that depend on insect pollination (Entwistle, 1972; Young, 1994). The cashew nut, native to Brazil, has both bisexual and male flowers on the same plant. This crop is frequently cultivated in the Neotropics (Roubik, 1995; Kevan and Imperatriz-Fonseca, 2002) and has two main pollinating species: the honey bee (*Apis mellifera* L.) and the native oil bee (*Centris tarsata* Smith) (Freitas and Paxton, 1998).

Many crops that depend on animal pollination are of high economic importance at a more local, country- or state-wide scale. For some of these crops, such as Brazil nut, melon, passion fruit, pumpkin, squash, vanilla and watermelon, animal pollination was found to be essential (Klein et al, 2007). Furthermore, a high number of crops depend partly (to certain degrees or under certain conditions) on animal pollination, such as agaves, annatto (or achiote), avocado, chayote, chilli pepper, common bean, dragon fruit, eggplant, guayule, jojoba, mesquite, papaya, peanut, pepper, pimento, rubber, quinine, sisal, soursop (or *guanábana*), star apple (or *caimito*), sunflower, tobacco and tomato (Roubik, 1995). Here we highlight two locally important native crops: passion fruit and avocado. Passion fruit (*Passiflora edulis* Sims) is cultivated throughout the Neotropics and has self-incompatible, large hermaphroditic flowers. It is mainly pollinated by large carpenter bees of the genus *Xylocopa*, as other frequent flower-visiting species are too small to touch the stigma during nectar and pollen collection (e.g. Benevides et al, 2009). Wind pollination is ineffective because pollen is heavy and sticky. Another important native crop is avocado (*Persea americana* Mill.), a variable and poorly understood species with respect to its pollination system. Avocado varieties vary between self-compatible to self-incompatible; but cross-pollination through bees, bats, flies and wasps improves fruit production

Table 5.1 *Pollinator dependence of the most cultivated crops in the Neotropics*

Species	*Crop*	*Pollinator dependence*	*Cultivated area (ha)*	*(%)*
Zea mays	Maize	None	26,314,959	24.6
Glycine max, G. soja	Soybean	Modest	24,124,332	22.6
Saccharum officinarum	Sugar cane	None	9,825,691	9.2
Phaseolus sp., *P. vulgaris, P. lunatus, P. angularis, P. aureus, P. mungo, P. coccineus, P. calcaratus, P. aconitifolius, P. acutifolius*	Bean dry like kidney bean, haricot bean, lima bean, azuki bean, mungo bean, string bean	Little	6,457,637	6.0
Coffea arabica, C. canephora (syn. *Coffea robusta*), *C. liberica*	Coffee	Modest	5,667,250	5.3
Oryza sp. (mainly *O. sativa*)	Rice, paddy	None	5,262,464	4.9
Triticum sp. (mainly *T. aestivum, T. durum, T. spelta*)	Wheat	None	3,236,071	3.0
Sorghum guineense, S. vulgare, S. dura	Sorghum	None	3,155,116	3.0
Manihot esculenta (syn. *M. utilissima, M. palmata*)	Cassava	Only breeding	2,791,040	2.6
Musa sapientum, M. cavendishii, M. nana, M. paradisiaca	Banana, plantain	Only breeding	2,128,586	2.0
Gossypium hirsutum, G. barbadense, G. arboreum, G. herbaceum	Cotton	Modest	1,735,189	1.6
Theobroma cacao	Cocoa	Essential	1,490,461	1.4
Citrus trifoliata	Orange	Little	1,442,261	1.4
Anacardium occidentale	Cashew nut, cashew-apple	High	1,354,993	1.3
Cocos nucifera	Coconut	Modest	672,713	0.6
Hordeum disticum, H. hexasticum, H. vulgare	Barley	None	667,234	0.6
Elaeis guineensis	Oil palm	Little	611,211	0.6
Nicotiana tabacum	Tobacco	Only sowing	545,856	0.5
Mangifera indica, Garcinia mangostana, Psidium spp.	Mango, mangostan, guava	High	458,435	0.4

Notes: Harvested area data given for each crop are extracted from the FAO dataset for the year 2007 (FAOSTAT, 2009). Argentina, Chile and Uruguay were excluded; but examples from these countries are discussed in the chapter when appropriate (e.g. Chacoff and Aizen, 2006). Listed crops accounted for 93 per cent of the total cultivated land in the Neotropics in 2007. Pollinator dependence data obtained from Klein et al (2007). Pollinator dependence: *none* = yield not dependent on animal pollination; *little* = yield reduction > 0 but < 10 per cent without pollinators; modest = 10–40 per cent reduction; *high* = 40–90 per cent reduction; *essential* = reduction >90 per cent; *only breeding* = pollinators increase seed production for breeding (in commercial farming, the plants are propagated from vegetative organs and the vegetative parts are harvested); *only sowing* = pollinators increase seed production to produce the vegetative parts that are harvested.

(Roubik, 1995). The flower is bisexual and opens twice; it functions as a female during the first opening, and functions usually as a male and releases pollen on the following day upon the second opening. Commercially grown avocado plantations are therefore planted with two complementary flowering groups to ensure the spatio-temporal availability of female and male openings for adequate pollination (Delaplane and Mayer, 2000).

In summary, 70 per cent of the leading crops in the Neotropics depend to some degree on animal pollination. This number is similar to that estimated for the global scale (74 per cent) (Klein et al, 2007), and also similar to tropical regions, in general (70 per cent: Roubik, 1995), to Argentina (74 per cent: Chacoff et al, 2010), Mexico (85 per cent: Ashworth et al, 2009) and the European Union (84 per cent: Williams, 1994). The latter two studies include many crops of minor importance in terms of production and total cultivated area, whereas the other calculations include major crops only. In general, however, few studies have evaluated pollination services in the Neotropics (Freitas et al, 2009); consequently, we know little about the pollinator relevance for many widely cultivated crops or about the variability of pollinator requirements among varieties.

Determination of Crop Pollination Services

Pollination can be important for agricultural and non-domesticated plants; however, the actual impact of these services is difficult to estimate. To better understand pollination services, it is important not only to measure the interaction between a pollinator and a certain crop or plant species, but to identify biophysical and socio-economic drivers in an interdisciplinary approach (see Figure 5.3; Bayon and Jenkins, 2010).

Pollinators can provide direct benefits by increasing the amount and inter-annual stability of crop yield quantity (kg of product ha^{-1}) and quality (e.g. fruit size, shape, weight), and indirect effects such as maintaining plant and animal biodiversity and their associated benefits for human welfare. These services can be promoted by either pollinator abundance or diversity (Hoehn et al, 2008; Klein, 2009; Klein et al, 2009; Vergara and Badano, 2009). We would like to note that some flower visitations may be a disservice to crops, as has been demonstrated for flowers in the wild. This can occur in the form of nectar or pollen robbery where a 'pollinator species' takes nectar or pollen without pollinating the plant (Irwin et al, 2001; Thomson, 2003; Hargreaves et al, 2009). However, we have found no studies showing that the exclusion of flower visitors has positive effects on crop pollination. The exclusion of wild visitors commonly reduces or does not significantly affect pollination services (Klein et al, 2007).

Many studies have measured pollinator abundance, pollinator richness/diversity, flower visitation rates, pollen deposition, pollen tube growth, and/or seed/fruit set (Klein et al, 2007). Fewer studies, however, have determined direct production variables (e.g. yield quality or quantity) at a farm (plot) scale. These calculations are also relevant at the socio-economic scale where decisions on land use are made (see the section on 'Socio-economic drivers affecting pollination services') (Ghazoul, 2007; Klein et al, 2008b; Veddeler et al, 2008). When estimating pollination services, the following processes and methods should be considered:

- *Biotic pollination* can be evaluated by comparing crop yield of pollinator exclusion (only self + abiotic pollination) and free pollination (self + abiotic + biotic pollination) treatments (e.g. Klein, 2009; Vergara and Badano, 2009).
- *Abiotic pollination* can be estimated by comparing an abiotic plus biotic pollination exclosure treatment with a pollinator exclosure treatment.
- *Self-pollination:* by preventing any outcross pollen from reaching the flower (abiotic plus biotic pollination exclosure), the degree of self-pollination can be evaluated.
- *Pollen limitation:* pollen addition (hand pollination) and control treatments are useful to understand the degree of pollen limitation (see review by Wesselingh, 2007).
- *Self-incompatibility:* the addition of pollen from the same individual versus addition from other individuals (out-crossing) can be used to quantify the degree of self-incompatibility.

Other considerations when studying the above processes are:

- *Natural history and field censuses:* knowledge on pollinators' natural history and censuses of flower visitation helps to understand plant–pollinator interactions and to identify key pollinator species and their requirements (e.g. for habitat) (Kevan and Imperatriz-Fonseca, 2002).
- *Number of replicates:* the estimation of the number of (independent) replicates needed given an expected variability and a required precision is critical for obtaining useful information from experiments.
- *Relevant production variables:* from an applied perspective, it is important to measure the quantity and quality of yield, and the spatial and temporal stability in both variables (Ghazoul, 2007; Klein et al, 2008b).
- *Spatial and temporal scale:* when possible, treatments should be applied to plots, which are usually the scale of interest when measuring pollination services (or sometimes entire plants). Special attention should be given to perennial plants, in which plant resource allocation strategies can involve years (e.g. high allocation to vegetative growth during one year, but higher allocation for reproduction in the following year). Therefore, experiments should ideally be followed during the whole plant productive cycle and over consecutive years.
- *Variability in pollen and pollinator limitation:* the impact of pollen limitation on crop production may vary greatly depending on other environmental factors such as resource availability (water, nutrients and radiation), abiotic conditions (e.g. frosts) and pests (Bos et al, 2007; Ghazoul, 2007; Klein et al, 2008b). Pollen limitation may also vary with crop variety, and the magnitude of pollinators' exclusion effects may greatly depend on the resident pollinator community. Studies over multiple seasons and years are useful to account for periodic weather perturbations and temporal variation in pollinator communities (Klein, 2009).

Box 5.2 Pollinators and coffee production in the Neotropics

Coffee is one of the most important cash crops in the Neotropics. It is traded at the global market and accounts for nearly 5.7 million hectares of land in 2007 (see Table 5.1). For many years, coffee has been the second leading export product of developing countries (ICO, 2009), providing income and employment for millions of people.

Since the 1950s, many studies have shown that pollinators promote coffee yield (production per plant or hectare) by increasing fruit set and/or berry weight (see reviews by Free, 1993; Klein et al, 2007). Pollinators have also been shown to reduce the frequency of 'peaberries' – that is, small misshapen seeds (Free, 1993; Ricketts et al, 2004). The magnitude of the positive effects on yield can vary greatly, between 10 and 40 per cent among studies using different methodologies and environmental conditions (see Table 5.1) (Klein et al, 2007). Studies finding positive effects on coffee yield include those performed at the plant scale (therefore not biased by resource allocation patterns within the plant) (Free, 1993) and those performed for more than one year (Ricketts et al, 2004). Positive effects of pollinators on both seed number and weight have also been found simultaneously, without the confounding effects of seed number versus size compensation (Ricketts et al, 2004). In most studies, the honey bee was found to be the most frequent visitor to coffee flowers, followed by stingless bees, and some semi-social and solitary bee species (see the previous section on 'Major pollinator taxa').

Research addressing the effects of habitat and landscape scale on coffee pollination began only during the last decade. They include studies in Panama (Roubik, 2002a), Venezuela (Manrique and Thimann, 2002), Costa Rica (Ricketts, 2004), Brazil (De Marco and Coelho, 2004), Ecuador (Veddeler et al, 2006) and Mexico (Vergara and Badano, 2009). These studies considered variables such as distance between coffee plants and adjacent forests or cultivation variables such as shade versus sun coffee. All studies found more bee species, higher visitation frequency, higher fruit set and/or higher berry weight on coffee plants bordering forests.

Other studies highlighted the monetary value of coffee pollination services, such as Roubik (2002b) in Panama; Ricketts et al (2004) in Costa Rica, and Benitez et al (2006), Olschewski et al (2006) and Veddeler et al (2008) in Ecuador. For example, the extrapolation of data gained from pollination experiments in Costa Rica estimated that the value of pollination services for two forest fragments (46ha and 111ha) in a single farm (480ha) was US$60,000 annually (Ricketts et al, 2004). Veddeler et al (2008) calculated that a fourfold increase in bee density would translate to an 800 per cent increase in net revenues for coffee farms in Ecuador. Certainly, wild habitats are providing important pollination services for this crop.

Figure 5.2 *Coffee production in Manabi, coastal Ecuador: From left to right are the Africanized honey bee,* Apis mellifera scutellata *(Lepeletier), foraging on coffee flowers; ripe coffee berries at harvest; traditional harvest with mules*

Source: D. Veddeler (bee and berries); A. M. Klein (traditional harvest)

- *Socio-economic assessments:* it is important to understand the value of pollination services for different aspects of a society (e.g. cultural and economic; see the section on 'Socio-economic drivers affecting pollination services').

Depending on the focus, other measurements and treatments can be included. Examples are the exclusion of vertebrate but not invertebrate pollinators to understand their interactions and relative contribution to pollination, or the study of niche complementarity among invertebrate pollinator species (Hoehn et al, 2008). Here, we emphasize methods to quantify the degree of overall pollination and pollinator limitation on crop production at a farm (plot) scale.

Management to Improve Pollination Services at the Landscape and Farm Scale

In the previous sections we described how lack of animal pollination can limit the yield of certain crops. There is also evidence that wild pollinator species are decreasing locally (Ricketts et al, 2008) and regionally (Biesmeijer et al, 2006; Brown and Paxton, 2009; Freitas et al, 2009) due to land-use changes and the application of agrochemicals, among other factors. A recent review suggested that the effects of habitat loss on flower visitation rates should be higher in the tropics compared to temperate zones (Ricketts et al, 2008). Therefore, it is increasingly important to understand the drivers affecting pollinator abundance and diversity for adequate pollinator management and conservation. Management for wild pollinators usually implies decisions at the landscape and farm level to provide floral resources, breeding areas and nesting habitats within the flying range of pollinators (Kevan and Imperatriz-Fonseca, 2002; Kremen, 2008).

Landscape and habitat management

Pollination services can vary widely depending on the quantity, quality and spatial arrangements of habitat types in the landscape. Because flying has energy costs and many pollinators have fixed nest sites, pollinators prefer flower visits close to their habitat. Recently, Ricketts et al (2008) reviewed 23 studies representing 16 crops on 5 continents to evaluate the effects of distance from natural or semi-natural habitats on pollination services. They found that visitation rates by wild pollinators and pollinator richness decreased exponentially with distance from natural habitat, reaching half of its maximum at 0.6km and 1.5km, respectively. However, they found no evidence of effects on fruit and seed set, although such effects were only measured by half of the studies and most of them did not measure the size, quality or stability of yield (see the previous section on 'Determination of crop pollination services'). Among the reviewed studies, only four were performed in the Neotropics. Decreases in native visitation rates with distance from natural habitat was observed for highland coffee in Costa Rica (Ricketts, 2004; Ricketts et al, 2004) and grapefruit in northwestern Argentina (Chacoff and Aizen, 2006); but no effect was found for passion fruit in eastern Brazil (Ricketts et al, 2008) or for oil palm in southern Costa Rica (Mayfield, 2005; Ricketts et al, 2008). Overall, these studies suggest that the conservation of natural habitats close to agriculture can be important to enhance wild pollinator diversity and flower-visiting frequency, although the effects of habitat conservation on pollination services needs to be further evaluated. Moreover, most of these studies use either distance or proportional area of natural habitat as the landscape variable; future studies should also consider the effects of the spatial arrangement of habitat patches in terms of distance, number, size and quality (Olschewski et al, 2010).

The magnitude of positive effects from natural habitat proximity can vary greatly among pollinator species. It is proposed that species with high dispersal abilities will be less affected by habitat degradation at relatively short distances. For example, a review concerning tropical crops found that small cavity-nesting bees and generalist beetles required natural forest near their foraging areas, whereas insects with large body sizes explored larger areas and were therefore less sensitive to isolation from forest (Klein et al, 2008a). Overall, taking into account the biology of species and considering different spatial scales will improve our understanding of the effects that land-use change has on habitat quality for pollinator species (Steffan-Dewenter et al, 2002; Tscharntke et al, 2005).

Habitat quality involves the abundance of appropriate floral resources, nesting places and the possibility to escape from natural enemies and diseases. Managing habitat quality requires detailed knowledge of the species' natural history. When the habitat is highly degraded, active management may be required (e.g. sowing or transplanting native species as well as constructing suitable habitat).

A matrix of agricultural and natural patches can be beneficial to pollinators because of a higher diversity of resources (Tscharntke et al, 2005; Winfree et al, 2007, 2008). Enhanced diversity and abundance of pollinators in these complex landscapes may also provide services to a wider spectrum of crops (Kremen, 2008). Pollination services should be greater when agricultural field sizes are smaller because of greater habitat complexity within the flying range of pollinators. Unfortunately, there is a trend towards increasing field size and homogenization of agricultural landscapes in the Neotropics and many other regions (Tscharntke et al, 2005; Aizen et al, 2009b). These landscape variables also interact with decisions at the farm scale because crop management influences the quality of habitat for wild pollinators.

Farm and pollination management

There are several agricultural practices that can improve the visitation of wild pollinators to flowers, such as small-scale farming, polycultures, sowing of diverse flower resources in edge habitats (e.g. field boundaries) and reduced use of agrochemicals (Tscharntke et al, 2005, Brosi et al, 2008). In general, farming practices that increase habitat diversity (and, thus, pollinator diversity) should promote pollination services because of:

- species complementarity, when species use different resource parts or promote positive intra-guild interactions;
- sampling effects, when higher biodiversity increases the probability of including species that provide important services; and
- redundancy, when different species provide a similar pollination service in highly diverse habitats, which is important for reorganization after disturbance (insurance hypothesis) (see reviews by Tscharntke et al, 2005; Klein et al, 2009).

For example, rustic shade coffee managed under native forest in Veracruz (Mexico) showed higher pollinator diversity and fruit production than less diverse sun coffee systems where native forest was removed (Vergara and Badano, 2009).

Although several thousand species contribute to pollination, only a few are managed. Examples include stingless bees as pollinators for tomatoes in Mexico (Cauich et al, 2004) and Brazil (Del Sarto et al, 2005) greenhouse production, and also for other crops such as cucumber and sweet pepper in the Neotropics (see reviews by Cortopassi-Laurino et al, 2006; Slaa et al, 2006). However, most managed pollinators are honey bees (Kevan and Imperatriz-Fonseca, 2002). This reliance on a single pollinator species seriously threatens the stability of pollination services. Indeed, higher incidence of pests and diseases in the US decreased the number of managed honey bee colonies during the past years, and raised several problems for the pollination of important crops such as almond in California (Oldroyd, 2007). In the Neotropics, for example, there has been an increase in the reproductive ability of the mite *Varroa destructor* Anderson & Truemann in the widely spread Africanized honey bee in southern Brazil (Carneiro et al, 2007). Furthermore, *Apis mellifera* L. is not the most efficient pollinator species for many crops (Freitas and Paxton, 1998; Greenleaf and Kremen, 2006). The temporal and spatial stability, as well as the rate of pollination services, can be improved by pollinator diversity (Greenleaf and Kremen, 2006; Klein, 2009; Klein et al, 2009).

Hand pollination is a difficult and laborious task that is currently performed only in expensive crops under intensive farming. This is the case for vanilla (*Vanilla planifolia* L.), an orchid native to Mexico and a highly pollinator-dependent crop species (Davis, 1983; Klein et al, 2007). In general, species with large flowers are easier to hand pollinate than species with small flowers. However, pollinators provide not only quantity of pollen, but also pollen quality (e.g. cross-pollination) and special techniques of pollen transfer (e.g. vibration). Performing such tasks by hand at the proper production scale is both challenging and expensive.

For many pollinator-dependent crops, there are some varieties that are non-dependent so that farmers have the choice to choose varieties that do not need pollinators. However, despite genetic engineering and crop breeding advances, many of the most important crop species depend on pollinating animals (see Table 5.1) (Klein et al, 2007).

Socio-Economic Drivers Affecting Pollination Services

Land-use decisions affecting pollination services are made at the household or farm scale in response to several environmental and socio-economic variables (see Figure 5.3) (Lambin et al, 2001). Crop production often depends on environmental drivers such as resource availability (e.g. water and radiation), abiotic conditions (e.g. temperature), incidence of pests and weeds, and pollination services (see Figure 5.3). Several socio-economic drivers interact with environmental variables to affect land-use decisions, such as markets, demog-

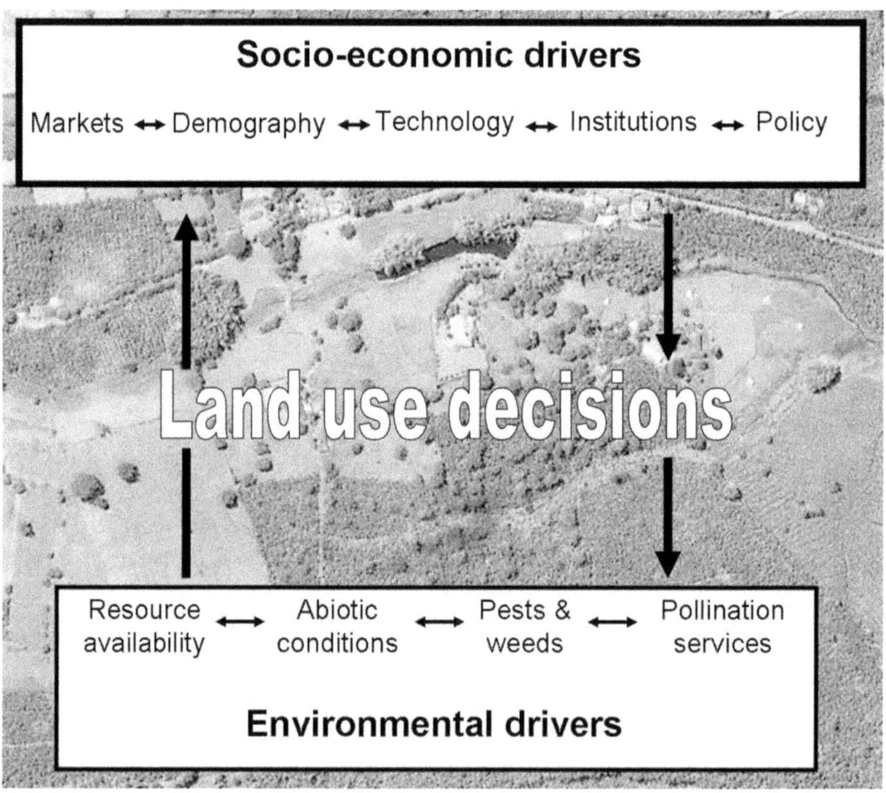

Figure 5.3 *Socio-economic and environmental drivers of land-use decisions and crop productivity*

Note: See explanations in the chapter.
Source: chapter authors

raphy, technology, institutional settings and public policy. This section briefly describes the socio-economic drivers of farmers' decision-making that influence the landscape and habitat management of pollination services (see previous section).

Relevant socio-economic drivers of land-use decisions are established markets, and the participation in trade. Besides income generation, participating in trade might have further advantages, such as access to credit, information, technology and urban centres. However, the importance of markets for small-scale producers depends on the type of land use considered, such as production of food (subsistence) versus cash crops (Cronon, 1985; Burgi and Turner, 2002; Black et al, 2003; Guhl, 2008). Supply and demand determine the market price and, thus, the profitability of crop species. The structure and functioning of markets allows us to understand how small-scale agricultural systems are connected to trade and market relationships. In general, pollinator-dependent crops achieve higher market prices (Gallai et al, 2009), thereby generating incentives to increase their production. However, a lack of pollinators might hinder the producers from doing so, and might force them to switch to less attractive non-pollination dependent crops.

Small-scale farmers' land-use decisions are often based on a comparison of net revenues. They depend on the product price, the quantity of the harvest and on the production costs. A case study in coastal Ecuador included these factors and assessed the impact of forest areas providing bee habitats and thereby enhancing pollination services for adjacent coffee production. It was shown that the impact on net revenues was significantly positive (Olschewski et al, 2006). However, alternative crops such as maize were more attractive from an economic point of view in that landowners had a strong incentive to convert forests into cropland. As a consequence, payments for single ecosystem services such as pollination are hardly sufficient to preserve bee habitats. Payment schemes should comprise further forest ecosystem services (e.g. carbon sequestration, soil and water conservation) in order to be effective.

Demography and other social criteria, such as gender, age and education, are common elements that influence land-use strategies (Mazvimavi and Twomlow, 2009). Institutional aspects such as landownership and tenure rights might be another powerful determinant (Wunder, 2000; Burgi and Turner, 2002; Black et al, 2003) – for example, owners are supposed to make different production decisions than tenants. Furthermore, it is important to consider underlying cultural beliefs and social perceptions of different land-use types (Nyerges and Green, 2000). Taking these into account might help to explain why families in the Neotropics often maintain small-scale farming despite modest income-generating opportunities.

Further important drivers of land-use change are agricultural knowledge and available technology (Angelsen and Kaimowitz, 1999; Burgi and Turner, 2002; Anastasopoulou et al, 2009). Among others, the inclusion of machinery may promote the cultivation of larger and more homogeneous fields. Additionally, the development of new crop varieties may affect pollination requirements and pollination service rewards. Little knowledge or experience on cultivation practices for a particular crop may also induce farmers to avoid the cultivation of that specific crop. Finally, public policies are crucial because they can establish incentives and recommendations regarding the adoption of specific land-use systems through their influence on several other drivers mentioned above (Burgi and Turner, 2002; Di Falco and Perrings, 2006; Anastasopoulou et al, 2009).

The variety of socio-economic drivers and their interactions show that simple explanations hardly provide adequate understanding of land-use change (Lambin et al, 2001). Various human and environmental conditions lead to specific land-use decisions, and policy recommendations aiming at habitat or landscape conservation should take these interactions into account.

Conclusions

We have shown that pollination services by wild pollinators are important for crop production in the Neotropics. However, our knowledge of the services that pollinators provide in terms of the amount, quality and stability of crop production is still deficient. It is also critical to understand how multiple socio-economic drivers influence the selection of particular management systems and, thus, the environmental services delivered. Land-use decisions based on short-term revenue calculations can lead to unsustainable results. Despite the high potential of social benefits, sometimes the net revenues obtained from pollination services through the preservation of a natural habitat are lower than other uses of that land, such as deforestation and crop cultivation. Future evaluations should also consider ecosystem services other than pollination and their interactions to reliably estimate the ecological and social benefits of conserving natural habitats. It is important to determine the value (monetary and non-monetary) of these services in order to raise awareness of ecosystem services when making decisions on particular land-use systems, and to support the design of appropriate conservation policies that benefit farmers and their environment.

Acknowledgements

We are grateful to C. L. Morales and J. Juhrbandt for their thoughtful comments on a previous draft of this chapter. Lucas A. Garibaldi and Iris Motzke were supported by the German Academic Exchange Programme (DAAD). Liliana Bravo-Monroy was supported by the University of Reading (Alumni Study Travel Fund) and the Society for Experimental Biology (the Company of Biologists Travel Fund).

References

Aizen, M. A., Garibaldi, L. A., Cunningham, S. A. and Klein, A. M. (2009a) 'How much does agriculture depend on pollinators? Lessons from long-term trends in crop production', *Annals of Botany*, vol 103, pp1579–1588

Aizen, M. A., Garibaldi, L. A. and Dondo, M. (2009b) 'Expansión de la soja y diversidad de la agricultura argentina', *Ecología Austral*, vol 19, pp45–54

Anastasopoulou, S., Chobotova, V., Dawson, T., Kluvankova-Oravska, T. and Rounsevell, M. (2009) 'Identifying and assessing socio-economic and environmental drivers that affect ecosystems and their services', *The Rubicode Project, Rationalising Biodiversity Conservation in Dynamic Ecosystems*, European Commission, Sixth Framework Programme

Anderson, E. F. (2001) *The Cactus Family*, Timber Press, Portland, OR

Angelsen, A. and Kaimowitz, D. (1999) 'Rethinking the causes of deforestation: Lessons from economic models', *The World Bank Research Observer*, vol 14, no 1 (February), pp73–98

Ashworth, L., Quesada, M., Casas, A., Aguilar, R. and Oyama, K. (2009) 'Pollinator-dependent food production in Mexico', *Biological Conservation*, vol 1424, pp1050–1057.

Bayon, R. and Jenkins, M. (2010) 'The business of biodiversity', *Nature*, vol 466, pp184–185

Bawa, K. S. (1990) 'Plant-pollinator interactions in tropical rain forests', *Annual Review of Ecology, Evolution, and Systematics*, vol 21, pp399–422

Benevides, C. R., Gaglianone, M. C. and Hoffmann, M. (2009) 'Yellow passion fruit (*Passiflora edulis* f. flavicarpa Deg. Passifloraceae) floral visitors in cultivated areas within different distances from forest remnants in north Río de Janeiro', *Revista Brasileira de Entomologia*, vol 53, pp415–421

Benitez, P. C., Kuosmanen, T., Olschewski, R. and Van Kooten, G. C. (2006) 'Conservation payments under risk: A stochastic dominance approach', *American Journal of Agricultural Economics*, vol 88, pp1–15

Biesmeijer, J. C., Roberts, S. P. M., Reemer, M., Ohlemuller, R., Edwards, M., Peeters, T., Schaffers, A. P., Potts, S. G., Kleukers, R., Thomas, C. D., Settele, J. and Kunin, W. E. (2006) 'Parallel declines in pollinators and insect-pollinated plants in Britain and the Netherlands', *Science*, vol 313, pp351–354

Black, A. E., Morgan, P. and Hessburg, P. F. (2003) 'Social and biophysical correlates of change in forest landscapes of the interior Columbia Basin, USA', *Ecological Applications*, vol 13, pp51–67

Bos, M. M., Veddeler, D., Bogdanski, A., Klein, A. M., Tscharntke, T., Steffan-Dewenter, I. and Tylianakis, J. M. (2007) 'Caveates to quantifying ecosystem services: Fruit abortion blurs benefits from crop pollination', *Ecological Applications*, vol 17, pp1841–1849

Brosi, B. J., Armsworth, P. R. and Daily, G. C. (2008) 'Optimal design of agricultural landscapes for pollination services', *Conservation Letters*, vol 1, pp26–27

Brown, M. J. F. and Paxton, R. J. (2009) 'The conservation of bees: A global perspective', *Apidologie*, vol 40, pp410–416

Burgi, M. and Turner, M. G. (2002) 'Factors and processes shaping land cover and land cover changes along the Wisconsin River', *Ecosystems*, vol 5, pp184–201

Carneiro, F. E., Torres, R. R., Strapazzon, R., Ramírez, S. A., Guerra Jr, J. C. V., Koling, D. F. and Moretto, G. (2007) 'Changes in the reproductive ability of the mite *Varroa destructor* (Anderson e Trueman) in Africanized honey bees (*Apis mellifera* L.) (Hymenoptera: Apidae) colonies in southern Brazil', *Neotropical Entomology*, vol 36, pp949–952

Cauich, O., Quezada-Euán, J. J. G., Macias-Macias, J. O., Reyes-Oregel, V., Medina-Peralta, S. and Parra-Tabla, V. (2004) 'Behavior and pollination efficiency of *Nannotrigona perilampoides* (Hymenoptera: Meliponini) on greenhouse tomatoes (*Lycopersicon esculentum*) in subtropical Mexico', *Journal of Economic Entomology*, vol 97, pp475–481

Chacoff, N. P. and Aizen, M. A. (2006) 'Edge effects on flower-visiting insects in grapefruit plantations bordering premontane subtropical forest', *Journal of Applied Ecology*, vol 43, pp18–27

Chacoff, N. P., Morales, C. L., Garibaldi, L. A., Ashworth, L. and Aizen, M. A. (2010) 'Pollinator dependence of Argentinean agriculture: Current status and temporal analysis', *The Americas Journal of Plant Science and Biotechnology*, vol 3 (Special Issue 1), pp106–116

Chiari, W. C., Arnaut de Toledo, V. D. A., Ruvolo-Takasusuki, M. C. C., De Oliveira, A. J. B., Sakaguti, E. S., Attencia, V. M., Costa, F. M. and Mitsui, M. H. (2005) 'Pollination of soybean (*Glycine max* L. Merril) by honey bees (*Apis mellifera* L.)', *Brazilian Archives of Biology and Technology*, vol 48, pp31–36

Chiari, W. C., Arnaut de Toledo, V. D. A., Hoffmann-Campo, C. B., Ruvolo-Takasusuki, M. C. C., de Oliveira Arnaut de Toledo, T. C. S. and Lopes, T. D. S. (2008) 'Pollination by *Apis mellifera* in transgenic soy (*Glycine max* (L.) Merrill) Roundup Ready (TM) cv. BRS 245 RR and conventional cv. BRS 133', *Acta Scientiarum-Agronomy*, vol 30, pp267–271

Cortopassi-Laurino, M., Imperatriz-Fonseca, V. L., Roubik, D. W., Dollin, A., Heard, T., Aguilar, I., Venturieri, G. C., Eardley, C. and Nogueira-Neto, P. (2006) 'Global meliponiculture: Challenges and opportunities', *Apidologie*, vol 37, pp275–292

Cronon, W. (1985) *Changes in the Land: Indians, Colonists, and the Ecology of New England*, Hill and Wang, New York, NY

Davis, E. W. (1983) 'Experiences with growing vanilla (*Vanilla planifolia*)', *Acta Horticulturae*, vol 132, pp23–29

De Marco, J. and Coelho, F. (2004) 'Services performed by the ecosystem: Forest remnants influence agricultural cultures', *Biodiversity and Conservation*, vol 13, pp1245–1255

Del Sarto, M. C. L., Peruquetti, R. C. and Campos, L. A. O. (2005) 'Evaluation of the Neotropical stingless bee *Melipona quadrifasciata* (Hymenoptera: Apidae) as pollinator of greenhouse tomatoes', *Journal of Economic Entomology*, vol 98, pp260–266

Delaplane, K. S. and Mayer, D. F. (2000) *Crop Pollination by Bees*, CABI Publishing, New York, NY

Di Falco, S. and Perrings, C. (2006) '16 cooperatives, wheat diversity and the crop productivity in southern Italy', in M.A. Smale (eds) *Valuing Crop Biodiversity on Farm Genetic Resources and Economic Change*, CABI Publishing, Oxfordshire, UK

Dick, C. W., Hardy, O. J., Jones, F. A. and Petit, R. J. (2008) 'Spatial scales of pollen and seed-mediated gene flow in tropical rain forest trees', *Tropical Plant Biology*, vol 1, pp20–33

Elmqvist, T., Cox, P. A., Rainey, W. E. and Pierson, E. D. (1992) 'Restricted pollination on oceanic islands – pollination of *Ceiba pentandra* by flying foxes in Samoa', *Biotropica*, vol 24, pp15–23

Entwistle, P. F. (1972) *Pests of Cocoa*, Longman, London
Falque, M., Lesdalons, C. and Eskes, A. B. (1996) 'Comparison of two cacao (*Theobroma cacao* L.) clones for the effect of pollination intensity on fruit set and seed content', *Sexual Plant Reproduction*, vol 9, pp221–227
FAOSTAT (2009) *ProdSTAT Database*, Food and Agriculture Organization of the United Nations, http://faostat.fao.org/site/567/default.aspx#ancor, last accessed October 2009
Fleming, T. H. and Muchhala, N. (2008) 'Nectar-feeding bird and bat niches in two worlds: Pantropical comparisons of vertebrate pollination systems', *Journal of Biogeography*, vol 35, pp764–780
Fleming, T. H., Geiselman, C. and Kress, W. J. (2009) 'The evolution of bat pollination: A phylogenetic perspective', *Annals of Botany*, vol 104, pp1017–1043
Free, J. B. (1993) *Insect Pollination of Crops*, Academic Press, London
Freitas, B. M. and Paxton, R. J. (1998) 'A comparison of two pollinators: The introduced honey bee *Apis mellifera* and an indigenous bee *Centris tarsata* on cashew *Anacardium occidentale* in its native range of NE Brazil', *Journal of Applied Ecology*, vol 35, pp109–121
Freitas, B. M., Imperatriz-Fonseca, V. L., Medina, L. M., Kleinert, A. D. M. P., Galetto, L., Nates-Parra, G. and Quezada-Euán, J. J. G. (2009) 'Diversity, threats and conservation of native bees in the Neotropics', *Apidologie*, vol 40, pp332–346
Gallai, N., Salles, J.-M., Settele, J. and Vaussuère, B. E. (2009) 'Economic valuation of the vulnerability of world agriculture confronted with pollinator decline', *Ecological Economics*, vol 68, pp810–821
Geiselman, C. K., Mori, S. A. and Blanchard, F. (2002) *Database of Neotropical Bat/Plant Interactions*, www.nybg.org/botany/tlobova/mori/batsplants/database/dbase_frameset.htm
Ghazoul, J. (2007) 'Recognising the complexities of ecosystem management and the ecosystem service concept', *GAIA*, vol 16, pp215–221
Greenleaf, S. S. and Kremen, C. (2006) 'Wild bees enhance honey bees' pollination of hybrid sunflower', *Proceedings of the National Academy of Sciences*, vol 103, pp13890–13895
Gribel, R., Gibbs, P. E. and Queiroz, A. L. (1999) 'Flowering phenology and pollination biology of *Ceiba pentandra* (Bombaceae) in Central Amazonia', *Journal of Tropical Ecology*, vol 15, pp247–263
Guhl, A. (2008) *Café y cambio del paisaje en Colombia 1970–2005*, Fondo Editorial Universidad EAFIT, Banco de la República, Medellín
Hamrick, J. L., Nason, J. D., Fleming, T. H. and Nassar, J. M. (2002) 'Genetic diversity in columnar cacti', in T. H. Fleming and A. Valiente-Banuet (eds) *Columnar Cacti and their Mutualists: Evolution, Ecology, and Conservation*, University of Arizona Press, Tucson, AZ, pp122–133
Hargreaves, A. L., Harder, L. D. and Johnson, S. D. (2009) 'Consumptive emasculation: The ecological and evolutionary consequences of pollen theft', *Biological Reviews*, vol 84, pp259–276
Helversen, O. V. (1993) 'Adaptations of flowers to the pollination by glossophagine bats', in W. Barthlott, C. M. Naumann, K. Schmidt-Loske and K. L. Schuchmann (eds) *Animal–Plant Interaction in Tropical Environments*, Museum Koenig, Bonn, Germany, pp41–59
Hoehn, P., Tscharntke, T., Tylianakis, J. M. and Steffan-Dewenter, I. (2008) 'Functional group diversity of bee pollinators increases crop yield', *Proceedings of the Royal*

Society B: Biological Sciences, vol 275, pp2283–2291

Howell, D. J. and Roth, B. S. (1981) 'Sexual reproduction in Agaves: The benefits of bats; the costs of semelparous advertising', *Ecology*, vol 62, pp1–7

ICO (International Coffee Organization) (2009) *International Coffee Organization Database*, www.ico.org, last accessed October 2009

Irwin, R. E., Brody, A. K. and Waser, N. M. (2001) 'The impact of floral larceny on individuals, populations, and communities', *Oecologia*, vol 129, pp161–168

Kearns, C. A., Inouye, D. W. and Waser, N. M. (1998) 'Endangered mutualisms: The conservation of plant–pollinator interactions', *Annual Review of Ecology and Systematics*, vol 29, pp83–112

Kevan, P. G. and Imperatriz-Fonseca, V. L. (2002) *Pollinating Bees: The Conservation Link between Agriculture and Nature*, Ministry of Environment, Brasília, Brazil

Klein, A.-M. (2009) 'Nearby rainforest promotes coffee pollination by increasing spatio-temporal stability in bee species richness', *Forest Ecology and Management*, vol 258, pp1838–1845

Klein, A.-M., Vaissière, B. E., Cane, J. H., Steffan-Dewenter, I., Cunningham, S. A., Kremen, C. and Tscharntke, T. (2007) 'Importance of pollinators in changing landscapes for world crops', *Proceedings of the Royal Society B: Biological Sciences*, vol 274, pp303–313

Klein, A.-M., Cunningham, S. A., Bos, M. and Steffan-Dewenter, I. (2008a) 'Advances in pollination ecology from tropical plantation crops', *Ecology*, vol 89, pp935–943

Klein, A.-M., Olschewski, R. and Kremen, C. (2008b) 'The ecosystem service controversy: Is there sufficient evidence for a "pollination paradox"?', *GAIA*, vol 17, pp12–16

Klein, A.-M., Müller, C., Hoehn, P. and Kremen, C. (2009) 'Understanding the role of species richness for crop pollination services', in S. Naeem, D. E. Bunker, A. Hector, M. Loreau and C. Perrings (eds) *Biodiversity, Ecosystem Functioning, and Human Wellbeing – an Ecological and Economic Perspective*, Oxford University Press, Oxford, pp195–208

Koopman, K. F. (1981) 'The distributional patterns of New World nectar-feeding bats', *Annals of the Missouri Botanical Gardens*, vol 68, pp352–369

Kremen, C. (2008) 'Crop pollination services from wild bees', in R. R. James and T. L. Pitts-Singer (eds) *Bee Pollination in Agricultural Ecosystems*, Oxford University Press, US, pp10–26

Kricher, J. (1999) *A Neotropical Companion: An Introduction to the Animals, Plants, and Ecosystems of the New World Tropics*, Princeton University Press, New Jersey

Lambin, E. F., Turner, B. L., Geist, H. J., Agbola, S. B., Angelsen, A., Bruce, J. W., Coomes, O. T., Dirzo, R., Fischer, G., Folke, C., George, P. S., Homewood, K., Imbernon, J., Leemans, R., Li, X., Moran, E. F., Mortimore, M., Ramakrishnan, P. S., Richards, J. F., Skanes, H., Steffen, W., Stone, G. D., Svedin, U., Veldkamp, T. A., Coleen, V. and Xu, J. (2001) 'The causes of land-use and land-cover change: Moving beyond the myths', *Global Environmental Change*, vol 11, pp261–269

Law, B. S. and Lean, M. (1999) 'Common blossom bats (*Syconycteris australis*) as pollinators in fragmented Australian tropical rainforest', *Biological Conservation*, vol 91, pp201–212

Manrique, A. J. and Thimann, R. E. (2002) 'Coffee (*Coffea arabica*) pollination with africanized honeybees in Venezuela', *Interciencia*, vol 27, pp414–416

Mayfield, M. M. (2005) 'The importance of nearby forest to known and potential pollinators of oil palm (*Elaeis guineënsis* Jacq.; Areceaceae) in southern Costa Rica',

Economic Botany, vol 59, pp190–196
Mazvimavi, K. and Twomlow, S. (2009) 'Socioeconomic and institutional factors influencing adoption of conservation farming by vulnerable households in Zimbabwe', *Agricultural Systems*, vol 101, pp20–29
McGregor, S. E. (1976) *Insect Pollination of Cultivated Crop Plants*, US Department of Agriculture, Washington, DC
Molina-Freaner, F. and Eguiarte, L. E. (2003) 'The pollination biology of two paniculate agaves (Agavaceae) from northwestern Mexico: Contrasting roles of bats as pollinators', *American Journal of Botany*, vol 90, pp1016–1024
Myers, N., Mittermeier, R. A., Mittermeier, C. G., da Fonseca, G. A. B. and Kent, J. (2000) 'Biodiversity hotspots for conservation priorities', *Nature*, vol 403, pp853–858
Muchhala, N. (2007) 'Adaptive trade-off in floral morphology mediates specialization for flowers pollinated by bats and hummingbirds', *The American Naturalist*, vol 169, pp494–504
Muchhala, N. and Thomson, J. D. (2010) 'Fur versus feathers: Pollen delivery by bats and hummingbirds, and consequences for pollen production', *American Naturalist*, vol 175, pp717–726
Nassar, J. M., Ramirez, N. and Linares, O. (1997) 'Comparative pollination biology of Venezuelan columnar cacti and the role of nectar-feeding bats in their sexual reproduction', *American Journal of Botany*, vol 84, pp918–927
Nathan, P. T., Raghuram, H., Elangovan, V., Karuppudurai, T. and Marimuthu, G. (2005) 'Bat pollination of kapok tree, *Ceiba pentandra*', *Current Science*, vol 88, pp1679–1681
Nyerges, A. E. and Green, G. M. (2000) 'The ethnography of landscape: GIS and remote sensing in the study of forest change in West African Guinea savanna', *American Anthropologist*, vol 102, pp271–289
Oldroyd, B. P. (2007) 'What's killing American honey bees?', *PLoS Biol*, vol 5, e168
Olschewski, R., Tscharntke, T., Benítez, P. C., Schwarze, S. and Klein, A. M. (2006) 'Economic evaluation of pollination services comparing coffee landscapes in Ecuador and Indonesia', *Ecology and Society*, vol 11, p7
Olschewski, R., Klein, A.-M. and Tscharntke, T. (2010) 'Economic trade-offs between carbon sequestration, timber production, and crop pollination in tropical forested landscapes', *Ecological Complexity*, vol 7, pp314–319
Ortega-Olivencia, A., Rodriguez-Riano, T., Valtuena, F. J., Lopez, J. and Devesa, J. A. (2005) 'First confirmation of a native bird-pollinated plant in Europe', *Oikos*, vol 110, pp578–590
Ricketts, T. H. (2004) 'Tropical forest fragments enhance pollinator activity in nearby coffee crops', *Conservation Biology*, vol 18, pp1262–1271
Ricketts, T. H., Daily, G. C., Ehrlich, P. R. and Michener, C. D. (2004) 'Economic value of tropical forest to coffee production', *Proceedings of the National Academy of Sciences of the United States of America*, vol 101, pp12579–12582
Ricketts, T. H., Regetz, J., Steffan-Dewenter, I., Cunningham, S. A., Kremen, C., Bogdanski, A., Gemmill-Herren, B., Greenleaf, S. S., Klein, A. M., Mayfield, M. M., Morandin, L. A., Ochieng, A. and Viana, B. F. (2008) 'Landscape effects on crop pollination services: Are there general patterns?', *Ecology Letters*, vol 11, pp499–515
Roesel, C. S., Kress, W. J. and Bowditch, B. M. (1996) 'Low levels of genetic variation in *Phenakospermum guyannense* (Strelitziaceae), a widespread bat-pollinated Amazonian herb', *Plant Systematics and Evolution*, vol 199, pp1–15

Roubik, D. W. (1995) *Pollination of Cultivated Plants in the Tropics*, Food and Agriculture Organization of the United Nations, Bulletin 118, Rome, Italy

Roubik, D. W. (2002a) 'Feral African bees augment Neotropical coffee yield', in P. Kevan and V. L. Fonseca (eds) *Pollinating Bees: The Conservation Link between Agriculture and Nature*, Ministry of Environment, Brazil

Roubik, D. W. (2002b) 'The value of bees to the coffee harvest', *Nature*, vol 417, p708

Roubik, D. W. (2009) 'Ecological impact on native bees by the invasive Africanized honey bee', *Acta Biológica Colombiana*, vol 14, pp115–124

Slaa, E. J., Sánchez Chaves, L. A., Malagodi-Braga, K. S. and Hofstede, F. E. (2006) 'Stingless bees in applied pollination: Practice and perspectives', *Apidologie*, vol 37, pp293–315

Steffan-Dewenter, I., Münzenberg, U., Bürger, C., Thies, C. and Tscharntke, T. (2002) 'Scale-dependent effects of landscape context on three pollinator guilds', *Ecology*, vol 83, pp1421–1432

Thomson, J. D. (2003) 'When is it mutualism?', *American Naturalist*, vol 162, ppS1–S9

Tscharntke, T., Klein, A. M., Kruess, A., Steffan-Dewenter, I. and Thies, C. (2005) 'Landscape perspectives on agricultural intensification and biodiversity – ecosystem service management', *Ecology Letters*, vol 8, pp857–874

Valiente-Banuet, A., Santos, R., Arizmendi, M. D. and Casas, A. (2007) 'Pollination biology of the hemiepiphytic cactus *Hylocereus undatus* in the Tehuacan Valley, Mexico', *Journal of Arid Environments*, vol 68, pp1–8

Veddeler, D., Klein, A. M. and Tscharntke, T. (2006) 'Contrasting responses of bee communities to coffee flowering at different spatial scales', *Oikos*, vol 112, pp594–601

Veddeler, D., Olschewski, R., Tscharntke, T. and Klein, A. M. (2008) 'The contribution of non-managed social bees to coffee production: New economic insights based on farm-scale yield data', *Agroforestry Systems*, vol 73, pp109–114

Vergara, C. H. and Badano, E. I. (2009) 'Pollinator diversity increases fruit production in Mexican coffee plantations: The importance of rustic management systems', *Agriculture, Ecosystems & Environment*, vol 129, pp117–123

Villalobos, S., Vargas, O. and Melo, S. (2007) 'Uso, manejo y conservación de 'yosú', *Stenocereus griseus* (Cactaceae) en la Alta Guajira colombiana', *Acta Biologica Colombiana*, vol 12, pp99–112

Vogel, S. (1969) 'Chiropterophilie in der neotropischen flora, neue mitteilungen II', *Flora*, vol 158, pp185–222

Wesselingh, R. A. (2007) 'Pollen limitation meets resource allocation: Towards a comprehensive methodology', *New Phytologist*, vol 174, pp26–37

Williams, I. H. (1994) 'The dependences of crop production within the European Union on pollination by honey bees', *Agricultural Science Reviews*, vol 6, pp229–257

Winfree, R., Griswold, T. and Kremen, C. (2007) 'Effect of human disturbance on bee communities in a forested ecosystem', *Conservation Biology*, vol 21, pp213–223

Winfree, R., Williams, N. M., Gaines, H., Ascher, J. S. and Kremen, C. (2008) 'Wild bee pollinators provide the majority of crop visitation across land-use gradients in New Jersey and Pennsylvania, USA', *Journal of Applied Ecology*, vol 45, pp793–802

Winter, Y. and Helversen, O. V. (2003) 'Operational tongue length in phyllostomid nectar-feeding bats', *Journal of Mammalogy*, vol 84, pp886–896

Wunder, S. (2000) *The Economic of Deforestation: The Example of Ecuador*, MacMillan Press Ltd, Hampshire, UK
Young, A. M. (1994) *The Chocolate Tree*, Smithsonian Institution Press, Washington, DC

6

Ecological Indexing as a Tool for the Payment for Ecosystem Services in Agricultural Landscapes

The Experience of the GEF-Silvopastoral Project in Costa Rica, Nicaragua and Colombia

Cristóbal Villanueva, Muhammad Ibrahim, Francisco Casasola and Claudia Sepúlveda

Introduction

In Latin America, pastures are the chief agricultural land use, comprising 30 per cent of the total land area (FAO, 2006). A significant portion of these pastures are mismanaged through reliance on monocultures of pasture species and through overgrazing. As a result of the combination of these factors, more that 50 per cent of the pastures in Central America are degraded (Szott et al, 2000), having lost both their productive capacity as well as their capacity to provide key ecosystem services. This situation leads to livestock systems based on pastures with low productivity and profitability, and negative externalities for the environment, such as soil erosion, contamination of water sources, biodiversity loss and high greenhouse gas emissions.

Alternative management practices exist to reduce the environmental impact of livestock systems, increase their productivity and reverse the environmental degradation associated with mismanaged pastures. These alternatives include silvopastoral systems, which are complex livestock production systems made up of combinations of trees and/or shrubs, with pasture species and livestock. Silvopastoral systems, together with other interventions (e.g. improved animal varieties, nutrition and veterinary practices), are a key strategy for the development of sustainable livestock farming models with the potential of improving animal production (Souza de Abreu, 2002; Betancourt et al, 2003), increasing farm profitability (Holmann et al, 1992; Alonzo et al, 2001) and providing ecosystem services (Chacón and Harvey, 2006; Ibrahim et al, 2007; Ríos et al, 2007; Sáenz et al, 2007).

However, despite the positive aspects associated with silvopastoral systems, their adoption has been low. This is, in part, because of the high investment cost, lack of capital and lack of technical knowledge for establishing and managing these systems (Alonzo et al, 2001). Moreover, they require a relatively long establishment period, in excess of six months, chiefly for the establishment of tree/shrub components. This long establishment period can result in reduced initial cash flow if the producer does not have local feed alternatives in order to maintain herd size during the establishment phase.

Nevertheless, there are several incentive mechanisms, such as payment for ecosystem services (PES), privileged or green credits, and product certification, to promote the adoption of silvopastoral systems. PES is a strategy to compensate rural landowners for at least part of the cost of adopting environmentally friendly land uses. One of the complex challenges for PES schemes is how to quantify the value of the service provided in order to make appropriate payments to the landowner in exchange for changes in management. For example, when quantifying the carbon in a pasture that has a high tree density, the carbon accumulated above, as well as below ground, has to be considered. The case of biodiversity is more complex than that of carbon, in part because it can be measured at various scales (e.g. at the level of a plot of land, a farm or a landscape) and no single unit of measure exists (see Chapters 3 and 10 in this volume). Furthermore, current techniques for the quantification and monitoring of ecosystem services have relatively high costs. This can affect the participation of providers of ecosystem services because of the high transaction costs; the payment received for ecosystem services may be less or only slightly superior to the monitoring costs.

In light of these limitations, there is a critical need for a tool that is precise and low cost in order to estimate the value of the ecosystem services provided by cattle farms. This chapter describes a methodology used by the Integrated Silvopastoral Systems for Ecosystem Management project, funded by the Global Environment Facility (GEF) and administered by the World Bank, as a tool for applying PES on cattle farms. The Ecological Index was employed by the GEF-Silvopastoral project to promote the adoption of silvopastoral systems on livestock farms located in pilot zones in Costa Rica, Nicaragua and Colombia.

How Indices Were Created for Carbon and Biodiversity Monitoring

The aim of the project was to use PES to promote the adoption of silvopastoral systems on cattle ranches in order to improve socio-economic conditions in the rural communities while simultaneously improving biodiversity conservation and carbon storage on the farms receiving payments.

One of the first steps of the Silvopastoral project was to develop an Ecological Index to estimate the quantity of additional ecosystem services provided by farms when a land use changed from A to B. It was important that the index could be employed to monitor ecosystem services, as well as to define how much a farm should receive in payments. The index was constructed using available secondary information (databases, technical and scientific publications) from studies completed by the Tropical Agricultural Research and Higher Education Center (CATIE) and other research centres in Latin America, as well as from estimates by experts.

As a starting point, we reviewed published literature about the categorization of different land uses in terms of their contribution to biodiversity conservation (species richness and other indices) and carbon capture (t C ha^{-1}). Forests (primary and secondary) provide the highest conservation values for multiple taxonomic groups, although we relied primarily on the case of birds. These same land uses provide the greatest potential for carbon fixation. Thus, in the development of our biodiversity and carbon indices, these reference land uses received the greatest possible value of 1.

In terms of carbon, primary or mature secondary forests may exhibit low fixation rates, but retain important carbon reserves (see Chapter 1). Nevertheless, to avoid creating perverse incentives, which could lead to clearing forests in order to replace them with land uses with greater fixation values, we gave forests high carbon values in our Ecological Index. Land uses that generally do not involve tree cover and which are highly vulnerable to soil erosion processes, such as annual crops, were located at the other extreme of our index. Pastures with a high tree density (>30 trees ha^{-1}), multi-strata live fences and mixed orchards received intermediate values. Other land uses were assigned ordinal values in relation to similar land uses and as a function of the tree cover that they retained. For those land uses not reported in the literature, relative values were developed by a panel of experts who also reviewed and validated the assignment of the indices for all the land uses of interest in the project.

Whereas our classification was specifically designed for PES on cattle farms, it is important to point out that these indices for the different land uses can be adjusted depending on the focus of a PES and new research results. The ordering of land uses, including agroforestry systems, can change in relation to the ecosystem service of interest. For example, the ranking can be quite different for soil erosion services, scenic value, pest control or provisioning services.

Carbon and Biodiversity Index for the Different Land Uses in the Agro-Landscape

The main land uses found within the pasture-dominated landscapes of this study were classified from 0 to 1 according to their biodiversity conservation value, and independently a value between 0 and 1 according to their carbon storage capacity. The two values can be summed to determine the total contribution of each land use to these services, providing a score between 0 and 2 for each land use. For example, primary forest has the greatest index value with 2 points: 1 point for carbon capture and 1 point for biodiversity conservation. On the other hand, degraded pastures were assigned a score equivalent to 0 because they make the smallest contribution to either carbon or biodiversity conservation (see Table 6.1).

Validation

The project monitored biodiversity (Sáenz et al, 2007) and carbon capture (Zamora, 2006) in the land uses predominating in the three countries involved in the project over a period of four years, in order to validate the literature-based Ecological Index used for the payment for ecosystem services. We used the results from more than four years of monitoring birds in Costa Rica to validate the relationship between the literature-based Ecological Index for Biodiversity and field data and to develop a refined index that can be used in place of the Ecological Index as a tool for PES for biodiversity. This index is known as the Biodiversity Index for Environmental Services (IBSA) and is calculated using the following equation developed by Sáenz (2005):

$$\text{IBSA} = \sum i(\text{IV}_i \times \text{Ab}_i/\text{abT}) \times (\text{S}_{\text{ha}}) + (\text{Va}_{\text{h}}) \qquad [6.1]$$

where:

IV_i = Importance Value, which measures two issues: the vulnerability of species 'i' with respect to the loss of tree cover, and its risk of extinction, as estimated by the International Union for Conservation of Nature (IUCN). Tree cover dependence of each species was graded as follows: 1 = open area species; 2 = species that require at least remnants of forest; and 3 = forest-dependent species. The IUCN status was graded as follows: 3 = in danger of extinction or on the IUCN Red List; 2 = threatened or small populations; and 1 = least conservation concern. The final value is the average value if the species category was known in both scales.

Ab_i = sum of all the individuals of species 'i' in a particular land use.

abT = total species richness in a landscape (Gamma diversity).

S_{ha} = proportion of richness in each land use with respect to the total for the landscape.

Va_{h} = value of the habitat variables that best explain species richness. In our case, we used tree cover, the variable that was best correlated to bird species

Table 6.1 *Ecological Index (ecological points per hectare) for the different land uses in the landscapes of the GEF-Silvopastoral project*

Land use	*Biodiversity Index (A)*	*Carbon Index (B)*	*Total index (A + B)*
Short rotation crops with a cycle less than 12 months, such as basic grains (corn and beans), vegetables and tubers.	0.0	0.0	0.0
Degraded pasture with less than 50% cover of grass and desirable forage; minimal presence of trees and/or shrubs. May show obvious signs of erosion.	0.0	0.0	0.0
Naturalized pasture without trees: pasture dominated by native and/or introduced species without trees and/or shrubs.	0.1	0.1	0.2
Improved pasture* without trees: pasture dominated by very hardy, high-productivity introduced species with a cover greater than 90% and without trees and/or shrubs.	0.1	0.4	0.5
Unshaded semi-perennial crops: plantain crops or coffee plantations with 2000 trees or more per hectare with full exposure to the sun. Minimal presence of shade or fruit trees.	0.3	0.2	0.5
Naturalized pasture with low tree density: pasture dominated by native and/or naturalized species with a density of < 30 trees ha^{-1}, DBH** > 5cm and > 2m height.	0.3	0.3	0.6
Enriched naturalized pasture with low tree density: pasture dominated by native and/or naturalized species, with recently planted trees of up to 5cm DBH, > 0.5m height and > 30 trees ha^{-1}.	0.3	0.3	0.6
Simple live fence:# linear system of trees that are pruned periodically (at least twice a year) for forage or fertilizer, or recently established live posts.	0.3	0.3	0.6
Enriched improved pasture with low tree density: pasture dominated by very hardy and productive improved species, with recently planted trees of up to 5cm DBH, > 0.5m height and > 30 trees ha^{-1}.	0.3	0.4	0.7
Orchard plantation (monoculture): plantations of shrubby perennial or semi-perennial tropical fruit trees or citrus grown homogeneously.	0.3	0.4	0.7
Grass-based forage bank: cut pasture or high-density forage sugar cane; with or without trees.	0.3	0.5	0.8
Improved pasture with low tree density: pasture dominated by very hardy and highly productive improved or introduced species with a density of < 30 trees ha^{-1}, DBH > 5cm and 2m height.	0.3	0.6	0.9
Shrub-based forage bank: high-density (> 10,000 plants ha^{-1}) shrubby forage plantation.	0.4	0.5	0.9
Naturalized pasture with high tree density: pasture dominated by native and/or naturalized species with a density > 30 trees ha^{-1}, DBH > 5cm and height of 2m.	0.5	0.5	1.0
Orchard plantation (polyculture): plantations of perennial or semi-perennial shrubby fruit trees or mixed citrus and/or in various strata.	0.6	0.5	1.1
Multi-strata live fence:§ linear system with trees growing freely in multiple strata or with at least one upper stratum at least 4m wide, 4m high or 4m crown.	0.6	0.5	1.1

Table 6.1 *continued*

Land use	*Biodiversity Index (A)*	*Carbon Index (B)*	*Total index (A + B)*
Diversified forage bank: pastures, forage sugar cane or cut shrubby plants in various strata (at least four species) with an upper stratum of trees of at least 4m.	0.6	0.6	1.2
Timber plantation (monoculture): homogeneous planting of trees of a single species, planted at high density, more than 500 trees ha^{-1}.	0.4	0.8	1.2
Coffee agroforestry system: coffee plantations with shade trees of various species, with at least 25% cover of the canopy or upper stratum.	0.6	0.7	1.3
Improved pasture with high tree density: pasture dominated by very hardy and productive improved or introduced species where the existing trees are mature and with a density > 30 trees ha^{-1}.	0.6	0.7	1.3
Forest or plantation of guadua (bamboo): forest or homogeneous or mixed (crop) plantation of guadua or other bamboos.	0.5	0.8	1.3
Timber plantation (polyculture): intensive plantation of timber-yielding trees of at least three native, naturalized or introduced species, planted with density > 500 trees ha^{-1}.	0.7	0.7	1.4
Successional shrublands: native vegetation in natural succession of at least 5m in height.	0.6	0.8	1.4
Riparian forest: natural vegetation of different strata on riverbanks or bodies of water with a minimum width of 4m.	0.8	0.7	1.5
Intensive silvopastoral system: very hardy and productive improved pastures associated with high-density forage shrubs, minimum 5000 trees ha^{-1}; also improved pastures with rows of high-density timber-yielding trees (pastures in alleys).	0.6	1.0	1.6
Secondary forest (intervened): native forest with base area greater than $10m^2$, intervened (high extraction of trees or non-timber resources, hunting, partial cutting); remnants of any size.	0.8	0.9	1.7
Secondary forest: native forest with moderate interventions in the last few decades; high biological diversity; base area greater than $10m^2$; remnants of any size.	0.9	1.0	1.9
Primary forest: native forest with no intervention in the last 30 years, more than 80% cover and high biodiversity; remnants of any size.	1.0	1.0	2.0

Notes: * Includes species such as *Brachiaria* spp., *Panicum maximum* and *Cynodon* spp., among others.
** DBH = diameter at breast height (1.3m height).
# A simple live fence has one or two dominant species managed by pruning to a similar height.
§ Live fence that has more than two shrubby species of different heights and uses for timber, orchard, forage, medicinal, ornamental purposes, etc.
Source: Murgueitio et al (2003)

Table 6.2 *Comparison between the Ecological Index for Biodiversity and the Biodiversity Index for Environmental Services (IBSA)*

Land use	*Ecological Index for Biodiversity*	*IBSA*
Primary forest	1.0	1.02
Secondary forest	0.9	0.93
Riparian forest	0.8	1.04
Secondary shrubby vegetation	0.6	0.82
Improved pasture with high tree density	0.6	0.40
Forage bank	0.6	0.05
Multi-strata live fence	0.6	0.57
Mixed species orchard	0.6	0.22
Naturalized pasture with high tree density	0.5	0.45
Timber plantation (monoculture)	0.4	0.39
Naturalized pasture with low tree density	0.3	0.13
Improved pasture with low tree density	0.3	0.16
Simple live fence	0.3	0.43
Naturalized pasture	0.1	0.07
Improved pasture	0.1	0.04
Degraded pasture	0.0	0.26

Note: Pastures with low tree density have less than 30 trees ha^{-1}; pastures with high tree density have more than 30 trees ha^{-1}.
Source: Sáenz (2005)

richness (Sáenz et al, 2007); but number of trees, diameter at breast height (DBH) or other variables could be used as well.

A comparison of the Ecological Index for Biodiversity and the IBSA showed a high positive correlation ($r = 0.78$; $p < 0.05$) between both indices. Forage banks, pastures with low tree density and mixed orchards had different values for these indices (see Table 6.2). In general, these results support the claim that the Ecological Index for Biodiversity assigned by the project was effective for the majority of the land uses. However, these examples demonstrate that it is necessary to revise and adjust index values according to local and context-specific data, particularly for biodiversity. The results also suggest how land uses with low conservation values can be improved to increase their conservation value. For example, in the case of forage banks, including at least one upper stratum of shrubby species in order to improve structural complexity increases avian diversity (Cárdenas, 1999).

Bird richness depends strongly on the complexity, composition and structure of the vegetation in the different land uses. The studies completed in the different pilot zones of the GEF-Silvopastoral project showed a similar pattern of bird richness – that is, the land uses with greater tree cover had the highest values (Milder et al, 2010). In forests, Costa Rica had a mean bird richness of between 6 and 8 species; Nicaragua, between 12 and 15; and Colombia, between 20 and 30 (10-minute monitoring). In contrast, the lowest bird richness values were found in the treeless pastures with 2, 7 and 10 species, respectively (Sáenz et al, 2007). These values do not take into consideration the influence of landscape context, such as distance to nearest live fence, and

Table 6.3 *Carbon stored in different land uses in Costa Rica*

Land use	*Carbon stored (t ha^{-1})*	*Ecological Index for carbon*
Degraded pasture without trees	32.2	0
Natural pasture without trees	136.2	0.1
Improved pasture without trees	156.3	0.4
Natural pasture with low tree density	139.0	0.3
Improved pasture with low tree density	140.5	0.6
Natural pasture with high tree density	164.6	0.5
Improved pastures with high tree density	164.9	0.7
Forest plantation, monoculture (teak)	202.8	0.8
Vegetable succession (tacotal)	171.5	0.8
Riparian forest	323.3	0.7
Secondary forest, intervened	127.6	0.9
Secondary forest	202.8	1.0

Note: Pastures with low tree density have less than 30 trees ha^{-1}; pastures with high tree density have more than 30 trees ha^{-1}.
Source: Zamora (2006)

secondary and riparian forests (Enríquez et al, 2007), which is inversely correlated with bird diversity in agricultural land uses.

We also validated our literature-based carbon values, using a modelling approach (CO2FIX model) (Masera et al, 2003), parameterized with our field data, and found that the land uses with high tree cover, such as forests, forest plantations and improved pastures with high tree density, had the highest values for carbon reserves in comparison with degraded pastures and land uses with reduced tree cover (Zamora, 2006) (see Table 6.3). The resulting values were strongly correlated with the literature-based Ecological Index defined by the project. It is important to bear in mind that the capacity of different land uses to store carbon will depend on local and regional variables, such as soil fertility, climate, system age, and composition and density of shrubby species.

Putting the Indices to Work: How Payment for Ecosystem Services (PES) Operates

The main guideline for the system of payments based on the Ecological Index was that the producer provides ecosystem services by means of changes in the land use on the farm from systems that provide minimal biodiversity or carbon storage services, such as degraded pastures, to systems with a greater capacity to provide these services. Temporal changes in the patterns of land use were taken as indicators of the amount of ecosystem services provided. Thus, payments were made in proportion to the total increase in ecosystem services provided by the farm, calculated by summing the results of multiplying the area under each land use by the corresponding Ecological Index for that land use, relative to a baseline sum established in year zero. The index was calculated at the farm level to avoid leakage problems (i.e. the score obtained by the farm as a whole determined the level of payment). The amount paid for the ecosystem services was calculated by multiplying the farm score by the monetary value assigned to each index point. In the case of Costa Rica, the monetary value of each index point was derived from the price paid by the National Fund for Forestry Funding (FONAFIFO), the institution legally designated by the government of Costa Rica to administer resources provided for the payment of environmental services (see Chapter 12 in this volume). In the case of the GEF-Silvopastoral project, the payment was made for the incremental points accumulated beyond those estimated for the base year. With this incremental system, the payment only provided an incentive for improvements, rather than a payment for services that were already provided prior to the project. This limitation was addressed during the implementation of the project.

From an operational point of view, the project established the following protocol to determine payment amounts (example of the project's PES scheme in Costa Rica):

- A first visit to the farm verifies the land-use map based on satellite images.
- Digitalization of the farm data using Arcview assigns each land use a carbon and biodiversity value based on the area under each land use and the corresponding Ecological Index. These values are then summed to provide a total score for each farm.
- During the first year, payments are made for the baseline value (US$10 per point).
- Payments during the subsequent years are made for incremental points in relation to the baseline (US$110 per point if the farmer has selected a two-year scheme; US$75 per point for a four-year scheme).

All payments were made by FONAFIFO, after the project reported the results of applying the above protocol. The maximum amount that a farm could receive was established at US$4500.

This methodology can create controversy or serve as a demotivating factor among the producers participating in the programme: the methodology favours farms that have poor and unproductive land use (high presence of degraded pastures or natural pastures) – that is, those which are not producing many ecosystem services in comparison with those that are already very developed in terms of land uses that favour conservation (forests) and production/conservation (silvopastoral systems) since these latter farms will not be able to offer more services than already provided. These farms only receive baseline payments for ecosystem services that result from their previous conservation efforts. For this reason, principles of justice, and the motivation to safeguard existing services, suggest that farms that have land uses exclusively for conservation should continue to be compensated economically – the reduce emissions from deforestation and degradation (REDD) approach. This approach would ensure better management and conservation of the forests in the territories.

Applying the Payment

The impact of the payments upon the dynamics of land-use change can be seen from the contrast of two Costa Rican farms, one of which was in a control group (no payment for ecosystem services) (see Tables 6.4 and 6.5).

Table 6.4 *Operation of PES on a farm participating in the GEF-Silvopastoral project located in the pilot zone of Esparza, Costa Rica*

Land use	*Index*	*Baseline*		*Year I*		*Year II*		*Year III*		*Year IV*	
	(points ha^{-1})	*Area (ha)*	*Points*	*Area (ha)*	*Points*	*Area (ha)*	*Points*	*Area (ha)*	*Points*	*Area (ha)*	*Points*
Degraded pasture	0	6.16	0	11.98	0	10.62	0	6.28	0	6.28	0
Natural pasture without trees	0.2	14.26	2.85	0	0	0	0	0	0	0	0
Improved pasture without trees	0.5	1.80	0.90	0.81	0.41	0.19	0.10	0	0	0	0
Semi-perennial agricultural crop without shade	0.5	0.55	0.28	0	0	0	0	0	0	0	0
Natural pasture with low tree density	0.6	1.62	0.97	5.97	3.58	3.90	2.34	4.22	2.53	4.22	2.53
Simple live fences*	0.6	2.24	1.34	2.87	1.72	3.45	2.07	3.45	2.07	3.45	2.07
Orchard monoculture	0.7	0.27	0.19	0.27	0.19	0.27	0.19	0.27	0.19	0.27	0.19
Grasses forage bank	0.8	0.10	0.08	0.24	0.19	0.55	0.44	0.76	0.61	0.76	0.61
Improved pasture with low tree density	0.9	0.20	0.18	6.19	5.57	8.88	7.99	8.26	7.43	8.26	7.43
Natural pasture with high tree density	1.0	0.70	0.70	0.20	0.20	0.20	0.20	2.74	2.74	2.74	2.74
Multi-strata live fence*	1.1	0	0	1.42	1.56	1.85	2.04	1.85	2.04	1.85	2.04
Improved pasture with high tree density	1.3	0.07	0.09	0.07	0.09	0.07	0.09	0.40	0.52	0.40	0.52
Secondary vegetable succession	1.4	1.45	2.03	1.86	2.60	2.91	4.07	2.91	4.07	2.91	4.07
Riparian forest	1.5	5.02	7.53	4.28	6.42	4.28	6.42	4.28	6.42	4.28	6.42
Intensive silvopastoral system	1.6	0	0	0	0	0	0	1.76	2.82	1.76	2.82
Secondary forest, intervened	1.7	2.38	4.05	2.71	4.61	2.71	4.61	2.71	4.61	2.71	4.61
Payment operations											
Total ecological points			21.19		27.14		30.56		36.05		36.05
Incremental ecological points					5.95		9.37		14.86		14.86
Payment for ecosystem services at baseline (US$10 per point)			211.9								
Payment for incremental ecosystem services (US$75 per point)					446.3		702.8		1114.5		1114.5

Notes: * Live fences are measured in kilometres.

Table 6.5 *Change in land use on a control group farm (no payment for ecosystem services) in the GEF-Silvopastoral project in Esparza, Costa Rica*

Land use	*Index (points ha⁻¹)*	*Baseline Area (ha)*	*Baseline Points*	*Year I Area (ha)*	*Year I Points*	*Year II Area (ha)*	*Year II Points*	*Year III Area (ha)*	*Year III Points*	*Year IV Area (ha)*	*Year IV Points*
Basic grains	0	0.57	0	0.57	0	0.57	0	0.57	0	0.57	0
Degraded pasture	0	0.56	0	0.56	0	0.55	0	1.57	0	1.57	0
Natural pasture with low tree density	0.6	2.45	1.47	0.14	0.08	0.14	0.08	0	0	0	0
Simple live fences*	0.6	1.55	0.93	2.4	1.44	2.33	1.4	2.39	1.43	2.39	1.43
Improved pasture with low tree density	0.9	2.06	1.85	3.91	3.52	1.43	1.29	1.43	1.29	1.43	1.29
Natural pasture with high tree density	1	1.06	1.06	1.52	1.52	0.46	0.46	0	0	0	0
Multi-strata live fence*	1.1	0	0	0	0	0.26	0.29	0.26	0.29	0.26	0.29
Improved pasture with high tree density	1.3	1.31	1.7	1.31	1.7	4.86	6.32	4.44	5.77	4.44	5.77
Secondary vegetable succession	1.4	0.16	0.22	0.16	0.22	0.16	0.22	0.16	0.22	0.16	0.22
Riparian forest	1.5	0.34	0.51	0.34	0.51	0.34	0.51	0.34	0.51	0.34	0.51
Total ecological points			7.74		8.99		10.57		9.51		9.51
Incremental ecological points					1.25		2.83		1.77		1.77

Notes: * Live fences are measured in kilometres.

The first farm was on the four-year scheme (i.e. it received a baseline payment of US$10 per point and incremental points after the baseline was paid at US$75 per point) (see Table 6.4). The Ecological Index of this farm for the base year was 21.19 (i.e. the farmer received US$211.9 corresponding to this baseline). Due to changes in land management, the farm scored 36.05 at the end of the contract period and received, over the four years, US$3378.1.

The second farm was part of the control group (i.e. it could not receive payment for any increased ecosystem services). In this case, the hypothesis that land-use changes on this type of farm would be fewer in comparison with the group of farms that received incremental payments was correct (see Table 6.5). From a score of 7.74 for the base year, the farm only increased its score by 1.77 at the end of the four-year period.

Land-use changes on the farm that participated in the payment for ecosystem services programme were primarily increased areas of natural pastures and improved pastures with trees (both high and low density), replacing degraded pastures; the establishment of the leucaena (*Leucaena leucocephala*) silvopastoral system in association with *Brachiaria brizantha*, also in degraded pasture areas; and an increase of simple and multi-strata live fences as new land divisions, replacing dead fence lines (see Figure 6.1).

On the other hand, on the farm that did not receive payments, the only positive changes were from natural pastures (of low and high tree density) to

Farm A, 2003

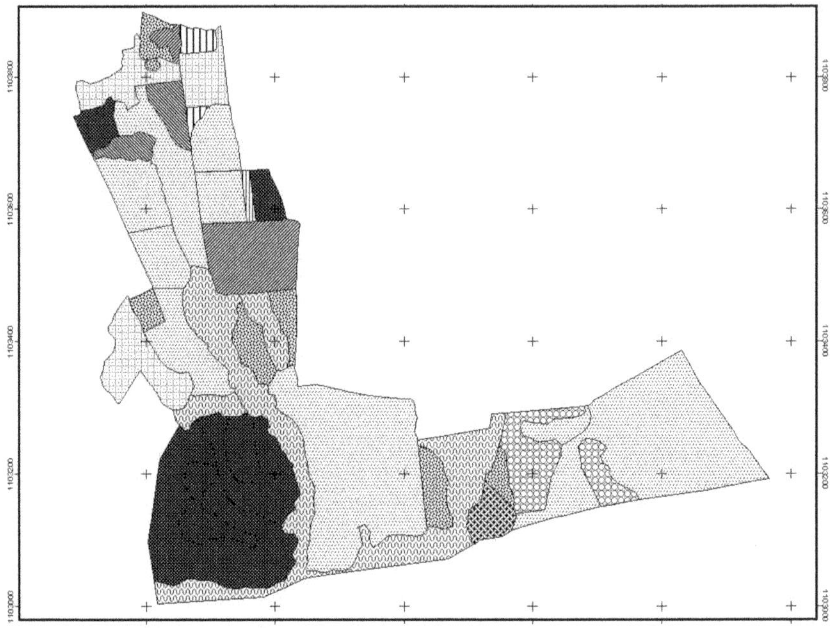

Farm A, 2007

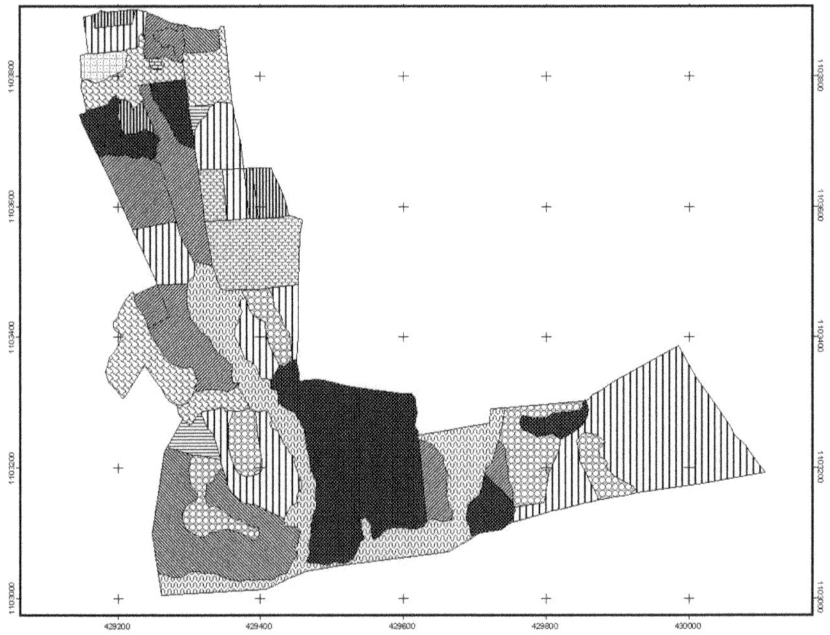

Figure 6.1 *Map of land uses during the years 2003 and 2007 on a farm under PES (Farm A) and on a control farm (Farm B) in Esparza, Costa Rica*

Source: chapter authors

Farm B, 2003

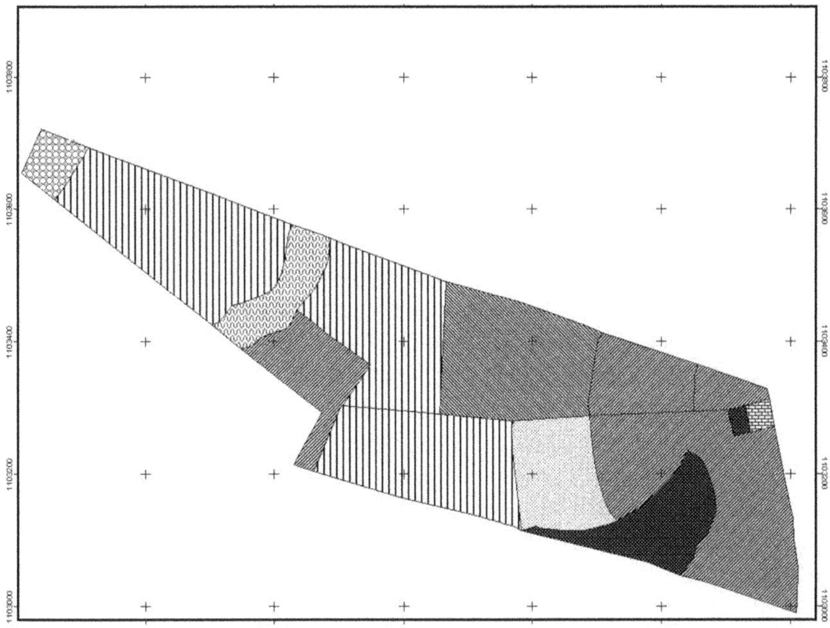

Farm B, 2007

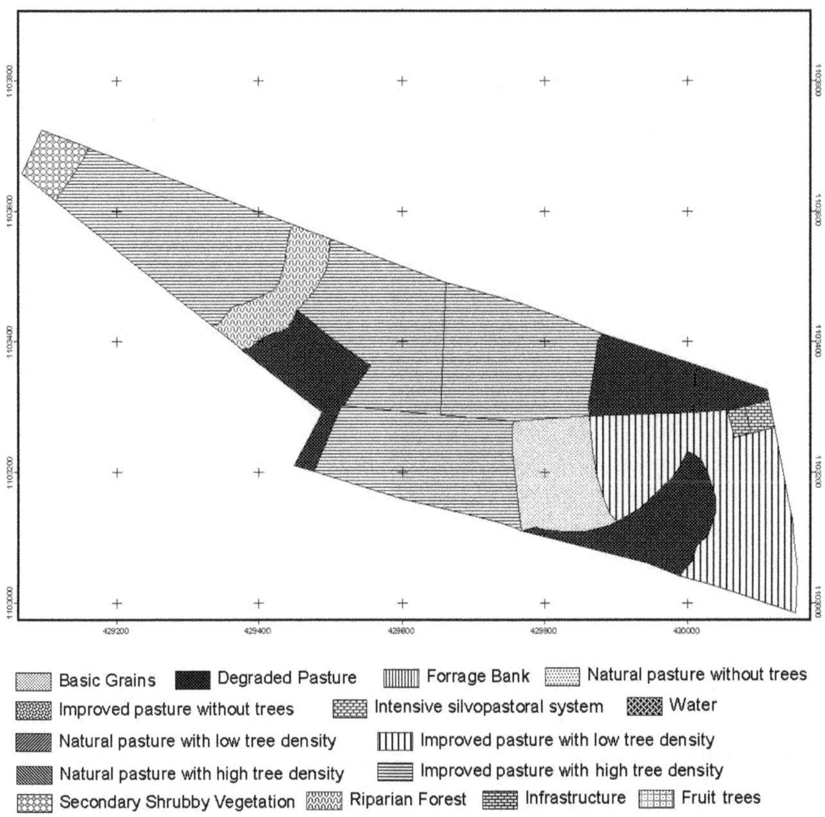

improved pastures (with low and high tree density). In this case, there were, in fact, some negative changes from natural pastures with low tree density to degraded pastures (see Figure 6.1). The results reflect the better response in changes in land uses (motivated by the accumulation of ecological points) of the farm that would receive payments for ecosystem services.

Cost of Monitoring

In many situations, the monitoring costs associated with PES have affected the viability of a programme since the payment received is less than or equal to the cost of implementing and monitoring conservation interventions. This can be a limitation for certification schemes as well as for accessing the Clean Development Mechanism (CDM) for carbon payments. The GEF-Silvopastoral project covered the transaction costs for monitoring and payment operations of the pilot schemes described in this chapter. Nonetheless, in order to quantify these costs, we selected the pilot zone of Costa Rica, where such information was available. Monitoring costs included the cost of the satellite image, drawing up maps of farms, acquiring field information (e.g. for ground-truthing), digitalization and data processing, and preparing reports for payment. Monitoring costs were directly related to the size of the farm, with greater operation costs per hectare associated with smaller farms, and the difficulties of monitoring certain land uses.

In general, land uses were measured using units of surface measurement (hectares), with the exception of live fences that were evaluated by length (kilometres). Thus, in order to make cost comparisons between land uses, live fence values were changed from a longitudinal unit to a surface unit (hectares); the length of a live fence was measured in the field with the help of the map of the farm, and the width corresponded to the average crown diameter of the trees included in the fence (Francesconi, 2006; Tobar and Ibrahim, 2010). Overall, the project's monitoring cost was US$4.18 per hectare. The breakdown of this amount showed that the office work associated with monitoring (preparing field equipment, and collecting and processing information using GIS) was the principal cost (46 per cent). Other costs were farm visits and monitoring of land uses (31 per cent), equipment and materials[1] (16 per cent), and transportation (7 per cent).

The unit area cost of monitoring land use decreased with farm size (i.e. it was less than US$2 per hectare on farms larger than 40ha in comparison to the US$4.18 per hectare average) (see Figure 6.2). This result illustrates that monitoring costs may be a barrier to the participation of small producers. Strategies need to be developed to reduce monitoring costs – for example, using the participation of organized groups as well as identifying alternative monitoring strategies or subsidies. For example, in one FONAFIFO scheme (FONAFIFO, 2010), small farms of some indigenous communities were organized as a single entity in order to receive agroforestry incentives (US$1.30 was paid for every tree planted in agricultural land), thereby reducing transaction costs.

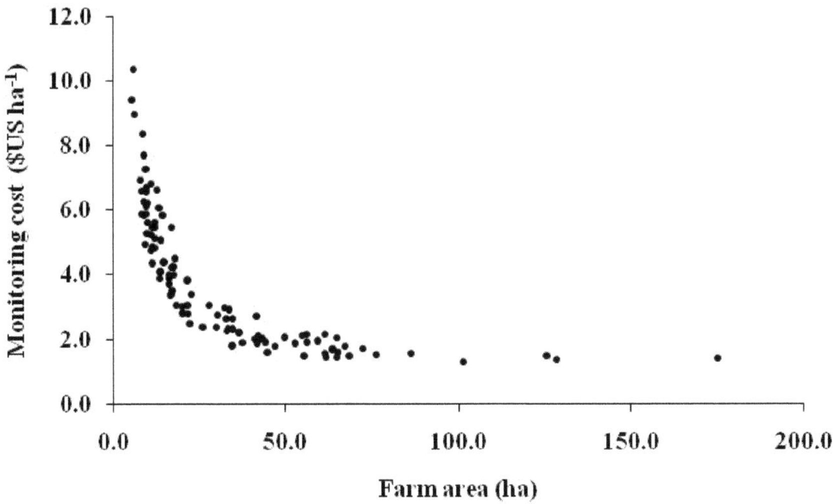

Figure 6.2 *Cost of monitoring land uses in relation to farm size*

Source: chapter authors

Monitoring costs can also vary depending on land use; this consideration is related to ease of verification in the field and in the office. This cost was less for forests, pastures, forage banks and intensive silvopastoral systems compared with simple and multi-strata live fences (see Table 6.6). This is attributed to the fact that live fences require detailed office work for their digitalization in satellite images and in databases. In terms of reducing the monitoring costs of the live fences, we recommend eliminating their consideration as an independent land use in favour of integrating them as a single land (e.g. consider pastures with live fences as a land use).

It is also important to analyse the relationship between the monitoring cost and the revenue that the PES generates for each land use so that the participating producers will be able to reduce transaction costs (this applies when they are responsible for covering that expense) and to implement land uses that make the greatest contributions to ecosystem services (a higher Ecological Index) in order to obtain a greater income from PES at the farm scale.

Conclusions

As a tool for monitoring and for payment for ecosystem services, the Ecological Index method costs less than current quantification techniques. Moreover, service providers find it easy to understand and it is flexible. For example, it can be modified according to the demands of ecosystem service markets or, depending on agroforestry and forestry system designs, adapted to a region's agroecological conditions, which define the potential for providing ecosystem services from different land uses.

Table 6.6 *Monitoring costs and income from ecosystem services in different land uses*

Land use	*Monitoring cost (US$ ha^{-1})*	*Ecological Index (points ha^{-1})*	*PES (US$ ha^{-1})**
Secondary forest	1.3	1.9	142.5
Improved pasture without trees	1.3	0.5	37.5
Improved pasture with low tree density	1.3	0.9	67.5
Improved pasture with high tree density	1.3	1.3	97.5
Natural pasture without trees	1.3	0.2	15.0
Natural pasture with low tree density	1.3	0.6	45.0
Natural pasture with high tree density	1.3	1.0	75.0
Grass forage bank	1.3	0.8	60.0
Intensive silvopastoral systems	1.3	1.6	120.0
Riparian forest	1.7	1.5	112.5
Multi-strata live fences	3.9	1.1	82.5
Simple live fences	6.6	0.6	45.0

Note: * Payment for ecosystem services, for each land use, was obtained by multiplying the index value by US$75 (payment per ecological point per hectare in a four-year scheme).

The Ecological Index for Biodiversity was validated using the IBSA resulting from the project's bird monitoring. This type of validation is necessary in order to make the appropriate adjustments in the indices of land uses. The Ecological Index for carbon might be improved, establishing a relationship between the ecological points and the carbon sequestered in an agroecosystem. By doing so, the environmental service per amount of carbon contributed by each land use might also be better evaluated.

The cost of monitoring falls as farm size increases, especially from 40ha upwards. This means that small farms should organize to reduce operating costs and thus achieve greater economic benefits for PES. The cost of monitoring land uses is less than the payment assigned for fostering ecosystem services. Surface area land uses (pastures with trees, agricultural crops and forests, among others) show a lower monitoring cost than linear systems (simple and multi-strata fences). This is due to the work of digitalizing the lines on the farm map in order to determine the length of the live fences.

To further simplify the methodology and achieve a higher cost–benefit ratio, the number of land uses might be reduced by combining pastures without trees with those that have a low tree density and by integrating pastures with live fences. The latter would help to reduce the transaction costs that the monitoring of individual live fences involves. In the same way, it would be important to define the minimum land-use area for monitoring.

Note

1 This heading includes the cost of purchase and ortho-rectification of a satellite image (500m^2), four global positioning systems (GPS), a computer, a printer, 16 ink cartridges and paper.

References

Alonzo, I., Ibrahim, M., Gómez, M. and Kees, P. (2001) 'Potencial y limitaciones para la adopción de los sistemas silvopastoriles para la producción de leche en Cayo, Belice', *Agroforesteria en las Américas*, vol 8, no 30, pp24–27

Betancourt, K., Ibrahim, M., Harvey, C. and Vargas, B. (2003) 'Efecto de la cobertura arbórea sobre el comportamiento animal en fincas ganaderas de doble propósito en Matiguás, Matagalpa, Nicaragua', *Agroforesteria en las Américas*, vol 10, no 39–40, pp47–51

Cárdenas, G. (1999) 'Composición y estructura aviar en sistemas de producción agropecuaria en el Valle del Cauca, Colombia', in *VI Seminario Internacional sobre Sistemas Agropecuarios Sostenibles*, Fundación CIPAV, Cali, COL

Chacón, M. and Harvey, C. A. (2006) 'Live fences and landscape connectivity in a Neotropical agricultural landscape', *Agroforestry Systems*, vol 68, pp15–26

Enríquez, M. L., Sáenz, J. and Ibrahim, M. (2007) 'Riqueza, abundancia y diversidad de aves y su relación con la cobertura arbórea en un agropaisaje dominado por la ganadería en el trópico subhúmedo de Costa Rica', *Agroforestería en las Américas*, vol 45, pp49–57

FAO (United Nations Food and Agriculture Organization) (2006) *FAO Statiscal Databases*, Rome, http://faostat.fao.org/default.aspx

FONAFIFO (Fondo Nacional de Financiamiento Forestal, CR) (2010) *Crédito Forestal*, www.fonafifo.go.cr/paginas_espanol/credito_forestal/e_cf_requisitos.htm, last accessed 10 March 2010

Francesconi, W. (2006) 'Bird composition in living fences: Potential of living fences to connect the fragmented landscape in Esparza, Costa Rica', *Tropical Resources Bulletin*, spring, vol 25, pp38–44

Holmann, F., Romero, F., Montenegro, J., Chana, C., Oviedo, E. and Baños, A. (1992) 'Rentabilidad de los sistemas silvopastoriles con pequeños productores de leche en Costa Rica: Primera aproximación', *Turrialba*, vol 42, no 1, pp79–89

Ibrahim, M., Chacón, M., Cuartas, C., Naranjo, J., Ponce, G., Vega, P., Casasola, F. and Rojas, J. (2007) 'Almacenamiento de carbono en el suelo y la biomasa arbórea en sistemas de uso de la tierra en paisajes ganaderos de Colombia, Costa Rica y Nicaragua', *Agroforestería en las Américas*, vol 45, pp27–36

Masera, O. R., Garza-Caligaris, J. F., Kanninen, M., Karjalainen, T., Liski, J., Nabuurs, G. J., Pussinen, A., de Jong, B. H. J. and Mohren, G. M. J. (2003) 'Modeling carbon sequestration in afforestation, agroforestry and forest management projects: The CO2FIX V.2 approach', *Ecological Modelling*, vol 164, no 2–3, pp177–199

Milder, J. C., DeClerck, F., Sanfiorenzo, A., Sanchez, D., Tobar, D. and Zuckerberg, B. (2010) 'Effects of tree cover, land use, and landscape context on biodiversity conservation in an agricultural landscape in western Honduras', *Ecosphere*, vol 1, no 1, pp1–22

Murgueitio, E., Ibrahim, M., Ramírez, E., Zapata, A., Mejía, C. and Casasola, F. (2003) *Usos de la tierra en fincas ganaderas*, first edition, Cali, Fundación Centro para la Investigación en Sistemas Sostenibles de Producción Agropecuaria, Colombia

Ríos, N., Cárdenas, A., Andradre, H., Ibrahim, M., Jiménez, F., Sancho, F., Ramírez, E., Reyes, B. and Woo, A. (2007) 'Estimación de la escorrentía superficial e infiltración en sistemas de ganadería convencional y en sistemas silvopastoriles en el trópico sub-húmedo de Nicaragua y Costa Rica', *Agroforestería en las Américas*, vol 45, pp66–71

Sáenz, J. (2005) *Índice de Biodiversidad para el Pago de Servicios Ambientales (IBSA)*, Informe de Consultoría para el Proyecto 'Enfoques Silvopastoriles Integrados para el Manejo de Ecosistemas', CATIE, Turrialba, Costa Rica

Sáenz, J. C., Villatoro, F., Ibrahim, M., Fajardo, D. and Pérez, M. (2007) 'Relación entre las comunidades de aves y la vegetación en agropaisajes dominados por la ganadería en Costa Rica, Nicaragua y Colombia', *Agroforestería en las Américas*, vol 45, pp37–48

Souza de Abreu, M. H. (2002) *Contribution of Trees to the Control of Heat Stress in Dairy Cows and the Financial Viability of Livestock Farms in Humid Tropics*, PhD thesis, CATIE, Turrialba, Costa Rica, CATIE

Szott, L., Ibrahim, M. and Beer, J. (2000) *The Hamburger Connection Hangover: Cattle Pasture Land Degradation and Alternative Land Use in Central America*, Serie Técnica no 313, CATIE, Turrialba, Costa Rica

Tobar, D. and Ibrahim, M. (2010) 'Las cercas vivas ayudan a la conservación de la diversidad de mariposas en paisajes agropecuarios?', *Rev. Biol. Trop.*, vol 58, no 1, pp447–463

Zamora, S. (2006) *Efecto de los pagos por servicios ambientales en la estructura, composición, conectividad y el stock de carbono presente en el paisaje ganadero de Esparza, Costa Rica*, MSc thesis, CATIE, Turrialba, Costa Rica

Part II

Marketing Ecosystem Services

7

Estimating the Cost and Benefits of Supplying Hydrological Ecosystem Services

An Application for Small-Scale Rural Drinking Water Organizations

Róger Madrigal-Ballestero

Introduction

There is a growing awareness about how land-use patterns in watersheds might affect the availability and quality of water needed to feed drinking water systems. Although there are several governmental regulations that aim to maintain and increase the attributes of different land uses to provide hydrological ecosystem services (HES),[1] these initiatives usually lack enforcement and financial resources. More importantly, these command-and-control practices fail to generate the appropriate economic incentives to guide private landowner decisions towards social well-being. Given these problems, payment for hydrological ecosystem services (PHES) has been proposed as a market-based approach that promises to overcome these deficiencies by means of direct payments to farmers who make environmentally friendly production decisions in recharge areas of watersheds of particular interest (Landell-Mills and Porras, 2002; Porras et al, 2008; Wunder et al, 2008; Boscolo et al, 2009). In order to guarantee financial self-sustainability, the funds to finance these schemes must come, ideally, from direct payments made by those who benefit from HES (e.g. consumers from drinking water systems).

In Central and South America, many local authorities of some municipalities are interested in developing PHES schemes as a component of their responsibility to provide drinking water to their communities (Alpízar and Madrigal, 2007a; Kosoy et al, 2007; Porras et al, 2008). However, the lack of an appropriate methodology to design, implement and evaluate these schemes might reduce its efficacy and financial sustainability. Furthermore, the absence of an adequate economic valuation of costs and benefits of providing HES reduces the possibility of better targeting of payments, limits the estimation of potential willingness to pay by consumers, and diminishes the likelihood that payments and charges reflect the true economic costs and benefits of decisions taken.

This chapter discusses the different economic valuation techniques that can be used to estimate the costs and benefits of providing HES and, hence, might be used to guide the decision-making process towards better watershed management. Specifically, these tools can be the building stones of an integrated approach to design PHES schemes in these settings. In addition, based on empirical data collected in four Latin American countries (Honduras, Nicaragua, Costa Rica and Colombia), this chapter presents a meta-analysis of findings related to the economic valuation of benefits and costs of proving hydrological ecosystem services, as well as how these findings could serve as an input to increase the performance of PHES schemes.

Context and Economic Valuation of HES

The analysed watersheds[2] have a clear human intervention and sustain the livelihoods of small rural communities dedicated to agricultural activities on hillsides (mainly coffee, vegetables, maize, beans and cattle). Even though there is high uncertainty about how land-use changes might affect the provision of water quality and availability (for further details, see Tognetti, 2000; Kaimowitz, 2001; Bruijnzeel, 2004), there are cases in which land-use and water conservation practices can be adopted in these settings to increase the potential of generating positive externalities in terms of HES (PASOLAC, 2000).

Unlike some other rural communities, the analysed cases have gravity-fed pipe-and-line systems that allow for carrying of water from springs or rivers in the upper watershed directly to households in lower parts of the watershed. Thus, there is a clear connection between land-use decisions taken upstream and the HES perceived by people downstream. This linkage is key to the estimation of the economic costs and benefits; however, the absence of a market price for the positive externalities[3] generated upstream entails important challenges to assess these values and how to incorporate them within current market transactions. These market failures are identified by the Millennium Ecosystem Assessment (MEA, 2005) as one of the principal causes of ecosystem degradation.

Table 7.1 *Some methods for the economic valuation of hydrological ecosystem services (HES)*

Method and description	*Potential use*
Replacement or avoided costs (Freeman, 1993). A decrease in the natural provision of environmental services might require investment in new technology or additional inputs to compensate for loss. The sum of all these expenditures is an approximation of the value of reinstalling the natural provision of environmental services.	This method can be employed either to measure the costs of HES provision (e.g. the implementation of soil conservation practices) or the benefits derived from it to downstream users (e.g. the avoided cost of buying bottled water to be 'defended' against water contamination).
Changes in productivity (Freeman, 1993). A decrease in the provision of HES might inevitably have an impact upon the production capacity of an economic agent, thereby reducing profits. This reduction is a measure of the damage caused by deteriorated environmental conditions or of the benefits to be attained if environmental conditions are improved. In other cases, the provision of HES requires the abandonment of an economic activity (cattle, etc.). In that case the estimation of the opportunity cost (the cost of the best possible alternative in which land could be employed) is needed.	This method requires real prior information on market prices. It is a particularly common practice to estimate the costs of land-use changes (opportunity costs) and the cost of conserving forests and other ecosystems. In some industrial or agricultural cases, changes in productivity can be used to estimate how variations in a baseline of HES affect private profitability.
Contingent valuation (Mitchel and Carson, 1989; Whittington, 2002). This is a survey-based method in which the respondent faces a hypothetical scenario describing a good or service and the particular setting in which it is to be provided. The respondent is asked to state his willingness to pay (WTP).	This method is widely used because it can deal with a broad range of situations, including those where no prior experience or information is available. In particular, it is well suited to estimate the benefits of potential watershed interventions.

The estimation of the economic costs and benefits of the provision of HES is anthropocentric in nature. This utilitarian approach implies that the values put on HES are defined by user groups and are relative in nature. That means that measures of economic benefits can be estimated by observing the changes in an individual's welfare due to variations in ecosystem services (ES). These changes in welfare (measured in any currency) reveal information about the value placed by individuals to any given level of ES.

The total value associated with the protection of any ecosystem can be divided into use values and non-use values (for details, see Freeman, 1993; Boardman et al, 2001). The former can be separated into direct and indirect consumption values. The latter are those values related to bequest, existence and altruistic concerns, among others. Even though these non-use values are of paramount importance, the estimation presented in this chapter does not aim to capture the economic value placed by people on them. In the case of PHES, the approach is simpler because the aim is to estimate a demand that reveals the economic value derived from direct use of the HES.

Economists have a menu of methods for assessing economic values of ecosystem services. In some cases, these methods use information revealed by people in different private markets (revealed preference methods); in other cases, the methods try to construct hypothetical markets for environmental goods (enunciated preference methods). For estimating the costs and benefits HES, we can use a combination of different methods. For example, different methodologies (contingent valuation, avoided costs, productivity changes) are available to the regulatory body in pursuit of demand information; already obtainable information, the type of users involved and the funds available to do research should guide the selection of the most suitable method. Irrespective of the method selected to estimate the demand for HES, particular attention must be paid to making sure that marginal or discrete change specifically associated with the planned intervention is evaluated. For example, a common mistake is to value all HES obtained from a given watershed, instead of the planned improvement from a well-defined baseline.

A list of potential methods that can be used is provided in Table 7.1. The application results of some of the methods presented in Table 7.1 for the estimation of HES and the design of PHES will be presented in the following sections.

An Integrated Approach for the Design of Payment for Hydrological Ecosystem Services (PHES)

This section briefly outlines the building pillars for the technical design of PHES. The reader must be aware that this is not a sufficient condition for PHES sustainability over time. The technical design must be part of a broader approach that includes an initial assessment of key political, legal and social conditions that might affect the feasibility of PHES; an adequate strategy for implementation; and, finally, an appropriate system for periodical evaluation of performance (Campos et al, 2006).

The technical design of PHES has five main components:

1 the construction of a dose–response relationship that captures the relationship between land uses and the provision of HES demanded by the potential beneficiaries;
2 the prioritization of areas with potential for HES provision;
3 an estimation of the costs associated with HES provision;
4 the identification and eventual measurement of a demand for HES; and
5 the creation or adaptation of an organizational framework suitable for the scale of intervention.

Figure 7.1 presents the main components of the methodology described above, framed into a market-based approach. An adequate estimation of the supply entails the first three components presented above. The demand curve is directly related to the identification and measurement of the benefits generated

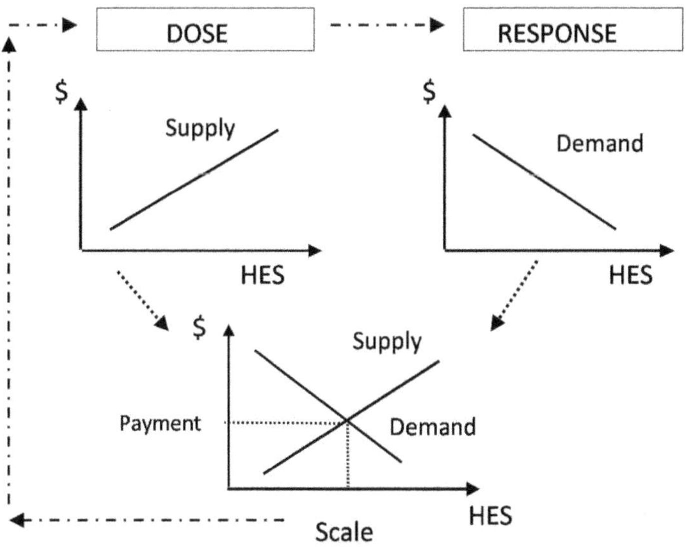

Figure 7.1 *Market approach to hydrological ecosystem services (HES) provision*

Source: Campos et al (2006)

by HES. Finally, the last component of the technical design is related to the organization of the market and the definition of rules that govern all transactions made under the PHES.

As mentioned in the previous section, we applied this framework to different sites. We concluded from these experiences that these five elements by no means imply a linear decision-making process, but rather an iterative approach in which the demand, the supply and the institutional framework affect and are affected by each other. As long as the availability and accuracy of information necessary to design PHES changes over time, as well as the social, economic and political settings, these schemes must evolve in order to be sustainable.

Valuing the costs of providing hydrological ecosystem services

The costs associated with maintaining or switching from one land use to another (as well as those associated with the modification of agricultural practices) depend on required changes from the baseline. Identification of these costs might help to design incentive programmes (such as PHES) that aim to reinforce certain activities or to persuade farmers to switch to more environmentally friendly practices. For this purpose, farmers in priority areas should be characterized according to type and profitability of their productive activity, type of property rights (including secure tenure), family size, and land availability, among other information required to understand landowner decisions and motivations (i.e. livelihood strategies). In other words, even if farmer

Table 7.2 *Estimation of the opportunity costs of natural regeneration*

Location	*Product*	*Net profitability (ha/year)*
Valle de Angeles, Honduras	Vegetables	US$1800
Copán Ruinas, Honduras	Coffee	US$357
Río Pasto, Colombia	Potatoes	US$2330
Jucuapa, Nicaragua	Maize, beans	US$300

participation in PHES is highly related to the amount of money paid, other factors might be crucial for the final decision.[4]

From a methodological perspective, if a certain land use needs to be maintained (e.g. conservation of forest) or completely abandoned (e.g. agricultural activities in highly vulnerable areas), then it is recommended that the opportunity cost is calculated by estimating the associated costs for this decision (excluding any transaction costs). This implies that opportunity costs depend on contextual factors that characterize areas with potential for provision of HES (as well as to other ES). In some areas, the opportunity costs might be relatively high due to highly profitable competing activities, while in others, the opportunity costs might be relatively low due to less profitable alternative uses. As the reader might deduce, areas with special attributes for HES provision and low opportunity costs are those in which PES might be easily implemented and which are cost effective.

From an empirical perspective, estimation of opportunity costs in different regions of Central America and Colombia entails serious problems due to the lack of accurate and easily accessible data. However, average estimations of the net profitability of some of the principal agricultural products in studied sites give us an idea of how costly it might be to induce some farmers to change from agricultural uses to natural regeneration. Alternatively, a general estimation is provided of how costly it is to keep actual areas (with agricultural potential, but based on natural regeneration) away from agricultural production.

The estimations presented in Table 7.2 might under- or over-state the real costs for some farmers. In small-scale settings, this problem might be minimal; but as geographical scope increases, it is unlikely that a single average estimation of opportunity costs will represent the majority of farmers effectively. This situation, exacerbated by asymmetries in information, imposes a serious challenge for fine-tuning PHES conservation incentive payments. Furthermore, worldwide PHES experience reveals a lack of opportunity cost measurements to define payments; nevertheless given the low additionality of some of these schemes, it seems that payments made are above opportunity land costs (Wunder et al, 2008).

Similarly, if a landowner decides to retain land forest cover, the forgone income derived by an alternative use represents the opportunity cost. Theoretically, under a PHES scheme, these costs must be covered by the payment; otherwise the farmer will be reluctant to accept the deal.[5] However, it is also true that it is very unlikely that farmers will receive the full estimated opportunity costs because legal barriers impede land use change in many settings. In this sense, payments can be viewed as a tool to partially reduce the costs that the law imposes on landowners. This opens the door to criticism that some of these schemes provide very low additionality (see Wunder et al, 2008, and Pattanayak et al, 2010, for further details on additionality).

On the other hand, it may be possible that HES could be generated in highly human-disturbed landscapes, such as agriculture in the Central American hills. The key requisite is that farmers adopt management practices that foster water and soil conservation. If improvements or changes in agricultural practices are needed, then the change in productivity is the suitable valuation method. Some common practices recommended for the provision of HES could be linked to productivity changes – for example, cultivation methods (e.g. improved drainage and soil conservation practices), reduced use of agrochemicals and fewer cultivated areas (e.g. from crops to secondary forest). The distinction described above is important because in many cases the generation of HES is carried out in human-altered landscapes, which, depending on current management practices, can already have favourable attributes for the generation of HES. A complete land-use change is not only expensive but may also be unnecessary.

The large variety of practices[6] that might be used to achieve this goal can have contrasting impacts upon private profitability, such as significant reduction (in the very first years or permanently), minimal impact and significant increase (e.g. agroforestry systems). The payments that could foster the adoption of these management practices under a PHES should reflect these diverse scenarios.

In cases where the impact upon private profitability is minimal, the need for funding becomes less acute. However, it calls attention to understanding what other factors (beyond financial aspects) might reduce the adoption of appropriate management practices by farmers. The experience in Honduras (mainly Copán) reveals that in cases where adoption minimally affects private profitability in the long run, a payment that covers the initial investment cost followed by additional yearly payments to cover maintenance of management practices is enough to promote adoption among farmers. However, it is necessary to add farmer training and support to speed up the adoption process and appropriate technical implementation. It is clear that in cases where the negative impact upon profitability is relatively high, the need for larger payments might create additional threats to the financial sustainability of PHES schemes.

The provision of HES is site and user specific, and therefore requires a very careful prioritization of intervention areas. However, this process might be cumbersome due to natural complexities (hydrological versus hydro-geological criteria) and general uncertainty in the relationship between land uses and HES.

Uncertainty and economic valuation of HES

Defining where to allocate PHES entails more technical challenges and costs compared to the selection of areas for providing other ES. Furthermore, the significant scientific gap on how land-use patterns affect water availability and quality undermines the capacity to define what types of conservation and land-use changes are more desirable.[7] As Wunder et al (2008) states: 'Additionality in land use still is not sufficient; however, we also need to know that the right land-use changes are being undertaken – that is, land uses that generate the desired ES.' This level of uncertainty affects the effectiveness of PHES schemes and, hence, the confidence of water users about the relationship of payments and water provision.

What types of land uses must be promoted is also important because water users might have different preferences about the actions promoted under PHES. Water quality and long-term availability are usually the main concern of local governments and water utilities, whereas stable and sustained water flows are more relevant to irrigation or hydropower generation. Land-use practices in a particular watershed can have a significant effect on water quality (reduced sediment and bacteriological content, less agricultural runoff). Although still relevant, the effect is less clear on average water flows.

Uncertainty can be further reduced by defining a dose–response relationship, leading, in turn, to increased robustness and effectiveness of PHES. This dose–response relationship approximates how a given land use (i.e. a dose) contributes to the generation of HES (i.e. a response). Unfortunately, the construction of this relationship is costly and troublesome. Still, at small scales of intervention (such as a watershed or a municipality in Central America), it is possible to reach a high level of precision by constructing a Land-Use Index based on dose–response relationships.

The term 'dose–response relationship' (as opposed to biophysical production function) is preferable in order to reflect the extreme level of simplicity used in this approach. For example, climatic and geographic variations, which are central elements in a biophysical production function, are not captured in the dose–response relationship. The degree of complexity in constructing a Land-Use Index obviously varies depending on the type of ecosystem service.[8]

A Land-Use Index is a list of land uses based on ordinal criteria, according to their contribution in the generation of HES. Land uses are subdivided into categories, and each level within a category is an incremental improvement compared to the previous level. Given the lack of information and the complexity of ecosystems, the construction of such an index based on exact quantitative information is unlikely. In other words, it is not possible to establish a cardinal classification of the land uses that reflects exact marginal changes in the provision of HES. The ordinal classification represents a less rigid and more pragmatic alternative; but, again, it is only an approximation of the true dose–response relationship.

Copán Ruinas is an interesting case in which efforts to reduce uncertainty have been made.[9] Given that water quality is the primary concern, the main points of water intake for the municipal water utility and the corresponding drainage areas were mapped and characterized based on current and potential land use. This allowed for high prioritization of 119ha within a relevant watershed of 71km². The exercise of prioritizing areas also includes criteria to weight the risk of land-use change. Previous policy instruments determine this. For example, the municipal government in Copán had already established an area of restricted use and had purchased land in the relevant drainage area, resulting in fewer high-priority areas that needed to be included in the PES scheme.

The Land-Use Index was used in the Copán Ruinas case to define a performance-based payments scheme. The index assigns a specific score to all land uses included. Scores range from 0, for presumed land uses that do not contribute at all to the provision of HES, to 1, for presumed land uses that contribute the most to HES provision. The Land-Use Index for HES was constructed so that it was general enough to cover most productive activities in the humid tropics, but was not particular to the Copán case study. It was based on secondary information, field validation and, most importantly, the active participation and consensus of a group of 30 international experts in the field who met in a two-day workshop organized for this purpose (Alpízar and Madrigal, 2005).

The design of a performance-based payment scheme combines the Land-Use Index and the associated costs of each land use. These costs and the methodology to estimate them depend on the required change from a baseline, as described earlier. Given the lack of information, it was assumed that the implementation of practices does not have any effect on in-farm productivity. The underlying logic is that incentives offered to the farmers to maintain or improve the environmental attributes of a particular land use should be positively correlated both to the magnitude of costs involved and, most importantly, to the generation of HES, as reflected on the Land-Use Index. The key is to define a payment per point for each category of the Land-Use Index and to pay according to the marginal contribution of each land use present in the category. Table 7.3 shows the performance-based payment scheme used in Copán.

There are two types of payments proposed:

1 payment for initial establishment (in relation to initial investments to adopt land conservation practices or to implement a forest surveillance plan); and
2 maintenance payment (in relation to all tasks necessary to properly maintain the initial investment).

The payment for the establishment of each land use is the result of multiplying the payment per point by the index value for the respective land use. The payment per point for each category results from dividing the total costs of

Table 7.3 *Performance-based payments based on the Land-Use Index (LUI)*

Score	*Use and/or land management*	*Establishment costs (US$/ha)*	*Maintenance costs (US$/ha)*	*Opportunity costs*	*Baseline costs (US$/ha)*	*Establishment payment (US$/ha)*	*Term (years)*	*Establishment costs (US$/ha/year)*	*Maintenance payment (US$/ha)*
Annual crops									
0.0	Annual crops –	–	–	–	–	–	–	–	
0.3	Annual crops with sustainable practices	46	16	0	46	80	1	16	14
0.3	Annual crops with physical soil investments	130	8	0	130	80	1	8	14
0.4	Annual crops with agroforestry practices	91	22	0	91	107	2	22	18
Payment per point					*267*			*46*	
Perennial crops									
0.2	Coffee, no shade, no soil cover	–	–	–	–	–	–	–	–
0.5	Coffee with shade, no soil cover	154	20	0	154	116	3	20	20
0.6	Coffee, no shade, with soil cover	34	12	0	34	139	3	12	24
0.7	Shaded coffee with soil cover	189	32	0	189	163	3	32	28
0.8	Certified organic coffee	226	38	0	226	186	3	38	31
Payment per point					*232*			*39*	
Forest and plantations									
0.2	Forest plantation with bare soil		–	–	–	–	–	–	– –
0.5	Isolated forest38	33	20	58	34	1	53	31	
0.8	Young secondary vegetation	38	33	20	58	54	2	53	49
1.0	Riparian forest	38	33	20	58	67	1	53	62
1.0	Secondary forest with surveillance	38	33	20	58	67	1	53	62
1.0	Primary forest with surveillance	38	33	20	58	67	1	53	62
Payment per point					*67*			*62*	

Source: Madrigal and Alpízar (2008)

adopting all the proposed practices in the category by the total number of points that can be achieved if all land-use management practices are implemented. Following Table 7.3, for example, the establishment payment per point (US$232) in the perennial crops category comes from the sum of the total baseline costs of the perennial crops category (US$603) divided by the total possible score of the category (i.e. 2.8). Note that the total baseline costs do not include opportunity costs due to the fact that farmers are not forced to set aside any portion of their land. Opportunity costs are only included for the case of forests and plantations.

For example, if a farmer owns a coffee farm with no shade and bare soil (payment equal to zero) and wants to add cover to the soil in his farm, he will be paid US$139 per hectare, distributed over three years. This is the result of multiplying the marginal contribution of that land use by the payment per point of adoption – that is, 0.6 × US$232. The payment of US$139 is a balance between costs and expected benefits. Adding soil cover to a coffee plantation is relatively cheap, but results in a high score compared to other practices. Hence, the high payment will clearly favour this practice.

In order to define maintenance payments, a similar estimation was used. Maintenance costs also include opportunity costs, if applicable. If the coffee farmer in the previous example satisfactorily installed soil cover, after the third year, he would receive an annual payment equivalent to US$24 per hectare. In addition, the farmer is free to adopt more improvements for his farm, which implies new payments.

The validations of this payment system carried out with farmers located in the priority protection zones of Copán shows promising preliminary results. The farmers understand that the payments will depend on the land uses that they have and the investments they carry out. In Copán, 150ha, which adds up to 24 farmers, many of whom are from a Mayan indigenous community (Chortí community), have been included in the project. These farmers receive nearly US$2000 in total annually (for further details, see Madrigal and Alpízar, 2008).

Valuing the benefits of providing HES

The creation of a PHES that could be financially sustainable in the long run depends on the existence of beneficiaries who are willing to pay the costs of these services. On many occasions, local governments or non-governmental organizations (NGOs) have tried to create markets for selling ES for which demand is non-existent (e.g. scenic beauty where there are no tourists), elusive (habitat for key species) or cumbersome (carbon credits). PHES are especially appealing precisely because the beneficiaries are, in most cases, clearly identifiable, whether water is used for drinking, irrigation or hydropower generation.[10] However, few efforts have been made to estimate how much the potential consumers are willing to pay and how to induce them to pay (Pattanayak et al, 2010).

People can value ES because they use them regularly (e.g. wood and drinking water) and also because of altruistic or ethical concerns, even when they do not use them regularly. The estimation of a demand for HES ultimately defines the temporal and spatial scale by determining the amount of money that can be raised through increased water fees. Different beneficiaries of HES require different valuation techniques (revealed preference methods or enunciated preference methods); but in all cases the aim is to estimate the maximum amount that they are willing to pay (WTP) for HES. The idea is to measure the amount of resources (measured in any currency) that users are willing to give up from other uses in order to have a certain level of HES. In other words, this trade-off reveals information about the value (benefit measure in monetary terms) that individuals place on any given level of HES.

One method that is particularly useful in estimating the WTP of beneficiaries of HES in drinking water systems is contingent valuation. A hypothetical

Table 7.4 *Estimation of maximum willingness to pay (WTP)*

Location	*Number of households with water connection*	*WTP (monthly average maximum per household)*	*Maximum annual budget*	*Determinants of WTP (statistically significant variables from probit regression)*
Valle de Ángeles, Honduras	579	US$0.98	US$6809	Family income (+) Use of bottled water (+) Weekly water shortages (+)
Copán Ruinas, Honduras	1190	US$0.94	US$13,423	Family income (+) Weekly water shortages (+) Women (+)
Quindío, Colombia	3750	US$0.66	US$29,700	Socio-economic stratus (+) Assets tenure (+)
Esparza, Costa Rica	4134	US$1.09	US$54,072	Family income (+) Women (+)

scenario is necessary for consumers to enunciate their WTP to a watershed management programme that promises improvements in water availability and/or water quality at home. The results of the application of this method (using similar surveys, data collection protocols and limited dependent variable models for econometric estimations) in the analysed communities are presented in Table 7.4.

The maximum WTP indicates the average cost that a family will be able and willing to pay each month as an extra fee to the actual water service bill. This measurement in itself constitutes an estimation of the benefits derived from a given programme and should therefore be considered the upper limit for any potential payment scheme. However, this amount is just a reference, since ultimately the price paid under a PHES scheme might be determined by other criteria. In particular, most regulations of public utilities and services use cost-based pricing as the norm, and HES are regarded as inputs into production and consumption that should therefore be paid at cost. In theory, payments should cover the costs to the supplier of adopting (maintaining) land uses that increase (preserve) the provision of HES in the relevant prioritized areas, and should be smaller than the aggregate WTP of the beneficiaries (this is equal to the maximum potential annual budget). In other words, measured economic benefits should be higher than the estimated costs.

Based on WTP and the number of households in each community, the maximum annual budget that can be raised in a particular PHES scheme can be easily calculated (see fourth column of Table 7.4). From this point, the analyst can perform different estimations to assess less ambitious scenarios – for example, what amount of money can be raised if only 50 per cent of the maximum WTP is charged to all households. These estimations of expected incomes can be compared to the expected costs related to payments to sellers of HES. For example, in Copán, the expected budget collected via user fees is barely enough to cover all the prioritized intervention scales, excluding transaction costs.

Finally, individual WTP depends on some observable characteristics (see last column of Table 7.4: these are independent variables in probit regression models used to estimate WTP). As expected, the probability of having higher WTP increases as family income or asset tenure increases. In addition, WTP tends to be higher in settings where people suffer water shortages and use bottled water to satisfy daily needs at home (cooking, washing and drinking). This might suggest that people value the provision of HES more in cases when an urgent need to improve or overcome a negative situation is present. Interestingly, women tend to value more the provision of HES. This might be due to cultural characteristics inherent to the studied sites that promote women's awareness of the necessity to invest in HES. In these locations it is very common to divide responsibilities according to gender; women stay at home in charge of cooking, cleaning and childcare (all highly dependent on water availability and quality) while men work all day in the field.

Institutional arrangements

The institutional framework should be adapted to the scale of the intervention (number of prioritized hectares and producers; funding) and should at the very least be capable of reaching agreements with suppliers of HES, monitoring contracts and managing funds. These contracts are key to the success of a HES market. They outline the Land-Use Index (when available) and the associated payments, and explain the conditionality of the payments and the sanctions for violating the rules.

PHES do not appear spontaneously,[11] but rather require the intervention of an agent who acts as mediator. For HES, this agent is frequently a representative of the beneficiaries (e.g. water utilities, irrigation board) or a local government (municipality). In some cases, a local NGO has taken the lead.

Given that transaction costs are likely to be high compared to payments to suppliers, the local institutional framework should be kept to a minimum. In our experience of successful case studies, most local governments, irrigation boards or water utilities already have the financial and technical capacity and the legal mandate to oversee the environmental management of their relevant catchments area, and this capacity should be used for the implementation of PES schemes. Where this capacity is not present, transaction costs are likely to severely hinder any implementation of locally financed PES schemes.

In Copán, transaction costs were estimated at roughly US$4000 per year, including 15 per cent of an accountant's time; 25 per cent of the time of a field operative with a degree in agronomy; transport; and materials. Happily, the environmental unit of the Municipality of Copán already had the technical capacity to manage payments and monitor compliance with PES contracts.

Conclusions

Economic valuation of HES can help us to discern the true costs and benefits generated by human decisions that affect ecosystems. While the theory behind these estimations is relatively straightforward, the inclusion of these results under effective market transactions offers greater challenges. Thus, the creation of technically sound PHES schemes is far from a simple task or a quick fix, and this point seems to be too easily overlooked.

Even though economic valuation of HES is a desirable task for decision-making, it is important to weight the expected gains in terms of efficiency versus the costs involved in the estimation. For example, in some small-scale scenarios, the implementation of a contingent valuation method to assess the WTP of 100 families for a PHES programme might be too costly in terms of the additional precision obtained. A simple estimation based on local criteria might be enough. Furthermore, in other contexts, politically viable actions instead of technically sound but cumbersome actions are sometimes required. However, it is also true that most policies (including PHES) need to evolve through a process of permanent learning and progressive incorporation of technical criteria that increases their effectiveness.

Given the current and ever-increasing interest in PHES, it is important to have a standardized best practice approach for evaluating the feasibility of this instrument for a particular site. Payments conditioned to performance can be a tool that improves the environmental and economic effectiveness of PHES schemes. However, these improvements imply additional transaction costs (implementation and monitoring) that can be significant, although this depends on the application scale and level of precision desired in the internalization of the externalities. If the transaction costs are relatively high, the viability of the PHES scheme can remain uncertain.

Local authorities in charge of the PHES scheme had the dual objective of increasing the provision of HES, while still paying farmers who, in the past, adopted environmentally friendly farming practices and land uses. Simply put, if payments are to successfully achieve the goal of *increasing* the provision of HES, they have to be targeted first to farmers who are not currently employing the environmentally friendly land uses and practices encouraged by the payments. A second objective, of *maintaining* the current provision of HES, is also valid and requires payments to farmers already using best practices (including forest). Focusing too narrowly on the first objective creates perverse incentives because it punishes 'good' past behaviour; therefore, a combination of the two objectives will probably have better results.

Notes

1 Ecosystem services are defined as the benefits that people perceive from the ecological processes that occur in ecosystems (MEA, 2005). Hydrological ecosystem services is a term adopted in this chapter to emphasize those ecosystem services that affect water quality and availability.

2 The communities studied are: in Honduras: Copán Ruinas and Valle de Ángeles; in Nicaragua: Jucuapa; in Colombia: Filandia and Cuenca Alta del Río Pasto; in Costa Rica: Esparza. For more details of sites as well as the economic valuation techniques employed, see Baltodano (2005); Cisneros (2005); Alvarado (2006); Retamal (2006); Tehelen (2006); Alpízar and Madrigal (2007b); Del Castillo (2008).
3 Externalities are defined as unintended side effects of the consumption or production decisions of an economic agent with characteristics of public goods (i.e. low excludability and low consumption rivalry). The benefits and costs derived from externalities are not reflected in regular market transactions.
4 For more details on factors that might influence the final decision of landowners to participate in PES schemes, see Zbinden and Lee (2005) and Wünscher (2008).
5 From a theoretical point of view, the willingness to accept estimation (the amount requested by a farmer to accept the change) might be identical to this opportunity cost. However, in practice there are incentives for farmers to overestimate this value. Some other mechanisms to reveal the true opportunity costs, such as auctions, have been explored. For more details, see Ferraro (2008) and Wünscher (2008).
6 For a detailed description of practices for soil and water conservation in Central American hillsides, see PASOLAC (2000).
7 Marchamalo et al in Chapter 2 of this volume discuss these issues further and provide an overview of methods and tools that may be applied to the quantification of soil and water services in the Neotropics, depending on the objectives and data availability.
8 CATIE, in collaboration with partners (notably those that were part of the RISEMP–GEF project) has built indices for carbon sequestration, biodiversity and HES. For more details, see Murgueitio et al (2003); Pagiola et al (2004); Alpízar and Madrigal (2005).
9 In Chapter 16 of this volume, Prins and León further discuss this initiative and how it is complemented with other mechanisms to guarantee drinking water provision.
10 There is sufficient evidence to suggest that the financial sustainability of PES depends on a demand-driven approach (Wunder et al, 2008; Boscolo et al, 2009) in which local beneficiaries finance the payments that landowners receive for the provision of ES. In other words, initiatives that pay insufficient attention to local demand or that rely heavily on external funds (such as subsidies from international organizations) tend to run into problems.
11 A notable exception is the bilateral 'Coasian' agreements between a group of suppliers and one large beneficiary – for example, hydropower plants, bottling companies and the like.

References

Alpízar, F. and Madrigal, R. (2005) 'Construcción de un índice de usos del suelo relacionados con la provisión hídrica: Insumo para una propuesta integral de PSE hídrico', SEBSA–GAMMA–CATIE, Turrialba, Costa Rica

Alpízar, F. and Madrigal, R. (2007a) *Bienes y servicios ecosistémicos en América Latina y el Caribe: Buenas prácticas, mecanismos de financiamiento y rol del Estado*, BID–CATIE, Costa Rica

Alpízar, F. and Madrigal, R. (2007b) *Valoración económica de servicios ambientales hídricos en paisajes intervenidos, cantón de Esparza, Costa Rica*, SEBSA, CATIE, Turrialba, Costa Rica

Alvarado, M. (2006) *Elementos claves para el diseño e implementación de un pago por el servicio ecosistémico de protección del recurso hídrico en el municipio de Valle de Ángeles, Honduras*, MSc thesis, CATIE, Turrialba, Costa Rica

Baltodano, M. E. (2005) *Valoración económica de la oferta del servicio ambiental hídrico en las subcuencas de los ríos Calico y Jucuapa, Nicaragua*, MSc thesis, CATIE, Turrialba, Costa Rica

Boardman, A. E., Greenberg, D. H., Vining, A. R. and Weimer, D. L. (2001) *Cost-Benefit Analysis: Concepts and Practice*, second edition, Prentice Hall, Upper Saddle River, NJ

Boscolo, M., Eckelmann, C., Madrigal, R., Mendez, B., Paveri, M. and Zapata, J. (2009) *Practical Experiences of Compensation Mechanisms for Water Services Provided by Forests in Central America and the Caribbean*, FAO-Facility, Rome, Italy

Bruijnzeel, L. A. (2004) 'Hydrological functions of tropical forests: Not seeing the soil for the trees?', *Agriculture, Ecosystems & Environment*, vol 104, pp185–228

Campos, J. J., Alpízar, F., Louman, B., Parrota, J. and Madrigal, R. (2006) *Enfoque integral para esquemas de pago por servicios ecosistémicos forestales*, Segundo Congreso Latinoamericano IUFRO-LAT, Costa Rica

Cisneros, J. (2005) *Valoración económica de los beneficios de la protección del recurso hídrico y propuesta de un marco operativo para el pago por servicios ambientales en Copán Ruinas, Honduras*, MSc thesis, CATIE, Turrialba, Costa Rica

Del Castillo, C. (2008) *Escenarios económicos para el manejo de la oferta del servicio ecosistémico de regulación y provisión del recurso hídrico para consumo humano, en la Subcuenca Alta Superior del Río Pasto, Colombia*, MSc thesis, CATIE, Turrialba, Costa Rica

Ferraro, P. (2008) 'Asymmetric information and contract design for payments for environmental services', *Ecological Economics*, vol 65, pp810–821

Freeman, A. M. (1993) *The Measurement of Environmental and Resource Values, Theory and Methods*, Resources for the Future, Washington, DC

Kaimowitz, D. (2001) 'Cuatro medias verdades: La relación bosques y agua en Centroamérica', *Revista Forestal Centroamericana*, vol 33, pp6–10

Kosoy, N., Martinez-Tuna, M., Muradian, R. and Martinez-Alier, J. (2007) 'Payments for environmental services in watersheds: Insights from a comparative study of three cases in Central America', *Ecological Economics*, vol 61, pp446–455

Landell-Mills, N. and Porras, L. (2002) *Silver Bullet or Fool's Gold? A Global Review of Markets for Forest Environmental Services and Their Impact on the Poor*, International Institute for Environment and Development, London

Madrigal, R. and Alpízar, F. (2008) 'Gestión y diseño adaptativo de un esquema de pago por servicios ecosistémicos en Copán Ruinas,' *Investigación Agraria: Sistemas y Recursos Forestales*, vol 17, no 1, pp79–90

MEA (Millenium Ecosystem Assessment) (2005) *Our Human Planet: Summary for Decision Makers*, Island Press, Washington, DC

Mitchel, R. and Carlson, R. (1989) *Using Surveys to Value Public Goods: The Contingent Valuation Method, Resources for the Future (RFF)*, Washington, DC

Murgueitio, E., Ibrahim, M., Ramírez, E., Zapata, A., Mejía, C. and Casasola, F. (2003) 'Usos de la tierra en fincas ganaderas: Guía para el pago de servicios ambientales en el Proyecto Enfoques Silvopastoriles Integrados para el Manejo de Ecosistemas', CIPAV, Cali, Colombia

Pagiola, S., Agostini, P., Gobbi, J., de Haan, C., Ibrahim, M., Murgueitio, E., Ramírez, E., Rosales, M. and Ruiz, J. P. (2004) *Paying for Biodiversity Conservation Services in Agricultural Landscapes*, Environment Department Paper no 96, FAO, Rome, Italy

PASOLAC (Programa para la Agricultura Sostenible en Laderas de América Central) (2000) *Guía técnica de conservación de suelos y agua*, PASOLAC, San Salvador, El Salvador

Pattanayak, S., Wunder, S. and Ferraro, P. (2010) 'Show me the money: Do payments supply environmental services in developing countries?', *Review of Environmental Economics and Policy*, vol 4, no 2, pp254–274

Porras, I., Grieg-Gran, M. and Neves, N. (2008) *All That Glitters: A Review of Payments for Watershed Services in Developing Countries*, Natural Resource Issues No 11, International Institute for Environment and Development, London

Retamal, R. (2006) *Valoración económica de la oferta del servicio ambiental hídrico para consumo humano en el Municipio de Copán Ruinas, Honduras*, MSc thesis, CATIE, Turrialba, Costa Rica

Tehelen, K. (2006) *Elementos principales de una propuesta de pago por servicios ambientales para el manejo de los recursos hídricos en la subcuenca del río Barbas, Quindío, Colombia*, MSc thesis, CATIE, Turrialba, Costa Rica

Tognetti, S. (2000) *Informe de síntesis: Relaciones tierra-agua en cuencas hidrográficas rurales, FAO*, Rome, Italy

Whittington, D. (2002) 'Improving the performance of contingent valuation studies in developing countries', *Environmental and Resource Economics*, vol 22, nos 1–2, p323–367

Wunder, S., Engel, S. and Pagiola, S. (2008) 'Taking stock: A comparative analysis of payments for environmental services programs in developed and developing countries', *Ecological Economics*, vol 65, pp834–852

Wünscher, T. (2008) *Spatial Targeting of Payments for Environmental Services in Costa Rica: A Site Selection Tool for Increasing Conservation Benefits*, PhD thesis, Bonn University, Germany

Zbinden, S. and Lee, D. (2005) 'Paying for environmental services: An analysis of participation in Costa Rica's PSA program', *World Development*, vol 33, no 2, pp255–272

8

Developing a Business Plan for Forestry and Other Land Use-Based Carbon Projects

Colin Moore and Till Neeff

Introduction

Carbon markets are currently the largest and most advanced outlet for valorizing and selling environmental services. The creation of international cap-and-trade schemes combined with an increasing public awareness of the need to tackle climate change has resulted in numerous markets that put a price on carbon. In this context, activities that reduce greenhouse gases (GHGs), including those in the forestry and land-use sector, create an environmental service that is increasingly sought after in the international carbon markets.

Accessing carbon markets can be a lengthy and costly process; to avoid unnecessary delays and costs, project developers should conduct a full project due diligence and outline a clear project business plan. This includes establishing a project's eligibility under the various carbon standards, assessing its carbon credit potential, determining the project's revenue and cost profile, and deciding on the best approach to commercialize the resulting credits.

To this end, we provide practical guidance in this chapter that will help project developers to understand the various steps required for establishing a business plan for forestry and land use-based carbon projects. The aim is to arm project developers with the necessary information that will allow them to determine whether, and under what conditions, pursuing carbon markets is a sensible proposition for their project idea.

A basic assumption of this approach is that project developers already have a detailed estimation of the GHG reduction potential of their project. This is a basic requirement upon which the various steps of the approach are based. Chapter 1 in this volume is strongly recommended as a precursor to this chapter as it discusses the various tools and methods available to quantify and monitor GHG emissions and balances from forestry and other land-use activities.

This chapter should also be read together with its companion chapter (Chapter 9), both of which stem from a prior publication entitled 'Making the step from carbon to cash: A systematic approach to accessing carbon finance in the forest sector' (Neeff et al, 2011). While this chapter focuses on the practical steps required in bringing a carbon project to market, in Chapter 9, Neeff and Fehse explain the options that projects have to commercialize their credits and what the roles are of the different actors involved in the sales process.

Background on Carbon Finance in the Forestry and Land-Use Sector

Types of carbon markets

Carbon markets can be broadly divided into regulatory and voluntary carbon markets. Regulatory carbon markets emerge in response to a regulatory commitment for emission reductions that a government or other regulatory body imposes on the emitters. The market players are therefore primarily those entities that are legally obliged to reduce emissions. Participants in voluntary markets, on the other hand, voluntarily decide to offset their own emissions – in particular, concerns about individual air travel and a growing sense of corporate social responsibility motivate voluntary market actors.

Current regulatory markets under the Kyoto Protocol

The Kyoto Protocol has generated several subsidiary carbon markets. The largest of these at present is the European Union Emissions Trading Scheme (EU ETS), established in 2005 as a means to help EU member states meet their Kyoto Protocol targets, where allowance trading values totalled US$100 billion in 2008 and despite the economic recession increased to US$118 billion in 2009 (Kossoy and Ambrosi, 2010). In conjunction with the EU ETS, the Clean Development Mechanism (CDM) has experienced significant investment. The value of the CDM market, both the primary production of credits and subsequent trading on secondary markets, reached nearly US$33 billion in 2008, although this was reduced in 2009 to about US$20 billion (Kossoy and Ambrosi, 2010).

Projects in the forest sector are, however, under-represented in the Kyoto markets. As of June 2010, there were only 15 registered afforestation and reforestation (A/R) CDM projects, and 39 in the validation stage. At present, forestry projects constitute about 1 per cent of the CDM pipeline (Kossoy and Ambrosi, 2010). Besides A/R, no other land use-based project types are permitted under the CDM.

In parallel, there is a series of other regulatory carbon markets that are not linked to the Kyoto Protocol but to alternative compliance regimes. These include initiatives in Australia, New Zealand and the US. None of these currently offers a market for forestry and land use-based activities in developing countries and we therefore do not discuss them further.

Future regulatory markets

Discussions are under way at the international level that could result in potentially new large markets for forestry and other land use-based offsets. Under the auspices of the United Nations Framework Convention on Climate Change (UNFCCC), parties are currently negotiating a successor agreement to the Kyoto Protocol. An important component of this agreement is whether and how developing countries should be compensated for reducing emissions from deforestation and forest degradation, conservation, sustainable management of forests and the enhancement of forest carbon stocks (REDD+). At the UNFCCC Cancun Conference in December 2010, the international community adopted the proposed text of the Long-term Cooperative Action (LCA) working group, which includes text on the development of a REDD+ mechanism. This sends the strongest signal yet of a future international agreement on REDD+ and therefore the possibility for developing countries to generate forest carbon offsets at a scale much larger than the current CDM.

During 2009–2010, the US took its boldest steps yet towards establishing a carbon cap-and-trade scheme. The US House of Representatives passed a bill in June 2009 that would limit emissions from certain high emitting sectors with the option to trade allowances and offset credits to help meet reduction obligations. The bill allowed for up to 1 billion tonnes of international offsets to be used for compliance purposes, with the expectation that a majority of these would come from REDD (reducing emissions from deforestation and degradation) activities. In the US Senate, a similar bill was promoted by Senators John Kerry and Barbara Boxer as the Clean Energy Jobs and American Power Act, but was not brought to a vote. More recently, in May 2010, Senators Kerry and Lieberman unveiled the American Power Act that proposed to establish a cap-and-trade scheme that would cover utilities in 2013 and expand to manufacturing in 2016. Under this scheme up to 1.5 billion domestic offsets and 0.5 billion international offsets could be used annually to meet compliance. Avoiding tropical deforestation was specifically mentioned as an eligible activity to generate international offsets. This piece of legislation, however, was also never brought to a vote leading to much uncertainty as to whether and when a cap-and-trade scheme will be introduced in the US.

Voluntary carbon markets

The total value transacted in the voluntary carbon markets in 2009 reached US$387 million (Hamilton et al, 2010). Although the voluntary markets grew by various orders of magnitude during the last several years, they represent only 1 per cent of the regulatory markets. Almost 84 per cent of the value in the voluntary carbon markets was transacted in direct sales between buyers and sellers in 2009, also known as over-the-counter (OTC) deals (Hamilton et al, 2010). Besides OTC transactions, there are also voluntary exchange-based markets under the Chicago Climate Exchange (CCX), an exchange for voluntary compliance in the US. Our analysis is concerned with the voluntary OTC markets only, and we do not further discuss the CCX due to the relatively small market share that it represents and the fact that the second commitment period of the CCX ended in 2010.

Forestry and other land use-based projects constitute a much higher voluntary market share by number of projects (24 per cent) than the regulated markets (less than 1 per cent) (Hamilton et al, 2010).

Forestry and other land use-based carbon project activities

A carbon project in the forestry and land-use sector is a land-use activity that is eligible under one of the existing carbon standards to generate carbon credits for trading on international carbon markets. Such carbon projects, however, can be classified in various ways. Figure 8.1 provides an overview of project-level activities in the forestry and land-use sector, following the classification under the Voluntary Carbon Standard (VCS, 2008).

Reducing emissions from deforestation and degradation (REDD). These projects, also known as avoided deforestation projects, reduce the rate of land-use change from forests to other land uses. Activities would be classified as REDD if they avoid the conversion of forest to a non-forest land use, and as improved forest management if they slow down or prevent a process of forest degradation that would gradually reduce the carbon content contained in a specific forest area.

Improved forest management (IFM). Adapting management practices in existing forestry activities can contribute to avoiding emissions. Example activities could include the protection of forests that would otherwise be logged, the improvement of management approaches and the declaration of protected areas.

Afforestation, reforestation and revegetation (ARR). Afforestation, reforestation and revegetation projects create new forests or increase woody biomass through planting or assisted natural regeneration. Agroforestry activities would also be considered under this category.

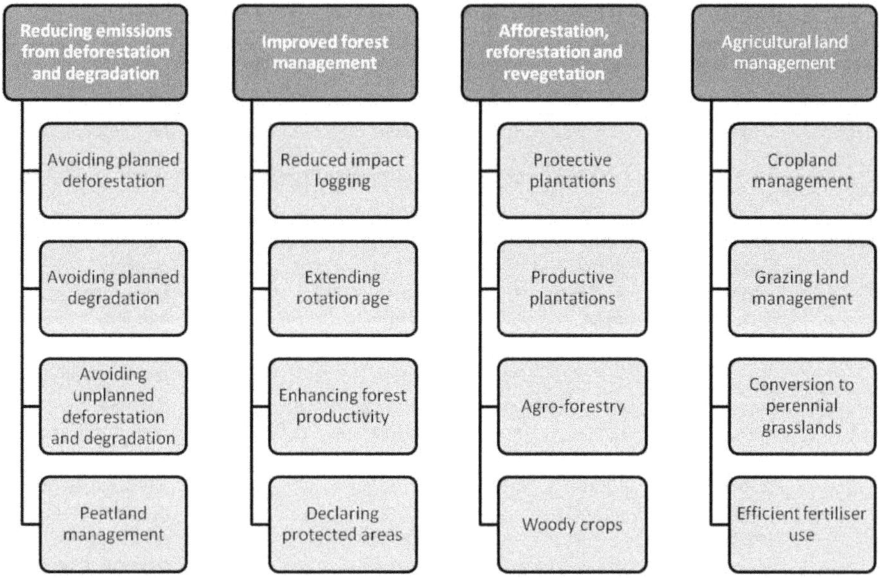

Figure 8.1 *Main types of forestry and land-use carbon projects with example activities as classified by the Voluntary Carbon Standard (VCS)*

Source: adapted from Neeff et al (2011)

Agricultural land management (ALM). A change in management practices on croplands or grazing lands can reduce emissions and enhance the carbon stock contained in the soil, as well as in the vegetation. ALM activities can consist of improved management of croplands and grasslands, or can also relate to the conversion of croplands and grasslands.

The carbon storage on agricultural lands can be increased by a variety of activities, including no-till agriculture, certain crop rotation schemes, and efficient management of nutrients, fertilization, irrigation and fallowing. Agricultural land can also be converted to perennial grassland (or, of course, be reforested as an A/R activity).

ALM is not eligible under the CDM; but various standards on the voluntary market (VCS, VER+ and CCX) include it in their scopes. So far, only a few ALM projects have been carried out for carbon finance (except under the CCX), and there still may be significant untapped potential in this sector.

Carbon standards available in the forestry and land-use sector

Each type of carbon market has evolved standards for project development and carbon accounting. These standards, which are often specific to a particular market, have been developed to ensure quality and credibility of any credit developed by an offset project. Table 8.1 lists the three principal standards that include forestry and other land-use activities in their scopes that are relevant to developing tropical countries.

Table 8.1 *Carbon standards that address forestry projects*

Standard	*Developed by*	*Scope*	*Features*	*Operational since*	*Carbon markets*
CDM	UNFCCC	A/R only	• Carbon accounting standard • Industry benchmark for carbon standards	2001 (2003 for forestry sector)	Regulatory Kyoto market
VCS	The Climate Group, IETA and WEF	ARR, REDD, IFM and ALM	• Carbon accounting standard • Backing of private-sector market players	2007	Voluntary OTC market Being used as industry benchmark for pre-compliance markets
CCB	Partnership of research institutions, corporation and NGOs	Land-based carbon projects	• Provides assurances of environmental and social co-benefits of forest carbon projects	2005	Regulatory and voluntary, although its use is purely voluntary on the project developer's part

Notes: ALM = agricultural land management; A/R = afforestation and reforestation; ARR = afforestation, reforestation and revegetation; CCB = Climate Community and Biodiversity Standards; CDM = Clean Development Mechanism; IETA = International Emissions Trading Association; IFM = improved forest management; NGO = non-governmental organization; OTC = over the counter; REDD = reducing emissions from deforestation and degradation; UNFCCC = United Nations Framework Convention on Climate Change; VCS = Voluntary Carbon Standard; WEF = World Economic Forum.
Source: adapted from Neeff et al (2011)

The VCS has also become the *de facto* standard for project developers wishing to develop projects for pre-compliance purposes (e.g. for a future international REDD+ mechanism). The VCS is one of the few standards that provide guidance for REDD project activities and is viewed within the voluntary space as the standard that generates the offsets of the highest quality. Adopting this high-quality standard most likely ensures that projects will meet all of the requirements of a future regulatory market and therefore be considered as eligible offset projects.

Other standards exist for carbon projects that do not specifically address land management projects or the forestry sector (e.g. Gold Standard), have limited market share (e.g. Voluntary Offset Standard (VOS), Plan Vivo Standard and VER+ Standard), have expired (e.g. CCX) or are limited to specific regions (e.g. the Climate Action Reserve (CAR) in the US); we therefore do not discuss them further.

In addition to the above carbon accounting standards are quality certificates for project development. The Climate Community and Biodiversity Standards (CCB Standards) have been widely used by proponents of land-use carbon projects. The CCB Standards aim to ensure high-quality design of land management projects that simultaneously minimize climate change impacts, support sustainable development and conserve biodiversity. However, unlike

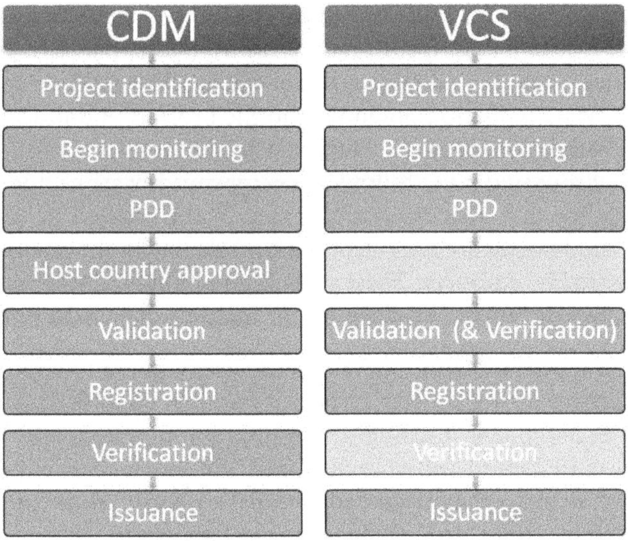

Figure 8.2 *Project development cycle for Clean Development Mechanism (CDM) and Voluntary Carbon Standard (VCS) projects*

the above-described carbon standards, the CCB Standards constitute a quality certificate related to the design, set-up and environmental and social co-benefits of a project, rather than a carbon standard that issues carbon credits. Potential credits must therefore be additionally verified through a carbon standard (e.g. VCS or CDM) in conjunction with the CCB Standards.

The carbon project development cycle

Carbon standards require projects to go through a number of defined stages before carbon credits are issued. The number of stages and requirements of each stage depend on the standard chosen. Although the CDM and VCS have quite similar project development cycles, some differences exist, as presented in Figure 8.2.

Project identification. This is the stage that this chapter aims to support. Projects are screened to identify those that are most feasible and are therefore good candidates to undergo further project development. The results of this due diligence can also be used to generate interest in the project and seek financing based on its business plan.

Monitoring. Project proponents must monitor and keep track of key data parameters that will allow them to calculate the project's actual GHG reduction benefits. Monitoring must be done according to the approach outlined in the methodology and project design document (PDD). Ideally, monitoring must begin together with the start of project implementation, often even before the PDD is developed.

Project design document (PDD). This is the most important carbon project document and includes all of the relevant information upon which the project can be assessed, including identification of the baseline, *ex-ante* calculation of the project's GHG reduction potential and the project's monitoring plan. The PDD is the basis for the project's validation audit.

Carbon accounting must follow a pre-approved baseline and monitoring methodology for that specific project type. In the context of carbon credit projects, methodologies are the rulebooks for calculating the amount of carbon credits generated by a project. These rules need to follow the larger framework of the respective carbon credit scheme – for instance, the CDM or the VCS.

Host country approval. Only CDM projects are required to seek official host country approval of the project's activities and its contribution to national sustainable development. Getting this approval is a prerequisite in order for CDM projects to pass to subsequent stages. Voluntary market projects may, however, wish to make the designated national authority aware of their project in order to receive some form of official project approval.

Validation. Third-party auditors assess the project documentation, including the PDD, against the criteria of the relevant standard. Successful validation results in a validation report that is submitted to the relevant standard's governing body.

Registration. Upon accepting the validation report, the standard's governing body will register the project in an official database. The CDM's executive board maintains the CDM database, while the VCS Association maintains the VCS's database.

Verification. Third-party auditors review projects' monitored data to verify the accuracy of the project proponents' GHG reduction claims. Successful verification results in a verification report that is submitted to the relevant standard's governing body. For the VCS, the verification can be carried out together with the validation.

Issuance. Upon receiving the verification report, the standard's governing body will authorize the relevant registry to issue credits to the project proponent's account, which can then be sold or traded.

As noted above, the project development cycle involves many steps and a variety of stakeholders, including project developers, external consultants, third-party auditors and the standard's governing body. For this reason, project identification to actual credits generation can easily take upwards of one year. The complexity of this cycle further strengthens the need to have a clear project idea and business plan at the project start to ensure that further delays during this project development cycle do not occur.

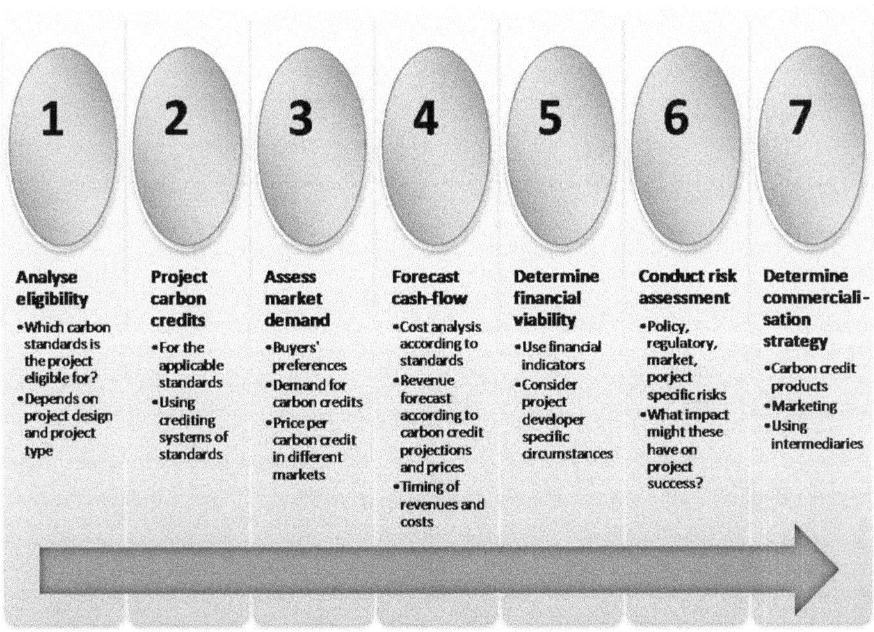

Figure 8.3 *Systematic approach to commercializing carbon credits*

Source: adapted from Neeff et al (2011)

The Carbon Project Pathway

This section outlines an approach that project developers may take to assess the eligibility and overall viability of their project under the two most relevant carbon standards for forestry and land-use projects in the Neotropics: the CDM and the VCS. As mentioned earlier, this approach assumes that a project is advanced far enough in its carbon project development that an initial GHG reduction projection has been performed. Once this projection is available, a series of steps need to be carried out to choose a standard and to optimize revenues (see Figure 8.3).

Step 1: Analyse eligibility

The CDM and VCS have different eligibility criteria and project scopes (as of the time of writing). Eligibility under future pre-compliance markets will depend on policy decisions, especially with regard to an international REDD+ mechanism. We do not, however, speculate as to the possible eligibility criteria under such schemes, but focus only on those where international forest carbon offsets are currently eligible.

Additionality. In order to qualify as an offset, the underlying activities to reduce greenhouse gases have to be additional (i.e. they need to demonstrate that the reduction would not have happened under a business-as-usual

Table 8.2 *Eligibility criteria and project scope under the CDM and VCS*

	CDM	*VCS*
Additionality required?	Yes	Yes
Project start date after ...	1 January 2000	1 January 2002
Project lifetime	Undefined	At least 20 years
Eligible host countries	Non-Annex I to Kyoto	In principle, any, although more useful for countries without cap-and-trade schemes
REDD	No	Yes
IFM	No	Yes
ALM	No	Yes
A/R	Yes	Yes
Land eligibility for REDD: forest ...	N/A	... for ten years before project start
Land eligibility for IFM: forest ...	N/A	... designated, sanctioned or approved by national or local regulatory bodies to be managed for wood products (e.g. sawn timber, fuelwood or pulpwood)
Land eligibility for ALM: no native ecosystems cleared...	N/A	... within ten years before project start
Land eligibility for A/R: no forest...	... since 1 January 1990	... for ten years before project start

Notes: ALM = agricultural land management; A/R = afforestation and reforestation; CDM = Clean Development Mechanism; IFM = improved forest management; N/A = not applicable; REDD = reducing emissions from deforestation and degradation; VCS = Voluntary Carbon Standard.
Source: adapted from Neeff et al (2011)

scenario). Additionality is widely accepted as a precondition for credible offsetting projects and, as such, is a basic quality requirement for almost all standards. The exact definition of additionality depends on the individual carbon standard, although, at present, the CDM's additionality tool is used as the industry benchmark and is also used by the VCS.

Project start date. Standards have different project starting dates, meaning the earliest possible date from which a project can claim credits.

Project lifetime. As an additional measure to ensure the permanence of GHG reductions, VCS projects need to demonstrate a minimum lifetime.

Host country. The VCS allows projects in all countries (although certain preconditions must be met for projects in Annex 1 countries), whereas the CDM is only applicable to non-Annex 1 countries. Annex 1 countries are developed nations and nations with economies in transition that are a party to the UNFCCC and have emission reduction targets assigned to them under the Kyoto Protocol. Non-Annex 1 countries are primarily developing countries and do not have any obligations to reduce their GHG emissions.

Project types. Standards limit the project scope to only certain types of projects. Most notably, the CDM is only applicable to A/R projects for which

Table 8.3 *Crediting schemes of the CDM and VCS standards*

Standard	*Carbon credits*	*Permanence approach*	*Fluctuations*	*Issuance interval*	*Crediting period*
CDM	tCER	Temporary credits that expire	Total GHG reductions and expiry	5 years	20 years twice renewable, or 30 years once
	lCER	Temporary credits that expire	Total GHG reductions and reversal	5 years	20 years twice renewable, or 30 years once
VCS	VCU	Risk-adjusted buffer	Long-term average GHG reductions	Up to annual	Equal to the lifetime of the project (20–100 years)

Notes: CDM = Clean Development Mechanism; GHG = greenhouse gas; lCER = long-term Certified Emission Reduction; tCER = temporary Certified Emission Reduction; VCS = Voluntary Carbon Standard; VCU = voluntary carbon unit.
Source: adapted from Neeff et al (2011)

several approved carbon accounting methodologies exist. The VCS has approved these same methodologies which allows for A/R projects to be developed under the same standard. While the VCS, in principle, allows for all land-use based projects, IFM and REDD carbon accounting methodologies only became available in 2010. At present, four REDD and two IFM methodologies have been approved by the VCS. The lack of any approved ALM methodologies means that this project type is currently impeded from registering with the VCS.

Land eligibility for A/R and ALM. In order to prevent the creation of so-called 'perverse incentives' (i.e. cutting down existing forests in order to reforest and claim carbon credits), restrictions have been established that allow afforestation and reforestation activities only on areas that have been without forest for a certain period of time, long enough to make it unlikely that deforestation took place in order to generate carbon credits.

Step 2: Project carbon credits

The crediting schemes of the standards describe how the amount of carbon credits can be calculated from the project's GHG reductions. In order to derive the amount of carbon credits, two issues need to be taken into account:

1 the approach to ensuring permanence of the GHG reductions; and
2 the rules on issuance and crediting periods (see Table 8.3).

Permanence approach. A fundamental challenge for forestry and other land-use projects is to ensure that the carbon stored in the biomass of trees or soils translates to permanent benefits, rather than eventually being released back to the atmosphere. Projects in the forest sector are prone to risks such as fire, pests or increased human pressures, which can cause previous GHG reductions to be lost. Several approaches have been developed to ensure permanence under the different carbon standards, such as issuing temporary carbon credits that need to be regularly reverified and replaced (CDM), or setting aside a buffer reserve of carbon credits that cannot be sold and are used to compensate for a potential loss in biomass (VCS).

Under the CDM, project developers can choose between two different types of temporary credits:

- Temporary Certified Emission Reductions (tCERs) correspond to the total amount of carbon removed since project start. They are valid until the end of the subsequent commitment period to the one in which they were issued, after which the credit buyer must replace these tCERs with other Kyoto units – for example, CERs, assigned amount units (AAUs), emission reduction units (ERUs), removal units (RMUs) or new tCERs. In practice, this means that credits for existing carbon stocks are reissued after each verification event (see Figure 8.4).
- Long-term Certified Emission Reductions (lCERs) correspond to the incremental amount of carbon removed by the trees since the last verification. If a reversal of GHG removals is observed from one verification event to another, then replacement AAUs, CERs, RMUs, ERUs or lCERs need to be made available from a lCER replacement account within the national registry (see Figure 8.4). At the end of the project crediting period, buyers of lCERs must replace these credits with permanent credits (e.g. CERs, AAUs, ERUs or RMUs). Unlike tCERs, a lCER cannot be replaced by another lCER.

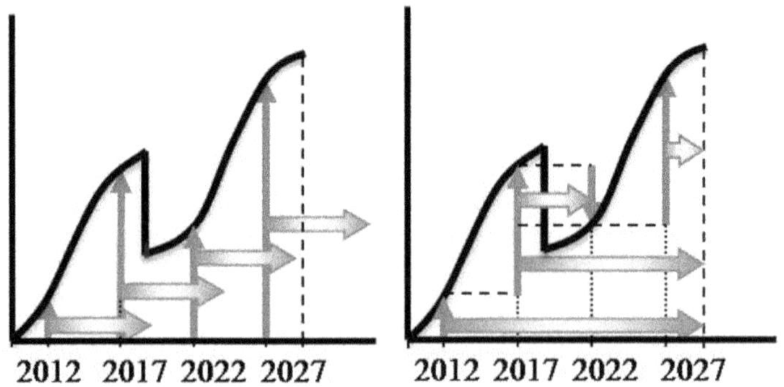

Figure 8.4 *Crediting schemes under the Clean Development Mechanism*

Note: The left graph shows temporary Certified Emission Reductions (tCERs) with carbon credits corresponding to the carbon increment since project start. The right graph shows long-term Certified Emission Reductions (lCERs) with carbon credits corresponding to the carbon increment since last verification.
Source: Locatelli and Pedroni (2004)

By issuing credits of temporary validity, the CDM reflects the fact that emission reductions in CDM forests may not be permanent. A tCER will ultimately be replaced by a permanent credit or an lCER. An lCER, on the other hand, can only ever be replaced by a permanent credit. In this regard, these forestry credits can only be used to show short-term compliance and ultimately will be replaced with a credit that does not carry the same impermanence risk.

In the VCS buffer approach, a certain amount of the project's carbon credits cannot be sold, but instead are set aside in order to buffer and ensure permanence. The VCS's registry includes a buffer account to hold buffer credits from all contributing projects, which is then used to replace carbon credits that become invalid due to biomass loss. The buffer size for each project is determined according to the risks that influence the probability of impermanence of the emission reductions. Third-party auditors assess projects against a set of predetermined impermanence risk criteria and, based on this assessment, rank projects as having an overall low, medium or high risk of impermanence. This ranking, in turn, corresponds to the buffer-withholding percentage that will be applied to the project.

Treatment of fluctuations. In the CDM, projects gain carbon credits according to total GHG reductions as measured at the point of monitoring and reflecting all coincidental peaks or troughs. In the VCS, carbon credits do not strictly reflect biomass on site, but represent the project's long-term average biomass increment. The long-term average biomass content in a plantation will be lower than its peak biomass content due to periods of harvesting and regrowth. By issuing credits that correspond to the long-term average biomass increment, rather than the actual biomass on site, fluctuations in the amount of issued carbon credits that are observed under the CDM are avoided.

Issuance interval. CDM forestry projects can issue carbon credits only every five years after the point of first verification, which can be chosen freely. There are no corresponding limitations under the VCS, which allows the issuance of annual carbon credits. It is important to consider, on a case-by-case basis, whether annual crediting makes economic sense in a land-use project, comparing the costs arising from monitoring, verification and issuance.

Crediting period. The time period during which GHG reductions can earn carbon credits is the crediting period. The crediting period for land-use projects is typically longer than for other sectors because of the longevity of land-use activities.

Step 3: Assess market demand

Carbon markets are driven by a variety of factors that are subject to changes: international and national regulations (regulatory markets), and corporate policies and consumer purchasing decisions (voluntary markets). In the carbon markets, it is therefore imperative that projects closely monitor market developments in order to ensure that carbon credits can be placed optimally. The need to monitor market developments will depend on the type of carbon project and the standards that it is eligible for. Most importantly, carbon credits from A/R projects can gain access to both voluntary and regulatory carbon markets; but carbon credits from REDD, IFM and ALM projects can currently only access voluntary carbon markets.

Regulatory markets

The users of carbon credits in the regulatory carbon markets are companies and governments that have to comply with their emission reduction targets.

Regulatory markets are large in size and would be able to absorb even substantial amounts of carbon credits. However, it should be reiterated that only A/R projects are eligible under the CDM and the temporary credits that these projects generate are unattractive to buyers in this market. Furthermore, because of the temporary nature of these credits, they are excluded from the EU ETS altogether. Therefore, while credits, in general, are in high demand, this demand does not extend to forestry credits.

Current Kyoto markets therefore offer limited opportunities for forest carbon projects; as a result, there is a risk for project developers that they may struggle to find interested buyers for their credits.

Pre-compliance markets

Future regulatory markets offer a potentially large market into which forest carbon credits could be sold. This is particularly important for REDD activities that can potentially generate substantial amounts of carbon credits. There is, however, uncertainty regarding various political decisions that would need to be taken to allow REDD carbon credits to access these regulatory markets, particularly with regard to negotiations under the UNFCCC. These relate to the overall role and architecture that a future REDD mechanism would take and whether credits from this mechanism will be eligible for use against a country's emission reduction target. Until there are clear indications of the treatment of REDD in a post-2012 climate change agreement, it will be risky to rely on selling large volumes of carbon credits from REDD in future regulatory carbon markets. There may, however, be opportunities for projects to benefit from early-mover advantage in these markets, securing purchase commitments, and potentially higher prices, more easily than projects developed at a later stage.

Voluntary markets

In the voluntary markets, buyers of carbon credits are looking to voluntarily offset the emissions that they cause. Although driven by a voluntary choice, there are often also considerations of public image and corporate social responsibility behind offsetting initiatives. Various project characteristics therefore greatly influence the attractiveness of projects, including technology, design and co-benefits (Hamilton et al, 2010). The attractiveness, in turn, is closely linked to the achievable carbon price.

A recent EcoSecurities survey (Neeff et al, 2010) identified clear voluntary market buyer preferences with regard to forest carbon projects (and closely reflects the results of the Ecosystem Marketplace's annual market report on voluntary markets) (Hamilton et al, 2010). The most important factor for choosing an offset project was the use of a respected standard, such as the VCS, CDM or CAR. Beyond this, the project types most sought after are reforestation projects with native species and REDD projects, while the least attractive is reforestation with commercial large-scale plantations. In terms of location, buyers prefer projects that are located in tropical developing countries, mostly because of the perceived benefits that additional finance can offer to communities in these countries.

Current voluntary carbon markets are still limited in size, and it is uncertain how large the forestry market share will be in these markets over the next few years. Market size is an issue for REDD projects, in particular, because some of these projects have the potential to generate very large amounts of carbon credits. Project developers of REDD projects are advised to check the size of their project against the total magnitude of the voluntary carbon markets before relying on sufficient market demand to absorb their credits.

Step 4: Forecast cash flow

Determining a project's cash flow depends upon a series of factors. These include the credit generation potential (see the section on 'Step 2: Project carbon credits' above), credit sales volumes (see the section on 'Step 3: Assess market demand' above), achievable credit sales prices, the magnitude of project development costs, and the timing of these costs and revenues. Combining all of these factors will allow project developers to model their project's cash flow and help towards assessing its overall financial viability.

Carbon prices

In the regulatory markets, carbon credits serve the purpose of compliance with emission reduction obligations, and all issued credits serve that purpose equally well. In this regard, the risk of credit delivery has been found to be the major determinant of prices. Carbon credits, particularly when purchased in a forward contract rather than on the spot market, carry a risk to the buyer because they have not yet been issued. Before issuance, there are risks relating to whether the project will actually achieve registration and if it will meet expectations. The lower prices paid for early-stage project-based carbon credits reflect these risks. The more risk mitigation mechanisms and guarantees that a project can offer, the higher the price for carbon credits.

At present, very few forestry credits exist in the CDM; therefore, accurate pricing for tCERs remains difficult. The World Bank's BioCarbon Fund currently provides the best price indicator for these credits, and is set up specifically to buy credits from emission reduction projects in the forest sector. Previous sales to the World Bank for tCERs have achieved prices of about

US$4 per tonnes CO_2 equivalent (t CO_2e) (Neeff et al, 2011). In addition to the credit delivery risks that affect forward sales, buyers' liability to continually replace their tCERs with replacement Kyoto units after the end of each subsequent commitment period explains the relatively low prices paid for these credits.

Carbon prices on the voluntary markets vary widely. For transactions in 2009, maximum prices ranged between US$0.10 to $111.00 per t CO_2e (Hamilton et al, 2010). The wide range in these estimates makes them useless for business planning; but it also shows how demand depends on the projects' characteristics. With regard to land-use projects, improved forest management projects received the highest weighted prices at US$7.3 per t CO_2e, followed by agroforestry (US$5.2 per t CO_2e), A/R (US$4.6 per t CO_2e), REDD (US$2.9 per t CO_2e) and, finally, agricultural soil conservation projects (US$1.2 per t CO_2e) (Hamilton et al, 2010). It would be prudent, however, to expect significantly lower prices if large quantities are sold to wholesalers for on-sale, as opposed to retailing small volumes directly to consumers. In turn, projects that sell small quantities and that have been certified under a well-known scheme boasting particular co-benefits, such as benefits for local communities and biodiversity conservation, may be able to capitalize on these aspects to obtain much higher prices.

Carbon project development costs

A project developed for carbon finance will face certain additional costs associated with the different steps of the project cycle. These carbon project development costs can be substantial and depend on both the carbon standard chosen and the project size (see Table 8.4). Projects need to be sufficiently large in order to justify the carbon project development costs since these are not related to project size (i.e. large projects have the same costs as small projects, which generate fewer revenues to cover the costs). Costs also depend on the amount of data that a project already has available, which is why we specify a large cost range for data collection.

Depending upon the commercialization model adopted by the project, some of these costs can be shared or transferred to other parties (see the section on 'Step 7: Determine commercialization strategy'). Investors may be willing to take on these costs or they can be shared with other project partners. However, in cases where the project developers choose to enter into late sales contracts, these costs will have to be entirely internalized.

The information displayed in Table 8.4 applies to large-scale projects and has only limited validity for those that are smaller in scale. Large-scale projects use the regular methodologies and registration processes without benefiting from simplification. Under the CDM and VCS, particular project categories have been designed for projects at a small scale that benefit from simplified rules. The simplifications of small-scale projects under these standards are designed to lower carbon project development costs and thus enable participation of projects with less access to funding.

Table 8.4 *Carbon project development costs of land-use projects and indicative timeline*

Activity	*Cost estimate*	*Type of cost*	*Timeline*
Planning stage			
Feasibility study	US$25,000–$35,000	Consultancy fee	Year 1
Project documentation	US$60,000–$125,000	Consultancy fee	Year 2
Data collection	US$5000–$30,000	Internal costs	Year 1
Validation	US$30,000–$55,000	Auditor fee	Year 1–2
Registration fee under the CDM	Same as issuance fee, but capped	Administrative fee	Year 2
Registration fee under the VCS	N/A	N/A	N/A
Operational stage			
Initial verification	US$25,000–$35,000	Auditor fee	Starting year 2
Ongoing monitoring	US$5000–$30,000	Internal costs	Starting year 2
Ongoing verification	US$15,000–$25,000	Auditor fee	Starting year 2
Issuance fee under the CDM	US$0.10 per t CO_2e for the first $15,000 per year; US$0.20 per t CO_2e beyond that	Administrative fee	Starting year 2
Issuance fee under the VCS	US$0.10 per VCU	Administrative fee to the VCS	Starting year 2
Registry fee under the VCS (market)	US$0.05 per VCU issued; US$0.02 per VCU transferred	Administrative fee to the registry	Starting year 2
Transaction stage			
Legal contracting	US$25,000–$60,000	Internal costs	Year 1
Marketing	Project specific	Internal costs	Starting year 1
Cost of sales	Percentage based upon negotiation with project partners	Agency fee	Starting year 2

Notes: CDM = Clean Development Mechanism; t CO_2e = tonnes CO_2 equivalent ; VCS = Voluntary Carbon Standard; VCU = voluntary carbon unit.
Source: adapted from Neeff et al (2011)

Cost and revenue accrual timing

The timing of carbon credit income differs according to project characteristics and carbon standards. Project developers tend to prefer revenues that accrue earlier rather than later in the project development process, and ideally costs for carbon project development would not have to be pre-financed. The timing at which revenues and costs occur may be a factor in choosing between carbon standards and working with different types of intermediaries. Typically, there are time lags for project development and operations, as shown in Figure 8.5.

Timing of costs. For practical reasons, we split costs between those that need to be invested upfront during carbon project development ('Planning stage' in Table 8.4), those that accrue on an ongoing basis ('Operational stage' in Table 8.4) and those that accrue when seeking to commercialize credits ('Transaction stage' in Table 8.4). The last can mostly be offset with revenues and are

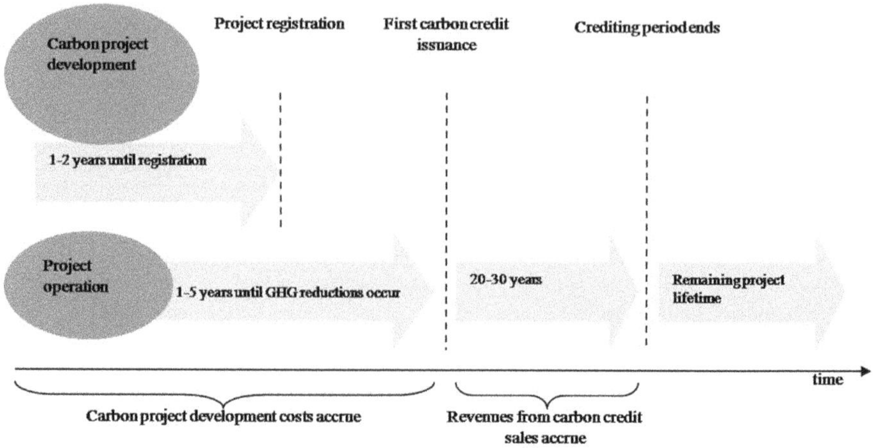

Figure 8.5 *Timing of accrual of carbon credits, costs and revenues*

Source: chapter authors

therefore less likely to generate a cash-flow bottleneck. The project development costs, however, require an initial investment that will only generate income after a certain period of time, and only if it is successful. Typically, project development may require about one to two years of lead time until registration (and usually longer until carbon credit issuance), and this delay should be factored into cash-flow projections.

Timing of GHG reductions. When GHG reductions physically occur depends on the project type and on project implementation. In some cases, REDD projects can reduce emissions very quickly (i.e. from the initial years onward) because reduced rates of deforestation directly translate into GHG reductions. A/R projects typically require more lead time before generating net carbon removals (e.g. two to five years after project start), while trees start growing and balance out emissions caused during initial site preparation.

Timing of carbon credit accruals. Carbon credits are issued once GHG reductions can be monitored and verified. Projects that issue credits under the CDM can only request issuance of carbon credits once in five years. It is advisable to postpone monitoring events until the end of commitment periods (Neeff and Henders, 2007); therefore, in forestry CDM projects, carbon credits will typically accrue during the corresponding intervals (i.e. 2012, 2017, etc.). Under VCS verification, intervals can be chosen freely by project proponents; however, verification intervals of more than five years for land-use projects will result in 50 per cent of the project's buffer being cancelled. If the verification interval is greater than 15 years, then the entire buffer is cancelled. This is to ensure that the environmental integrity of the buffer approach is credible. If a project does not verify its emission reductions periodically, the VCS assumes that this is because the project has lost carbon and therefore the buffer credits are cancelled to reflect this loss.

Timing of revenues. Typically, regardless of the date when a carbon credit buyer may have entered into a contract to purchase credits, they will pay for those credits upon delivery (i.e. after issuance). Nevertheless, some buyers may provide some upfront payment or other contributions to pre-financing, including paying for some or all of the project development costs.

Step 5: Financial viability

The overall financial viability of a project should be assessed with financial indicators such as net present value (NPV) or internal rate of return (IRR). These financial indicators will allow project developers to determine the financial viability of a project when modelling various project scenarios based on various input parameters, such as credit accrual, timing of costs and revenues, sales price and expected volumes of sales.

For example, although an A/R project may generate overall large amounts of credits and revenues, it is possible that there will be a long lead time before these accrue. In the meantime, project developers will incur carbon project development costs as well as the costs to implement the project activities. Depending on a project developer's individual situation, the delay before revenues accumulate may be too long to be financially viable unless additional sources of financing are found. In this case, the opportunity cost will be too large and the project developer will choose not to pursue carbon market certification.

Additionally, in many cases, the sale of carbon credits might be the only source of revenues for a land-use carbon project. It is therefore possible that these revenues, once discounted, are not attractive enough for project developers to risk the costs of going through project development.

Whether a project is financially attractive to an individual project developer will ultimately depend upon whether these financial indicators meet a project developer's requirements while taking into consideration the risks. By using financial indicators and modelling multiple scenarios, project developers can determine under which set of conditions a project is deemed attractive. It will then be up to the project developer to assess whether they can realistically develop a project under the parameters selected that makes the project attractive in their eyes.

Step 6: Risk assessment

There are several risks that may affect the ability of a project to successfully generate emission reductions or its long-term profitability. Some of the risks that need to be considered are outlined below. We distinguish between project risks, on the one hand, that relate to the project's ability to generate emission reductions, and policy, regulatory and market risks, on the other, that relate to external factors.

Project risks. These are risks specific to a project's overall performance and ability to successfully pass through the various stages of project development.

Because of these risks, the prices that credits can achieve in forward transactions are typically lower as the buyer carries the risk that these credits will, in fact, not materialize. Issued credits do not carry this risk profile and can therefore command higher prices.

Policy risks. Decisions taken by parties to the UNFCCC, national governments or a standard's governing body over the general direction of the markets can have a large impact upon land-use carbon projects. Specific examples include:

- Post-2012 risk: whether parties reach a global deal on a post-2012 successor to the Kyoto Protocol and the form that this takes is of fundamental importance to projects seeking access to carbon markets. A scaled-down deal or one that includes only a few countries would have obvious impacts upon the size and fluidity of a global market. Similarly, whether a REDD+ mechanism forms a part of this deal will greatly affect the role that forest carbon offset projects can play in a post-2012 agreement.
- US policy: climate change legislation in the US has suffered many setbacks in the US Senate since its initial approval in the House of Representatives. Further delays to pass such legislation will also postpone the start date of any cap-and-trade scheme and the opportunity for projects to sell into this market. Similarly, the final rules of such a scheme have not been determined and could limit the scope for forest carbon projects in tropical developing countries.
- EU ETS policy: the exclusion of forestry credits from the EU ETS has been a hindrance to the development of the forest carbon market. The continued exclusion of forestry in future phases of the EU ETS will sustain the current low demand for forestry credits in Europe.

Regulatory risks. These are risks primarily taken at the level of the standard's governing body or national government. A stable, predictable regulatory environment is preferred for the purposes of business planning. For example:

- National participation: a country may choose not to participate under a particular scheme and consequently negate the opportunity for a project to also participate. Projects that had counted on longer-term national commitments under a scheme might find themselves excluded from the opportunity to generate credits. There is much discussion around this topic, particularly in the context of project-based REDD interventions.
- Standard-specific regulations: a standard's governing body establishes the eligibility criteria under which projects can participate. Changes to eligibility criteria, methodological approaches to calculate emission reductions, or procedures to register projects can all affect a project's ability to successfully generate emission reductions.
- Country-specific regulations: national governments play a role in establishing the conditions under which projects can participate under various

schemes. For example, national governments establish forest definitions and CDM host country sustainable development criteria, and also determine tax and levy rates on project revenues. Changes to these conditions during the lifetime of a project could jeopardize either the eligibility of a project or its potential to generate attractive revenues.

Market risks. These are risks inherent to operating in a dynamic marketplace where supply and demand are fluid and prices fluctuate. The magnitude and predictability of these fluctuations is an important consideration that project developers should consider:

- Market demand: demand for forest carbon credits depends upon a variety of factors, such as the wider economic climate, the total volume of credits supplied from other projects, and the ability of buyers to reduce their own GHG emissions without the use of offsets. It is important that these factors combine to ensure a sustained demand for forestry credits to ensure a project's longevity.
- Price: price fluctuations and, most importantly, a sharp drop in prices can cause projects that were previously profitable to become unprofitable.

Step 7: Determine commercialization strategy

Determining how to commercialize the credits generated by a project is the final step in the process towards determining the best approach to bringing a project to market. Three main options exist, each of which carries its own risks and affects the achievable credit price.

Early sales contracts. Project developers sell the project's future credit stream to a dedicated buyer or number of buyers. Prices for these types of contracts are typically low; however, the project developer benefits from guaranteed future sales and the possibility of some upfront financing that can assist with project implementation.

Partnering with carbon specialists. Project developers partner with specialized carbon companies or consultancies that can assist with project development and credit sales. By relying on the specialized expertise of these actors, projects can increase their chances of successfully registering and achieving high sales prices. On the other hand, these specialized companies typically get paid on a success fee basis that equates to a percentage of the credits generated, thus reducing the overall number of credits that the project developer has available to sell.

Late sales contracts. By internalizing all costs and engaging directly with buyers only once, credits are issued at a higher price and all credit revenues accrue to the project developer. Under this approach, however, the project takes on all the project development costs and risks without any upfront payments or support from specialized carbon companies.

Conclusions

Despite some of the challenges for forestry and other land use-based projects under the carbon markets, these markets offer a new and potentially lucrative outlet to finance these activities. Successfully bringing a land-use carbon project to the markets, however, requires a significant amount of time, planning, upfront capital and risk mitigation. As such, moving ahead with project development needs to be considered as an investment decision and analysed as such. This means choosing which market to access, modelling financial returns, assessing the risks that could affect the project and determining a strategy to maximize sales. Combining these together will result in a business plan that will provide project developers with the best opportunity to enter the carbon markets and generate carbon finance.

References

Hamilton, K., Sjardin, M., Peters-Stanley, M. and Marcello, T. (2010) *Building Bridges: State of the Voluntary Carbon Markets 2010*, Ecosystem Marketplace and Bloomberg New Energy Finance, Washington, DC, and New York

Kossoy, A. and Ambrosi, P. (2010) *State and Trends of the Carbon Market 2010*, World Bank, Washington, DC

Locatelli, B. and Pedroni, L. (2004) 'Accounting methods for carbon credits: Impacts on the minimum area of forestry projects under the Clean Development Mechanism', *Climate Policy*, vol 4, pp193–204

Neeff, T. and Henders, S. (2007) *Guidebook to Markets and Commercialization of Forestry CDM Projects*, CATIE, Turrialba, Costa Rica

Neeff, T., Ashford, L., Davey, C., Durbin, J., Fehse, J., Hedges, A. et al (2010) *The Forest Carbon Offsetting Report 2010*, EcoSecurities, Oxford, UK

Neeff, T., Moore, C, Henders, S. and Ascui, F. (2011) *Making the Step from Carbon to Cash: A Systematic Approach to Accessing Carbon Finance in the Forest Sector*, Forests and Climate Change Working Paper, Food and Agriculture Organization of the United Nations, Rome, In press

VCS (Voluntary Carbon Standard) (2008) *Guidance for Agriculture, Forestry and Other Land Use Projects*, VCS, 18 November, www.v-c-s.org/afl.html, accessed 24 April 2010

9

A Functional Anatomy of the Project-Based Carbon Markets

Till Neeff and Jan Fehse

Introduction

Carbon markets have grown at exhilarating rates during the last few years. The World Bank estimates the volume of global carbon markets in 2009 to total US$144 billion (Kossoy and Ambrosi, 2010). These markets trade environmental commodities called carbon credits, which are associated with and, in many ways, similar to carbon ecosystem services. As such, carbon markets have, at least in theory, provided an opportunity for project developers to sell forest carbon ecosystem services. Although the role of forestry and agriculture in those markets has been limited so far, this is likely to change in the future as forestry (and also other types of land management) receives increasing interest from climate change policy-makers.

As a result of the success of these environmental commodities, investment bankers and the world of private capital have entered carbon markets. Some analysts have raised concerns that this has made carbon markets somewhat opaque and confusing for many potential ecosystem services vendors, who are not used to thinking in the mindset of private capital markets. In this chapter, we try to provide a functional anatomy of the carbon markets, and set out to provide potential ecosystem services vendors with an overview of how these carbon markets function. In particular, we discuss the roles and functions of different actors in the market.

It should first be pointed out that carbon credits as the carbon markets trade them are, in most cases, not the same as ecosystem services. The two largest carbon markets (that trade under the Kyoto Protocol and the European Union Emissions Trading Scheme, or EU ETS) build on a regulatory cap of emissions and on issuing tradable carbon credits ('allowances') to emitters ('cap and trade'). Project-based carbon credits ('offsets') can then be imported into the cap-and-trade scheme to help emitters comply with caps. Projects generate carbon credits for import through reducing greenhouse gases (GHGs) below a baseline scenario ('baseline and credit'). Most importantly, this happens under the Clean Development Mechanism (CDM), which applies to different sectors, including forestry.

The different types of allowances and offsets are environmental commodities and they share the same unit of trade (i.e. the ton of carbon dioxide equivalent). Only some of those carbon credits deserve the label of ecosystem services. In our intuitive understanding, only those carbon credits that are project based and where the projects are carried out in the land management sector can bear the label of ecosystem services. This distinction is important in order not to overrate the current carbon markets as tools for sustainable rural development. For instance, in 2009 only 3 per cent of the carbon markets by volume were project based and less than 1 per cent of the CDM came from land management activities (Kossoy and Ambrosi, 2010). The share of land management-related activities in voluntary carbon markets was larger; but voluntary markets have been tiny overall, compared to regulatory markets (Hamilton, et al, 2010). Carbon ecosystem services have only played a very minor role in carbon markets so far, although this may change in the future. A more complete discussion of carbon markets and their policy dynamics is provided in Chapter 8 of this volume.

In summary, this chapter explains how the carbon markets function as outlets where land management projects can sell carbon ecosystem services as carbon offsets. First, our analysis begins with a description of the role of carbon markets and how various actors function within carbon markets in relation to the CDM. We briefly discuss how role allocation changes within carbon markets other than the CDM. Second, we address carbon contracts and the business models of carbon market intermediaries. We discuss the drivers of choosing to work with particular types of carbon market intermediaries in a sound commercialization strategy. Finally, we analyse how the current limitation to voluntary markets affects the roles and actors in the trade of carbon ecosystem services, as well as how future developments may contribute to additional changes.

Parts of the description provided here stem from a prior publication (Neeff et al, 2011) and this chapter should be read together with its companion Chapter 8 in this volume. While this chapter addresses concepts of carbon markets at a somewhat theoretical level, Chapter 8 puts these concepts to work and elaborates upon how to develop business plans that (among others) rely on revenue streams from carbon markets.

Key Functions of Carbon Markets

Our functional description of carbon markets begins by looking at the Clean Development Mechanism as a classic case due to the fact that the CDM attracted the largest volume among all project-based carbon market mechanisms. The description is subsequently generalized in order to cover other types of carbon markets beyond the CDM.

Carbon market roles in the CDM

CDM markets are driven by a demand for compliance instruments to satisfy regulatory commitments for emission reductions. Users comprise those entities that need to comply with the emission reduction commitments, and generators are the projects that provide carbon credits to the markets. The transactions are typically not made directly between the generator and the user; but there are a series of intermediary roles that need to be filled so that the project can be implemented, the carbon credits can be issued and the transaction can occur (see Figure 9.1).

For instance, a carbon project could be the establishment of a state-of-the-art waste management system in India, and the generator of CDM carbon credits, therefore, could be the municipality running the landfills. The users for the carbon credits could be a medium-sized energy producer from Southern Europe that has emission reduction commitments under EU regulations.

In this example, there is a geographical and cultural disparity between the carbon credit buyer and the CDM project since they are based in different countries and belong to different industry sectors that do not otherwise work together. Therefore, for the Southern European energy company, effectively sourcing CDM projects is as challenging as it is for the Indian municipality to effectively place their carbon credits. In order to bridge this gap, various intermediaries take care of finding the activity and conduct due diligence (*origination*) and match carbon credit projects with their buyers (*sales*).

Regarding financial aspects, a financier provides capital (*financing*) and therefore enables the project owner to implement the activity (for clarification, we do not look at technical assistance, funding for feasibility studies, etc. here, but at much more significant financing for the underlying project). The municipality from India might not have cash available to improve systems on their landfills. Next to project financing, another important financial service is

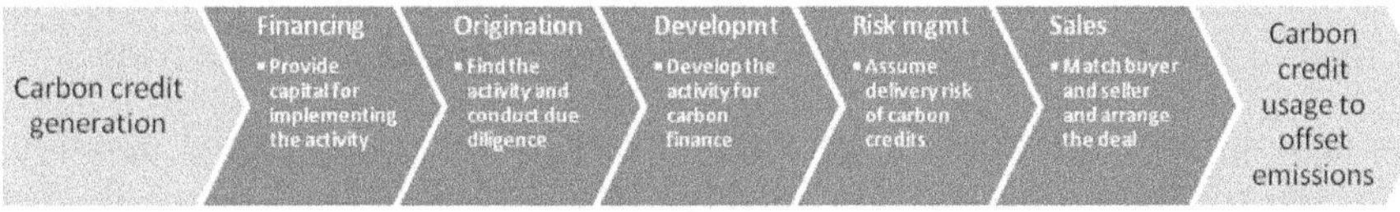

Figure 9.1 *Key roles of carbon markets in the Clean Development Mechanism (CDM)*

Source: chapter authors

risk management to guarantee delivery of carbon credits from projects with uncertain project success and according to changing CDM rules (we will elaborate further upon such services). Without a risk management scheme, the Southern European energy producer might find contracting with an Indian counterpart too risky.

Lastly, neither the seller nor the buyer of the carbon credits may have the technical expertise to conduct carbon accounting for the project and therefore might contract with an intermediary for *carbon project development* who registers the respective project under the CDM and takes care of all carbon technical and methodological aspects.

Intermediaries in the CDM carbon markets

Next to the users and generators in the CDM, there are additional actors who work as intermediaries to fill key roles in varying configurations (see Figure 9.2). Sometimes users and generators try to fill some of these key roles themselves – for example, when users invest directly in projects, or when a generator tries to carry projects by themselves through the project development cycle. We believe that the classic case involves series of intermediaries who fill the key roles and help users and generators to access the carbon markets. These intermediaries are classified as follows.

Carbon consultants can be hired on a daily basis, mainly for project development. Their business model relies on generating revenues through consulting fees. There is little involvement in financing, origination, risk management and sales.

Carbon brokers help with finding a buyer, negotiating and arranging a deal. The upside from carbon credit sales would then be shared between the project and the broker. Differently from traders, funds and banks, they do not take a principal position in the deal, but only match-make.

Carbon funds receive earmarked capital from their investors for sourcing carbon credits on their behalf. The funds aim for high volumes of carbon credits within the given investment with no delivery guarantee, but carbon funds capitalize on their ability to originate, conduct due diligence and implement the projects. Differently from traders, funds manage the capital of their investors. Apart from carbon funds' main goal of acquiring carbon credits, there are also capital market funds whose primary goal is to maximize returns for their investors. The latter is treated alongside financial service providers.

The aggregators and traders sell carbon credits with guaranteed delivery. They aim for high prices on so-called secondary carbon markets when also managing credit delivery risk through balancing a project portfolio. There is an incentive for efforts at origination since these efforts often result in principal positions.

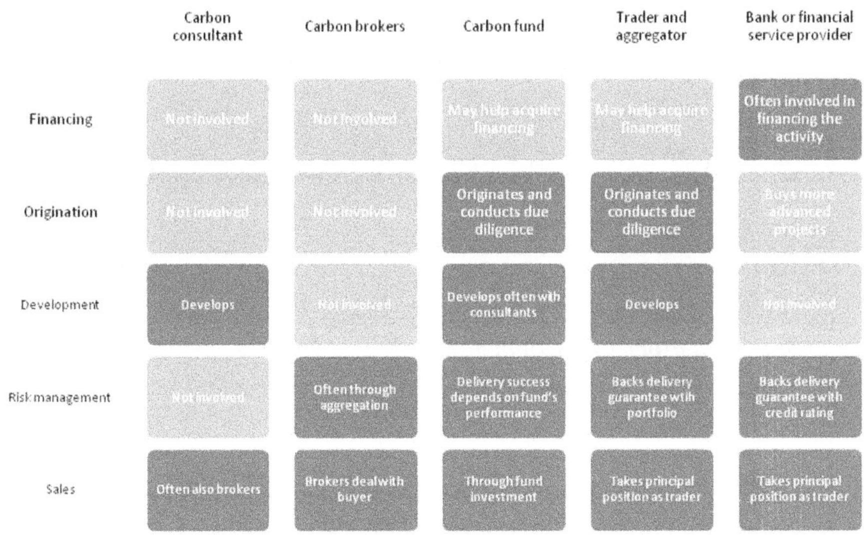

Figure 9.2 *Various kinds of intermediaries in the CDM carbon markets and how they fill key roles*

Note: Main roles are highlighted. The roles displayed here match those displayed in Figure 9.1.
Source: chapter authors

The banks and other financial service providers have started to treat carbon credits and the underlying activities as yet another asset class. They often invest in the underlying activities and leverage superb credit ratings to guarantee delivery and achieve high prices. Typically, financial service providers chase the more advanced projects and little effort is spent on origination of early-stage activities.

Risks, prices and risk mitigation in the CDM

In regulatory markets, including the CDM, carbon credits serve the purpose of compliance with emission reduction obligations, and all issued credits serve that purpose equally well. However, a carbon credit, particularly when purchased in a forward contract, carries a risk to the buyer because often it has not yet been issued (a forward contract is where the transaction terms are agreed upon early in the project and carbon credits are only exchanged for payments when the project is implemented and carbon credits issued). Thus, before issuance there are risks with regard to compliance and project performance.

The risk is significant because users might face hefty fines if they are unable to surrender enough carbon credits to regulators. The risk of credit delivery has therefore been found to be the major determinant of prices (Kossoy and Ambrosi, 2010). The lower prices paid for early-stage project-based carbon credits reflect the relevance of risks, while projects further advanced in the

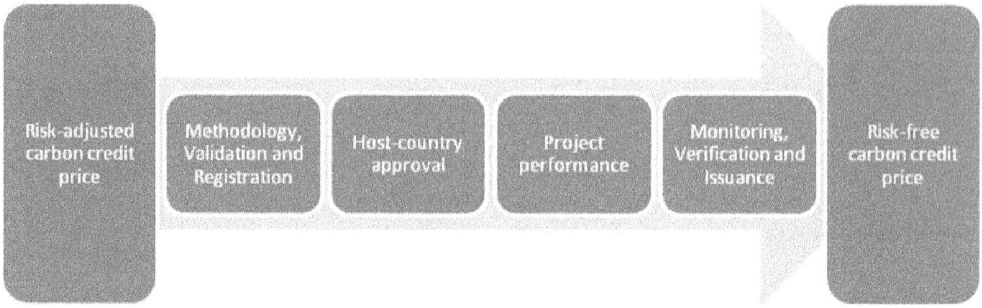

Figure 9.3 *Project-related risk factors as carbon projects move through the stages of carbon project development; these factors define the price differential between risk-adjusted and risk-free carbon credit prices*

Source: chapter authors

project development cycle may be able to command much better prices. The project-related risk factors (see Figure 9.3) explain why the risk-adjusted carbon price is lower than the risk-free carbon credit price. For instance, a project that has already dealt with two risk factors (methodological aspects, validation and registration, as well as host-country approval) carries less risk because only two risk factors remain: (i) project performance and (ii) monitoring, verification and issuance. Beyond the project-related risk factors, there are also risk factors surrounding the regulatory framework itself that are not related to project performance. Since these risk factors have no bearing upon the relative price differences between projects, the risk discussion does not include regulatory framework risks.

With the project-related risk factors, the more guarantees and risk mitigation that a project can offer, the better a price it can achieve in the market. This mechanism is being used for risk management that capitalizes on the ability of carbon intermediaries to assume the risk of carbon credit delivery in carbon transactions. Intermediaries use various common ways to mitigate risk of carbon credit delivery (see Figure 9.4). Most commonly, delivery guarantees

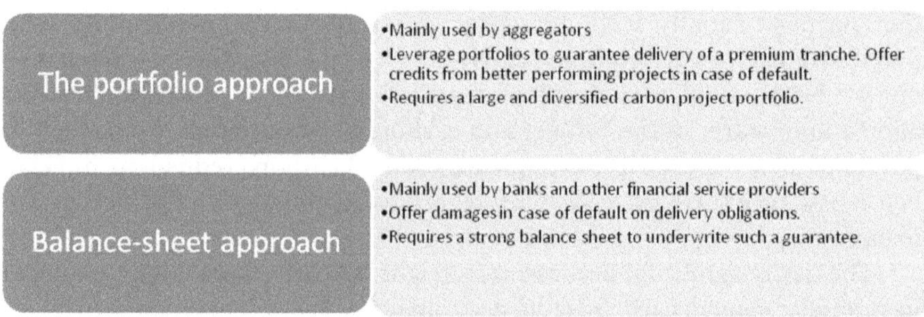

Figure 9.4 *Risk mitigation mechanism in carbon credit transactions as used by carbon intermediaries*

Source: chapter authors

are either backed by large and diversified carbon credit portfolios or, more simply, by financial guarantees and, thus, by the balance sheet of the intermediary.

Carbon market roles in carbon markets other than the CDM

Market design features will determine which key roles will occur in the ensuing carbon markets. The preceding sections looked at the markets shaped by the design of the CDM; here we broaden the analysis by considering carbon markets that are designed differently. Figure 9.5 summarizes how the market roles differ according to their design features.

The CDM, the Joint Implementation mechanism and most of the voluntary carbon markets function as *project-level markets* – for example, through specific projects that reduce emissions and whose carbon credits offset emissions elsewhere. Depending on the nature of the commitment to offset, we further distinguish between *project-level regulatory markets* and *project-level voluntary markets*.

With the CDM, as well as the Joint Implementation mechanism, the Kyoto Protocol has created a *project-level regulatory market*. We believe that in other markets similarly designed there would also be intermediaries with similar business models and therefore the full set of market roles would apply as presented in Figure 9.1.

In a *project-level voluntary market*, the driver of the buyers' interest in a given project is its quality when measured against the public image criteria (Hamilton et al, 2010). Since the carbon credit users do not face regulatory pressure, managing the risk of delivery is therefore much less important than in the CDM, and the corresponding role can be overlooked.

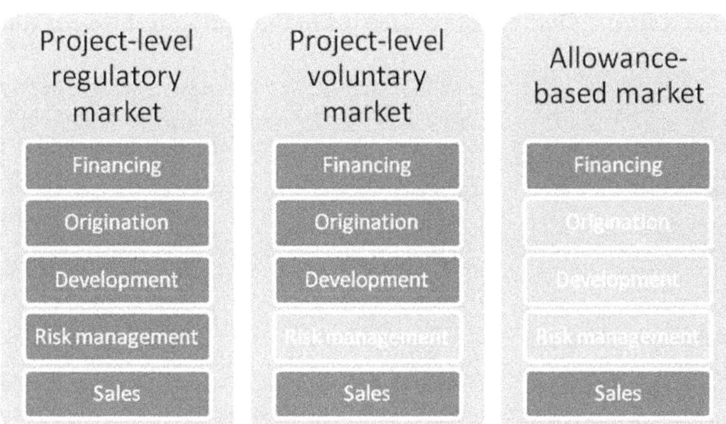

Figure 9.5 *Carbon market design features and the market roles to which they give origin*

Note: The roles match those displayed in Figure 9.1.
Source: chapter authors

The Kyoto Protocol's International Emissions Trading Mechanism, the EU ETS and the Chicago Climate Exchange market are *allowance-based markets*. In such cap-and-trade schemes (see explanation in the 'Introduction' to this chapter), emitters receive emission allowances as part of a compliance commitment. Since allowances do not need to be generated through project implementation, there is no need for origination and carbon project development; moreover, there is no delivery risk involved from project underperformance and, consequently, no need for risk management.

Some kinds of carbon finance instruments are not market based and do not consider carbon credits trading; as a result, they do not follow the mechanics explained here. For instance, incentive schemes, including the classical environmental services payment schemes, are not market based. Notably, the fund-based financing structures discussed in the context of REDD+ (reducing emissions from deforestation and degradation and enhancing carbon stocks) are not market based.

Transaction Structures in Carbon Markets

Carbon projects have the basic choice of selling their carbon credits directly to end-users or using the services of intermediaries. Intermediaries utilize different business models and offer different services, summarized here.

Carbon contracts

There are various types of carbon contracts that intermediaries typically offer. These contracts correspond to the carbon market intermediaries' different business models (see Figure 9.6). In all carbon transactions there needs to be an Emission Reduction Purchase Agreement (ERPA), which is the sales agreement for carbon credits. Owners of projects have the option to hire consultants to

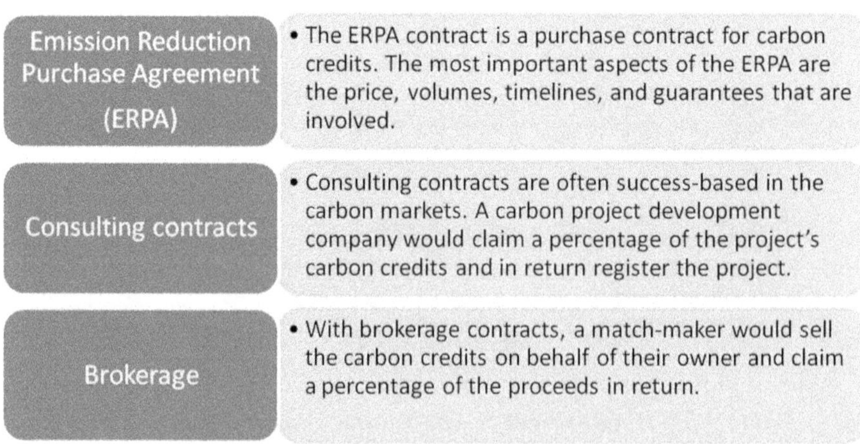

Figure 9.6 *Common carbon contract types*

Source: chapter authors

	Project stage of involvement	Source of profit for intermediary	Kind of carbon contract
Consultant	Early stage	Raises cash-fees or percentage of revenues	Consulting contract
Broker	Advanced stage	Levies percentage of revenues	Brokerage contract
Fund	Early stage	Levies overheads from fund investors	ERPA
Trader	Early and advanced stage	Mark-up through on-selling	ERPA
Bank	Advanced stage	Mark-up through on-selling	ERPA

Figure 9.7 *Overview of carbon market intermediaries' business models*

Note: The listed intermediaries match those displayed in Figure 9.2.
Source: chapter authors

prepare their projects through carbon project development for such a transaction. They also have the option to engage brokers for arranging an ERPA.

Business models of carbon market intermediaries

We summarize the intermediaries' business models according to the project development stage in which they get involved, the revenue-generation mechanism and the types of carbon contracts (see Figure 9.7).

Engaging with intermediaries

Working with different intermediaries and using different transaction structures are important components of a project's commercialization strategy. The choice of the best kind of intermediary depends on the project's circumstances (see Figure 9.8).

Engagement with *carbon consultants* is particularly interesting for projects at an early stage that require technical support for carbon project development, that are willing to assume the risk of changing carbon prices, and that can wait for carbon revenues to accrue. Many consultants work together with brokers. *Carbon brokers* are useful for projects that may be unable to achieve the highest prices on carbon markets since they lack the necessary visibility and experience in negotiating carbon deals.

Engaging with *carbon traders, aggregators and carbon funds* is an option for projects that would rather not assume the risk of changing carbon prices, that require technical support and that do not want to pay consultants. In some cases, projects may receive revenues earlier by using this strategy. Selling to

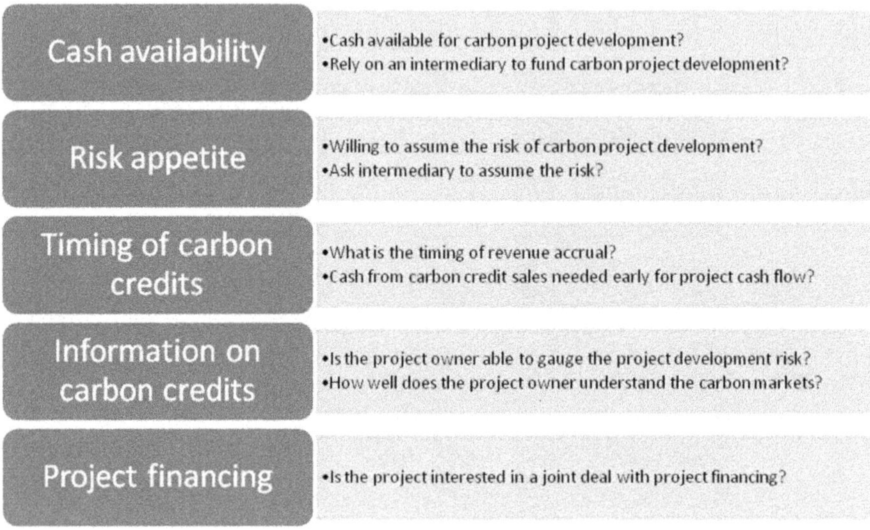

Figure 9.8 *Factors contributing to the best choice of intermediaries for selling carbon credits*

Source: chapter authors

banks and carbon investment funds is an option if projects want to negotiate joint deals between carbon credit sales and project financing.

Commercialization strategies

By choosing different kinds of carbon intermediaries (according to the factors displayed in Figure 9.8), we believe that projects essentially have three options for defining their commercialization strategies (see Figure 9.9).

Project owners can decide to sell their carbon credits at an *early stage in project development.* In doing so, the buyer would usually assume the subsequent project development costs. Since the project development risk would still be high at the initial stages, the carbon price would naturally be lower. Through funding the carbon project development, the carbon credit buyer would also assume the project development risks.

On the other hand, project owners could also enter into *success-based contracts that often include a brokerage component.* The intermediary would then fund part of the carbon project development and in turn claim a percentage of the carbon credits. Effectively, the carbon project development costs, as well as the revenues, would be shared between the project owner and the intermediary. Sales of carbon credits would only occur once the carbon credits are issued.

Lastly, project owners could decide to fund carbon project development entirely themselves, either using internal capacity or third-party consultants. Thus, in waiting for carbon credits to accrue, a better carbon credit price could be achieved. Naturally, the project owner would need to assume the risk of

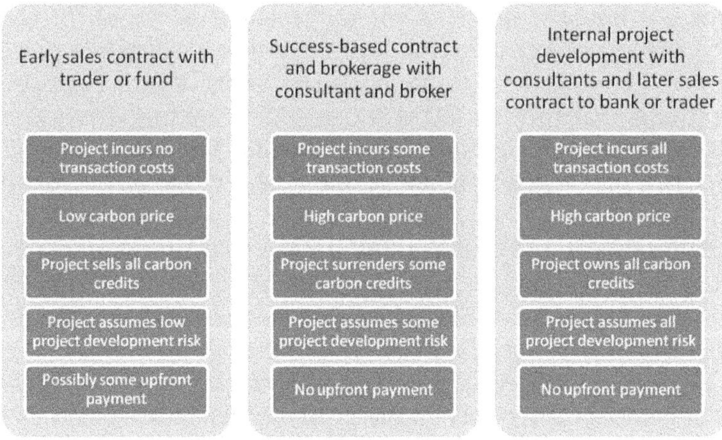

Figure 9.9 *Common commercialization structures for using carbon credit intermediaries*

Source: chapter authors

investing in carbon credit project development without being certain about the amount of carbon credits that the project could potentially deliver.

Carbon Ecosystem Services Markets

Having analysed the general aspects of market actors and their relationships in carbon markets, in general, we now turn to forestry and agriculture and discuss how the general functioning of carbon markets applies to carbon from land management activities in tropical countries. The mechanisms used in carbon credits trading from forestry and agriculture projects are analogous to those used for other types of projects. Agriculture has so far been excluded and forestry has been quite limited in the regulatory carbon markets; forestry has, however, been rather successful in voluntary carbon markets. Therefore, the case of carbon forestry and agriculture in the description of voluntary carbon markets provided in this chapter may be even more relevant. Nevertheless, since our policy outlook expects forestry to gain importance on the regulatory carbon markets as well, we also expect regulatory carbon market structures in the fashion of the CDM to soon become real for carbon ecosystem services.

Current role of forestry and agriculture in carbon markets

At this point in time, there is no regulatory carbon market where carbon ecosystem services from tropical forestry and agriculture projects can be sold effectively. One reason is that the majority of regulatory carbon markets are currently closed to land-based carbon ecosystem services from projects in tropical countries. This applies to the EU Emissions Trading Scheme (EU ETS), the New Zealand ETS, the New South Wales scheme in Australia, the Regional

Greenhouse Gas Initiative in the US and the Japanese industry ETS. The only regulatory carbon markets currently open to forestry CDM are those that occur directly between the member countries of the Kyoto Protocol. The second reason is that forestry CDM is being hampered by overly complex modalities and a disadvantageous crediting system. The most severe limitation is that the CDM only issues temporary carbon credits to forestry projects that meet little interest from carbon credit buyers who typically require permanent carbon credits (Neeff and Ebeling, 2007). The CDM has so far failed to provide an effective incentive scheme for tropical forestry, and in 2009 only less than 1 per cent of the CDM came from forestry projects (Kossoy and Ambrosi, 2010).

Next to these regulatory markets are the voluntary carbon markets, and the share of land-based projects on these voluntary markets has been much larger. For instance, in 2009, land-based projects accounted for as much as 24 per cent of the offsets sold (Hamilton et al, 2010). Although voluntary carbon markets are much smaller (in 2009, voluntary markets turned over approximately US$387 million compared to US$144 billion for the regulatory carbon markets) (Hamilton et al, 2010) because of the much larger market share, they provide an alternative to regulatory carbon markets as an outlet to carbon ecosystem services.

Outlook on carbon markets for tropical forestry and agriculture

A trend towards increased recognition of the importance of the land management sector in climate change mitigation is driving discussions about including it more fully in carbon markets. Limitations to the carbon markets of being an outlet for selling carbon ecosystem services may therefore be alleviated within the next three to five years. For example, a mechanism for avoided deforestation (REDD+) could be included in the UNFCCC's post-2012 agreement. In addition or alternatively, tropical forestry could gain access to regulatory carbon markets when included in climate change legislation at the level of individual countries. For instance, in the US, discussions on domestic emission caps could potentially be met by importing carbon credits from international forestry activities.

In spite of a clear expectation that more carbon financing will become available for forestry and other land-based activities, so far it is anyone's guess as to what extent the financial modalities will be market based and project based. Some of the relevant carbon finance instruments could be quite different from the CDM: first, because intervention could occur at a level going beyond the project and address the entire forestry and agricultural sectors; and, second, because the financial instruments are not necessarily market based but could also rely on funding in the style of development aid.

On the other hand, it is also widely expected that new instruments will be set up that function at a project level and where performance would be rewarded through tradable carbon credits. Such new instruments could be directly part of a new multilateral or bilateral climate change agreement or even be set up domestically in host countries. Under any of those scenarios, carbon ecosystem services could eventually find a liquid market to help fund tropical forestry and agriculture.

References

Hamilton, K., Sjardin, M., Peters-Stanley, M. and Marcello, T. (2010) *State and Trends of the Voluntary Carbon Markets 2010: Building Bridges*, Ecosystem Marketplace and Bloomberg New Energy Finance, Washington, DC

Kossoy, A. and Ambrosi, P. (2010) *State and Trends of the Carbon Market 2010*, World Bank, Washington, DC

Neeff, T. and Ebeling, J. (2007) 'The future of forestry offsets – will voluntary markets overtake the CDM?', in D. Lunsford (ed) *Greenhouse Gas Market 2007*, IETA, Geneva, Switzerland, pp132–135

Neeff, T., Moore, C., Henders, S. and Ascui, F. (2011) *Making the Step from Carbon to Cash: A Systematic Approach to Accessing Carbon Finance in the Forest Sector*, Forests and Climate Change Working Paper 8, FAO, Rome, In press

10

The Value of Biodiversity in Agricultural Landscapes

Fabrice A. J. DeClerck and Jean-François Le Coq

Introduction

What is the value of biodiversity? How much are people willing to pay for it and how is this value determined? These questions continue to be debated by ecologists and economists alike. One measure is the willingness to pay for the conservation of a single species or an ecosystem. Brooklyn Zoo has an interesting example in their gorilla exhibit, which costs US$3 more per person to visit than the normal entrance fee. The visitor is left with the question: is it worth an additional US$3 to see the gorillas? After walking through the exhibit, however, the visitor is invited to decide how to invest this money: which of the five species presented in the exhibit does he or she want to protect? After selecting a species, the visitor is further asked to decide how to spend the funds, with options such as paying for park guards, buying habitat, funding conservation science, or funding development projects for the local communities surrounding the reserve. It's a provocative exhibit and a means of valuing biodiversity. Since the opening of the exhibit in 1999, more than US$1 million have been raised to support the zoo's conservation efforts in Africa.

However, beyond the dollar amount that individuals invest through donations to conservation groups such as The Nature Conservancy, the World Wide Fund for Nature and others, or the additional amount we are willing to pay for Rainforest Alliance-certified coffee, this is a pretty limited means of deriving the economic value of biodiversity since it hinges on the popularity of singular species, and the ability of individual conservation groups pleading the cases of specific species or regions of conservation concern. Recently, a new focus has emerged for valuing biodiversity, which places emphasis on the functional role that biodiversity plays in the provisioning of ecosystem services (Daily, 1997; Costanza et al, 1997; Loreau et al, 2002; Perrings et al, 2009; TEEB, 2010).

Although biodiversity has often been lumped in with payment for ecosystem service (PES) programmes, we suggest that biodiversity in and of itself is not a service and, therefore, any payment for biodiversity is bound to be small and unsustainable. We also demonstrate that biodiversity, both at plot and landscape scales, plays a critical role in determining the quantity, quality and flow of most ecosystem services. The value of biodiversity should therefore not be based on the value of individual species *per se*, but on the services provided by those species within the ecological community.

Biodiversity: Intrinsic Value or Economic Value?

In Chapter 3, DeClerck and Martínez Salinas discussed means of measuring and evaluating ecological communities associated with agricultural systems of the Neotropics. They placed particular emphasis on measures that permitted understanding whether the species being protected in agricultural landscapes were of conservation concern or not, since the main goal of ecologically based certification schemes is to protect wildlife or wild nature to the extent possible. In terms of specific schemes, the authors targeted certification, or eco-labelling, of agricultural crops as the primary tool for promoting such conservation. The primary goal of certification is protecting threatened species and habitat, with the primary source of funding from the scheme derived from the coffee buyer who is, in essence, purchasing the right to drink a cup of good coffee with a free conscience regarding the impacts of coffee production upon rainforest destruction. The payment is motivated by a philosophy of conservation for conservation's sake, or the notion that we share this planet with other species, and it is our moral obligation to ensure that our actions do no harm to the myriad of species with whom we cohabit the planet.

Many have argued that biodiversity has an intrinsic value or a value that is independent from human well-being. Barbier et al (2009) explain that this value is used as a means of trumping arguments based on human values, or denying the trade-offs that people are willing to make between conservation and other objectives. Followers of this school would argue that biodiversity has no measurable economic value and worry that attempts to assign values to species would only justify the extinction of species when the economic value of the cause of extinction is greater than the value of the species themselves.

Let's take a step back. What is biodiversity? According to the Convention on Biological Diversity, biodiversity is 'the variability among living organisms from all sources, including terrestrial, marine, and other aquatic ecosystems, and the ecological complexes of which they are part. This includes the diversity within species, between species, and/or ecosystems' (CBD, 1992). Ecologists traditionally considered biodiversity as the passive recipient of environmental variation – that is, that biodiversity was distributed across landscapes according to variables such as temperature, precipitation, soils, elevation and other topographic elements (Holdridge, 1967). While this is true, more recently, ecologists have recognized that much more than passively responding to environmental gradients, biodiversity is an active player in shaping ecosystems and the services that they provide (Naeem, 2002).

In turn, ecosystem services are defined by the Millennium Ecosystem Assessment (MEA, 2005) as the benefits that people obtain from ecosystems, including provisioning services such as food, water, timber and fibre; regulating services that affect climate, floods, disease, wastes and water quality; cultural services that provide recreational, aesthetic and spiritual benefits; and supporting services such as soil formation, photosynthesis and nutrient cycling. Alternatively, Daily (1997) defines ecosystem services as 'the conditions and processes through which natural ecosystems, and the species that make them up, sustain and fulfil human life. This includes both goods and functions.' Note that biodiversity and species, the major currency of biodiversity, are missing from both definitions. However, in the ecological definition of ecosystems is the collection of living organisms (species) in a specific physical space interacting with each other and with the environment. It is precisely the interactions between individuals of the same species, interactions between species, and interactions with the environment over a broad range of timescales (daily to evolutionary) from which ecosystem services are derived. The concept of ecosystem services implies that by changing the conditions or species composition of an ecosystem, we also change the capacity of that ecosystem to provide certain services. Agricultural intensification tends to reduce species richness, reducing the capacity of the system to provide ecosystem services (Flynn et al, 2009; Laliberte et al, 2010). Payments for ecosystem services in agricultural landscapes, in contrast, provide incentives for farmers to increase species richness in these same landscapes to maintain or restore specific services.

This concept has evolved to the point that the Millennium Ecosystem Assessment (MEA, 2005) asserts that biodiversity is the backbone upon which all ecosystem services rest, bestowing a supporting system without which their production would be impossible. Others have suggested other analogies: that biodiversity comprises the global operating systems or the infrastructure upon which ecosystems services are provided. In this sense, *biodiversity is not an ecosystem service* that can or should be paid for, but rather is the basic building block of ecosystem services. The species composition and diversity of a system, including its spatial location and temporal dynamics, are, in effect, what will determine the capacity of the system to provide ecosystem services as it will determine the stability of the provisioning of these services. Biodiversity is not only critical in terms of providing essential services such as decomposition of waste, nutrient cycling and primary production, to name just a few; it also plays a key role in the stability of ecosystem services – that is, stabilizing the flow of services under variable environmental conditions. As such, biodiversity has been considered the 'life insurance for life itself' (see www.cbd.int).

For these reasons, we feel that payments specifically for biodiversity conservation will be difficult to rationalize as an ecosystem service. Rather, most payments of biodiversity will be through evolving markets for carbon (see Chapter 9, this volume), hydrological services (see Chapter 7) or scenic value that explicitly recognize that biodiversity is the key ecosystem component responsible for providing these services. In addition to these rather large global or landscape-scale services, with payments given to landowners from third parties, there is growing interest in managing ecosystem services that directly benefit the farmer, such as pest control (see Chapter 4), pollination (see Chapter 5) and soil fertility, whose values are measured in terms of decreasing cost of inputs, decreasing losses of crop to pests and diseases, and increasing productivity. What is essential to understand is that biodiversity is behind each of these services.

Methods to Value Biodiversity

We have argued that biodiversity is not an ecosystem service, but rather that changes in species richness, composition or abundance in an ecosystem alter the capacity of the system to provide certain services. In this chapter we focus on this functional approach since it is most applicable to the valuation of biodiversity *vis-à-vis* its contribution to providing ecosystem services. It is our opinion that developing strong linkages between biodiversity, the services that it provides and the valuation of these services is both one of the most promising, as well as one of the more sustainable, tools for biodiversity conservation. Thus, in this chapter we do not consider methods used to capture the *intrinsic value* of biodiversity, or a value independent of human well-being, since these values are not used to design PES mechanisms. In line with Barbier et al (2009), we will reflect on the value of ecosystem services embedded in human contexts, considering biodiversity through its contribution to ecosystem services and their instrumental role in meeting human needs whatever they are (hedonic, economic, etc.). Below, we review the most common methods used to value ecosystem services in the Neotropics, their limits and advantages, and discuss their possible applications to assigning economic value to biodiversity.

Economic valuation of ecosystem services is a tricky issue since they are not tradable goods or commodities and thus have no existing market value. The literature characterizes different goods according to their property right in order to define the possibility of becoming marketable products (commodities). Since ecosystem services are difficult to qualify as private goods and generally fall under the classification of public goods with no excludability or rivalry, they are subject to market failures *per se* (i.e. market transactions cannot ensure an optimal allocation of these types of goods). Ecosystem services such as carbon fixation and water quality share this characteristic, as does biological connectivity: the beneficiary of biological connectivity cannot ban access to other beneficiaries (no excludability). The use of biodiversity by a beneficiary does not impede others from using it (no rivalry). As such, biodiversity itself (abundance and richness of species, or flagship species) has no theoretically available market value.

In order to overcome these difficulties, different methods have been developed to derive an economic value for biodiversity's assets or services provided. From the perspective of establishing instruments that reward or pay for ecosystem services, various valuation methods have been applied (see Table 10.1). Two primary criteria can be used to classify ecosystem services (ES) valuation techniques. First, as PES mechanisms rely on the idea of creating a market, current valuation techniques can be classified into two groups by whether they consider the rationality of offer side (providers of ES) or demand side (buyers of ES). Second, valuation techniques can be divided into two groups according to their measurement basis: the methods based on tangible economic value, such as opportunity costs or avoided costs; and the methods that measure the preference of the actors on a declarative basis, also known as 'stated preference methods' or contingent valuation methods, such as willingness to accept (WTA) or willingness to pay (WTP).

Authors also consider in their classification of valuation methods the type of value estimated. They currently consider the no-use value, which represents an individual valuing of the pure existence of a natural habitat or ecosystem, and the use value, which involves some human interaction with the environment. The use value is generally broken down into two categories: the direct use value, which refers to consumptive and non-consumptive uses that involve some form of direct interaction with environmental goods or services (such as recreational activities, resources harvesting, drinking clean water, breathing unpolluted air, etc.); and indirect use values that refer to ecosystem services that can only be measured indirectly, such as supporting and protecting activities (Goulder and Kennedy, 1997; Barbier et al, 2009).

Aside from the methods presented here, other methods for valuing ecosystems services, especially developed for the regulating services, could be useful in estimating biodiversity value. This is the case of the 'option-insurance value' that refers to the capacity of agricultural systems to adapt to economic and environmental external shocks. This value depends on the ES contribution to the resilience of the system and the level of risk acceptance by a society, which can be captured by the WTP to reduce a risk or the empirical assessment of individual risk aversion. However, we will not develop it further since it has been poorly used to design PES schemes.

Valuing opportunity costs from the offer side is very common (see Chapter 13 in this volume). This valuation consists of establishing the cost to halt practices that are detrimental to the provisioning of a specific service, while promoting the adoption of new practices that favour the provisioning of ecosystem services. The value of the ecosystem services is determined by comparing the potential incomes generated by both practices. In Costa Rica, for example, the government established a national system of payment for

Table 10.1 *Main methods used to derive the value of ecosystem services*

	Offer side-oriented valuation	*Demand side-oriented valuation*
Evaluation based on tangible economic value or avoided costs	Opportunity costs	Replacement, maintenance
'Stated preference methods' (contingent valuation methods)	Willingness to accept (WTA)	Willingness to pay (WTP)

ecosystem services (PES) managed by the National Forestry Financing Fund (Fundo Nacional de Financiamiento Forestal, or FONAFIFO) as a means of reducing deforestation and land conversion to pastures by paying farmers to maintain land under forest cover. In order to determine how much they would have to pay landowners, they determined the opportunity cost to halt deforestation as the cost of abandoning extensive cattle ranching equal to the annual income generated by extensive cattle ranching on that land. The advantage of this method is that the valuation is based on measurable existing economic activities, such as agricultural production. It corresponds to the economic and financial logic of the provider of the ecosystem service in a cost–benefit decision process. Nevertheless, it has some limitations since it does not value a specific service in and of itself and does not discriminate the value of each type of ecosystem service. A second limitation is that providers may not follow a pure cost–benefit pattern in their decision-making, but may include other variables such as social and cultural preferences, or their aversion to risk in their decisions. Methods founded on opportunity costs may not sufficiently capture these other dimensions and may not correctly determine the actual cost incurred by the provider by adopting a new practice.

In order to overcome these shortcomings in the design of payment for ecosystem service mechanisms, other methods of valuation have been developed, such as willingness to accept (WTA). The principle of this valuation method is that a declaration of value by the landowner can be used to determine the financial amount that a provider considers an acceptable incentive to change their current practices. This value is based on a declaration by the providers using a method similar to willingness to pay (WTP), which we cover below.

Two methods are currently used to capture value from the demand side. The first one is WTP, which is similar to WTA; but rather than determine the payment that a landowner would accept to change their practices, it aims to determine how much the user or beneficiary of the service is willing to pay in order to receive a specific service. As a contingent valuation method (Mitchel and Carson, 1989), WTP tends to capture the value of the ecosystem service according to a declarative statement. The basic method to estimate WTP consists of individual declarations collected through surveys given to potential beneficiaries of an ecosystem service. This simple method has some shortcomings, such as over- or under-estimation of the value of a service due to lack of information and/or misunderstandings. For example, beneficiaries may be prone to making overly optimistic statements (I would pay US$50 a month for clean water) in a survey environment that do not accurately reflect how they would truly feel if they received that US$50 bill in their mailbox.

However, this method is particularly useful in providing a value for conservation since it is able to capture no-use as well as indirect and direct use value. For example, a survey applied to 240 participants in Costa Rica demonstrated that residents and foreigners were willing to pay a contribution between US$63 to $78 in order to protect 356ha (46 per cent) of the Manuel Antonio National Park in Costa Rica that was in danger of being converted to other land uses due to a 1991 constitutional court decision (Adamson, 2001, cited by Moreno, 2005). This technique has also been used to measure the WTP for voluntary contributions to increase forest protection (Hearne and Motte, 2001). The authors showed that Costa Rican citizens were willing to pay US$0.32 and $0.24 per month to protect an additional 1ha of forest with the goal of conserving the biodiversity and the scenic beauty of both remote areas and accessible areas. In contrast, foreign tourists were willing to pay up to US$6.41 in a one-shot contribution to increase conservation for biodiversity and US$3.10 for scenic beauty.

Many other technical designs have been developed to fine-tune this contingent valuation method and to limit possible biases (Carlsson, 2010). These include experimental techniques that enable researchers to put the survey recipients in more realistic conditions and settings to state their preferences. For example, this technique has been used to measure WTP for voluntary contributions to support national parks in Costa Rica (Alpízar et al, 2008). The authors paid particular attention to putting the respondents in realistic and credible conditions in order to avoid behaviour biases. Interviewers were selected and trained for credibility; they wore uniforms similar to those from park authorities. Tourists were approached at the station after they visited the main park attractions. Respondents were chosen randomly, with only one person per group invited to participate in the survey. Some respondents made actual contributions while others simply stated their hypothetical contribution. Various treatments were repeated to test the influence of the degree of anonymity and information provided about the contributions of others on the contributions stated by the respondent. They showed a substantial bias favouring hypothetical contributions over actual contributions; however, the influence of the social contexts was about the same when the subjects made actual monetary contributions as when they stated their hypothetical contributions.

WTP is a flexible tool that can be applied in a variety of ways to value biodiversity. It can be used to assess the value of specific characteristics of biodiversity, such as the greater willingness to pay for the conservation of 'charismatic mega-fauna' (pandas, whales and elephants) over less charismatic species (the giant water bug). However, it is more difficult to apply the concept to ecological communities comprised of multiple species in large part because the means of quantifying ecological communities are complex and difficult to grasp (see Chapter 3). WTP is more easily applied to determining the economic value of ecosystem services derived from biodiversity, such as carbon for climate change reduction, water (e.g. water tariffs) or landscape beauty (e.g. national park entry fees). Willingness to pay methods are also inherent in green certifications such as Rainforest Alliance or Bird Friendly coffee, where the additional cost of the certified commodity is indicative of the consumer's willingness to pay extra for a certified product. In some cases, it is the seller who determines this value, which is expressed not as an additional cost passed on to the consumer, but in a form of advertisement. For example, several chains such as McDonald's and American Airlines offer Rainforest Alliance-certified coffee, but at the same cost as their other coffee offerings, suggesting that these businesses perceived the added value in terms of increased customer satisfaction, rather than as increased dollar value paid for the product. In contrast, Café Britt in Costa Rica packages their organic coffee in a beautiful bag depicting a coffee system immersed in a forest ecosystem replete with toucans and capuchin monkeys. This imagery of a biodiverse coffee plantation increases the beneficiaries' willingness to pay US$1 more per 12oz bag of their organic coffee than for their Tarrazú blend.

The second demand-side method for valuing biodiversity is to calculate the replacement or avoidance costs (Freeman, 1993). As with the notion of opportunity costs, the principle behind the replacement costs valuation is that the value of the service is determined by the cost to the beneficiary to replace a service in its absence. For example, this cost could be the cost of construction and maintenance of a water treatment facility to replace the hydrological services that are no longer provided by forested watersheds in the case of water quality. Probably one of the more famous examples, though not Neotropical, is of the water quality services received by the residents of Manhattan in New York. According to a news piece by the Rand Corporation (see www.rand.org):

> *The clean, plentiful water that New York City residents drink isn't the result of a technological miracle. The natural systems of upstate New York's Catskill/Delaware watershed provide most of the City's drinking water. While other cities spend billions of dollars on filtration systems, the New York City water supply predominantly depends on the natural landscape to filter water for its 1.4 billion gallons of water each day. To fend off the $6 billion dollar price tag for construction of a new filtration facility and the associated $300 million per year for operating costs, New York City is protecting its vital water-filtering ecosystems by investing up-front in nature's services.*

In the case of the New York watershed example, preserving this watershed is valued in terms of cost of replacement of the services provided by the protected watershed (see Table 10.1), where the cost of building a water treatment plant (US$6 billion) serves as the comparative value.

Likewise, the value of soil conservation as an ecosystem service could equal the cost of removing the silt from reservoirs behind hydroelectric plants in addition to the loss of potential energy production due to the volume of the reservoir occupied by silt, rather than water. This method is also useful in evaluating ecosystem services associated with agricultural productivity: the value of maintaining soil fertility can be evaluated as the cost of chemical fertilizers applied to sustain the crop productivity; pest regulating services can be evaluated through the cost of pesticide applications in the absence of these services from agroecosystems, as proposed by Avelino et al (see Chapter 4).

Since no perfect method exists, and to avoid the shortcomings of the different methods, some designers of payment schemes have taken to using combinations of different valuation methods as a means of facilitating negotiations between stakeholders. In Costa Rica, for example, the city of Heredia's Public Service Enterprise (ESPH), which is in charge of supplying clean water for the city, combined various methods to define payments (i.e. tariff) levels that residents should be charged to maintain forest cover and thus contribute to the sustainability of clean water from the catchment (Villalobos and Solano Valverde, 2007). For the offer side, they determined the opportunity costs of the existing forest area at US$104 per hectare per year (ha^{-1} y^{-1}) by using pasture for cattle breeding as the reference land use. The recovery cost (cost to recover the loss of the ecosystem service) evaluation was based on the investment costs for reforestation of US$577 ha^{-1} y^{-1}. Taking into account the amount of water used from the ecosystem, they established a total water cost of US$0.015 per cubic metre. From the demand side, they established that residents of Heredia were willing to pay an additional US$0.03 per cubic metre of water in order to maintain and restore forests in the watershed that contribute to the water supply, which was higher than the opportunity and replacement costs. Following the demand of the public services regulating agency, the WTA value was used to define the minimum amount that a property owner requires to receive to maintain forest. Through interviews, this value was determined to be approximately US$44 ha^{-1} y^{-1}. Finally, the tariff was set by the regulating agency at US$0.0035 per cubic metre of water in 2001 and re-evaluated at US$0.0073 per cubic metre in 2004, which is still below the WTP value designated by the users. With this low level, which is less than 1 per cent of the total value paid by an average family using 25 cubic metres per year (m^3 y^{-1}), the additional tariff collected for conservation of ecosystems amounts to US$570,000 collected since the beginning of the programme in 2000.

Our discussion of these common methods does not preclude the development of other methods. The valuation of biodiversity associated with specific ecosystems can be derived from the direct and indirect activities generated by the ecosystem as well. This type of valuation method using the global chain concept is particularly useful for evaluating the importance of conservation within the regional or national economy. In Costa Rica, a recent news report on the funding needed by national parks argued that parks have been undervalued. The report integrated direct and indirect added value generated by national parks (including tourism, transport, etc.) and suggested that Costa Rica's national parks generate US$1.5 billion in revenues per year (Moreno Diaz et al, 2010). Although this kind of method is not specifically directed at the creation of a PES mechanism, it supports greater investment in conservation mechanisms and protected areas, in general. In Costa Rica's experience, such economic valuation of natural resources and ecosystems demonstrates the importance of investing in protected areas and contributes to the political justification of maintaining and improving the national PES mechanism.

Finally, the value of ecological assets is not universal since it is anthropocentric, individual based, subjective, context dependent, marginal and state dependent (Goulder and Kennedy, 1997). Thus, results of any economic valuation method are site specific and subject to local factors. The opportunity costs or replacement costs can vary according to the prices of the country (wages, infrastructures). Similarly, evaluations of willingness to pay depend not only on objective (physical or ecological) properties of the assets, but also on socioeconomic contexts such as human preferences, local institutions, and culture (Barbier et al, 2009).

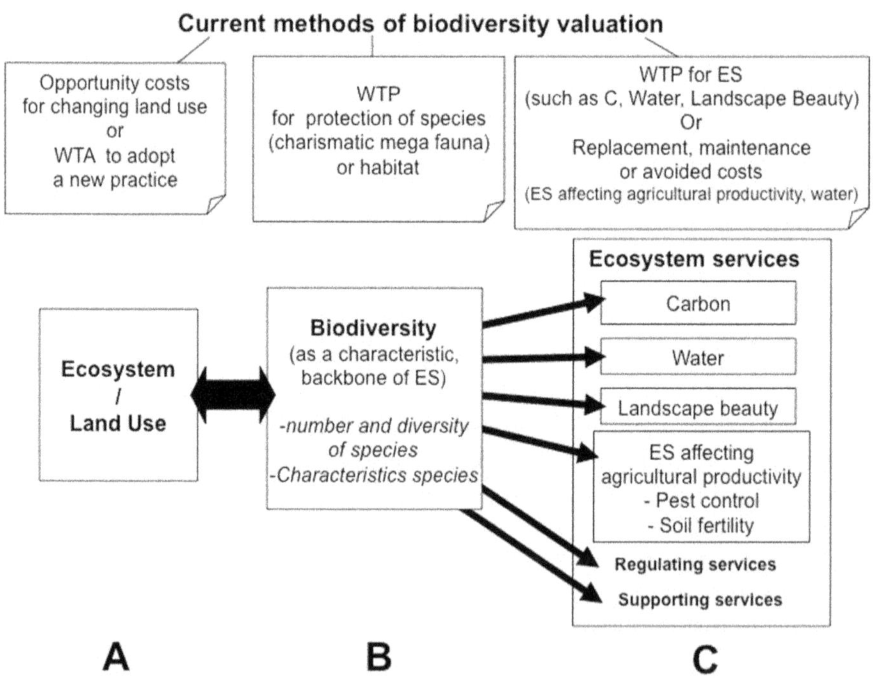

Figure 10.1 *The three primary methods for estimating the value of biodiversity discussed in this chapter*

Note: We also highlight the relationship between the ecological characteristics of the land use (the number, abundance and composition of species) and their impact upon the delivery of certain ecosystem services.
Source: chapter authors

In summary, since biodiversity, determined as the number and diversity of species and their functions, is incredibly complex to measure, there is no absolute method for determining its economic value. Thus, a surrogate measure, such as land-use type, can be used to approximate the economic value of biodiversity (see Figure 10.1). Biodiversity, considered as a key characteristic of an ecosystem, can be comprehensively captured at the ecosystem or land-use level using mainly 'offer side' valuation methods, such as opportunity cost or WTA (A); or at a more specific level considering the individual species, especially 'charismatic mega-fauna', the diversity of species or specific habitat using demand-side methods such as WTP (B). Finally, biodiversity's value can be indirectly quantified through the main ecosystem services produced (C), with demand-side methods such as the replacement cost method or WTP (see Figure 10.1). In this last case, the challenge is the lack of existing information that links biodiversity and ecosystem functioning to the provisioning of specific ecosystem services (Daily, 1997; Barbier et al, 2009).

Payment Options for Biodiversity

Valuing the services provided by biodiversity is just the second step in establishing payments (the first step is quantifying the service provided). In addition to these two steps, different payment options can also be considered. In this section we explore the existing payment options relating to biodiversity.

First, payment for biodiversity could be accomplished through specific ecosystem services derived from biodiversity, as we have suggested above. In some of these cases, no specific payment is made for biodiversity *per se*; rather, the payment is made for the service provided by biodiversity with no explicit acknowledgement of the conservation value obtained. This is often the case in water- or carbon-oriented PES. In other cases, payments can be made for specific ecosystem services; carbon and water again serve as good examples, where the added conservation value is recognized and areas that have a greater biodiversity value are either prioritized or receive a larger payment. For example, FONAFIFO's PES in Costa Rica encompasses payments of four services: carbon, water, biodiversity and scenic value. This PES prioritizes payments for conservation, natural regeneration and reforestation within the 37 nationally recognized biological corridors. The PES recognizes a specific PES modality for forest conservation within 'biodiversity vacuums' or areas that are strategic for maintaining biological connectivity.

In addition to this example, it is also worth mentioning a local pilot PES scheme being developed by a non-governmental organization (NGO) to support forest conservation in Costa Rica's central valley (Fundación de Desarrollo de la Cordillera Volcánica Central, or FUNDECOR) and the foundation of cooperation Costa Rica–USA (Crusa). This innovative PES mechanism captures voluntary contributions from local car rental companies (which derive much of their revenue from tourism dollars) that want to be carbon neutral. It then directs these carbon payments towards forest conservation or reforestation projects in specific locations where biodiversity conservation is important (national parks, biological corridors, private reserves, etc.). In this way, carbon services are received through biodiversity conservation, and biodiversity conservation is achieved through carbon payments.

Eco-labelling, green stickers or certification for conservation serves as a second payment option for biodiversity. Here the payment is based on the consumer's willingness to pay for conservation. Although eco-label requisites regarding biodiversity can be discussed (see the comparison between Smithsonian Bird Friendly and Rainforest Alliance certification described in Chapter 3), for some promoters or adopters of certification, the motivation behind certification responds to a market strategy in addition to environmental consciousness (Faure and Le Coq, 2009). For both, certification can serve as a means of securing a specific market share. For adopters, many of whom have multiple seals (e.g. organic + Rainforest Alliance + Starbucks CAFE practices) the multiple seals ensure access to a diversity of markets, ensuring multiple options in case of price variations. Certification is a valuable and promising mechanism to protect biodiversity and to provide incentives for producers to adopt or maintain environmentally friendly practices (see also Chapter 15). Recent years have demonstrated a steadily increasing market share of certified products. It is also worth noting that certification schemes that target social justice, such as Fair Trade (the most popular certification standard in Central America) now include environmental criteria in their norms because consumers prefer a product that encompasses social and environmental values. Note, however, that with the exception of Smithsonian's Bird Friendly certification, most labels recognize 'environmental values' in the form of reduced impacts regarding water contamination and soil erosion rather than species conservation *per se*.

A third option for paying for biodiversity is through direct financial support to protected areas and biodiversity. In Costa Rica, some actors of the national tourism industry created private foundations in order to provide financial support to the administration and maintenance of national parks. Proparques, which is largely comprised of private businesses, including the coffee sector, recognizes that the allure of tropical forests, howler monkeys, toucans and tree frogs serves as an important tourism draw to the country. Raising awareness of tourism actors has been translated into specific incentive programmes that promote the adoption of sustainable practices such as the Certificate for Sustainable Tourism developed by the Costa Rican Institute for Tourism, which is highly sought after by hotels and guide groups.

Ensuring that Critical Species Are Conserved: Intrinsic Value Piggybacking on Economic Value

Valuing the contribution of biodiversity to multiple ecosystem services is discussed in other chapters of Part II. With the exception of cases where the ecosystem service in question is provided by intact natural ecosystems (hydrological services offer the best example, as does carbon under REDD+ rules), most land uses managed for ecosystem services in agricultural landscapes will fail to conserve species of particular conservation concern. These are species at risk of extinction from deforestation, agricultural expansion or other human interventions. This implies that achieving greater conservation value, in terms of protecting species of conservation concern (see Chapter 3) must be explicitly planned for. Venter et al (2009) describe this as an opportunity in relation to carbon storage under new REDD++ rules by mapping priority areas for the proportional allocation of REDD funds:

- to forest-losing countries to minimize carbon emissions only;
- to minimize the loss of forest vertebrates; and
- to minimize carbon emissions while simultaneously doubling benefits to biodiversity.

The three scenarios would reduce deforestation by up to 20 per cent; however, the first scenario would only protect 9.6 species, whereas the second would protect 35.7 species, and the third would protect 19.2 species. This example demonstrates that in addition to often supporting the provisioning of services, integrating biodiversity conservation (species) with ecosystem services (here, forest cover) can provide added value, including the conservation of wild biodiversity, which may not directly contribute to the service being paid for, even though diversified tropical forests have been found to be a more stable form of carbon storage than most other land uses (Bunker et al, 2005).

Costa Rica has enacted this sort of thinking by prioritizing payments for ecosystem services in biological corridors, as previously discussed. FONAFIFO has developed a differential payment for these services. In FONAFIFO conservation contracts, which are five years in duration, farmers who protect and restore forest systems are eligible to receive US$320 per hectare in payments. However, if the conserved forest is found in an area that is critical for conservation, such as biological corridors, the landowner is eligible for a payment of US$375 per hectare. In this sense, biodiversity has greater value (+US$55 ha^{-1}) than 'regular' forest conservation (mainly consisting of carbon sequestration). In addition to the added dollar value given for forests that preserve carbon plus biodiversity, payment schemes are prioritized within the nationally recognized biological corridors, an important consideration in a country where the demand for payments exceeds currently available funds.

The Chorotega Biological Corridor, a subunit of the larger Mesoamerican Biological Corridor (MBC), is a second example of where the service provided by biodiversity has greater value than the biodiversity itself in terms of willingness to pay. The MBC is a multinational conservation effort established as the Wildlife Conservation Society's *Paseo Pantera* ('the panther's path') in 1990. The project's goal is to ensure biological connectivity between southern Mexico and northern Colombia. In theory, the MBC would link the protected areas of Mesoamerica, permitting a jaguar (the flagship species of the programme) to walk the length of Mesoamerica without leaving forest cover. Although the programme was widely popular initially, particularly with international conservation groups, it was little implemented in large part because it focused primarily on the intrinsic value of biodiversity and failed to take into consideration the needs of the local communities. These needs include agriculture production, access to clean and regular sources of water, and protection from natural disasters that were readily perceived as more important than connectivity by local communities and their governmental representatives. In the Chorotega Biological Corridor, the main motivating force for reforestation in the region is ensuring a perpetual and stable water supply to the local community, and that in the eyes of the members of this biological corridor, this hydrological stability is achieved by protecting biodiversity in the form of natural forest cover in the uppermost stretches of the watershed. To date, 60 per cent of the upper part of the Nosara River watershed has been recovered through land purchase, reforestation projects and negotiation of PES in the region. We feel that this case shows that the value of biodiversity here is equal to its capacity to provide a clean and regular source of water.

Location, Location, Location: Spatial Targeting of Ecosystem Services

In the previous section, we mentioned that FONAFIFO prioritizes payments in specific locations depending on their capacity to provide specific services (for another example of prioritization, see Chapter 13 on a Mexican PES). This brings up an important point: the capacity of biodiversity to provide certain specific services may be spatially explicit and therefore location can affect the value of biodiversity. The value of biodiversity is not universal across a landscape; therefore, understanding the spatial patterns and processes related to biodiversity and ecosystem services will affect the value. Much as in real estate, location matters. For example, carbon is an example of a relatively spatially independent ecosystem service at landscape scales. 1ha of forest in one portion of a landscape will store a similar amount of carbon to 1ha in a second portion of the landscape. However, other services such as pollination, scenic value and hydrological services are very spatially explicit. Pollinators typically do not move very far (< 1km), so interventions that are intended to promote pollination should occur either on the parcel where the service is desired or in close proximity (Ricketts, 2004; Ricketts et al, 2004). The scenic value of a landscape is also quite spatially explicit. This is frequently seen in real estate brochures or hotel advertisements in Costa Rica where the value of the property or room increases with the splendour of the view. In agricultural systems, well-managed coffee agroforests contribute to the scenic value (see Figure 10.2); however, this value only exists if the farm or region derives value from it through economic activities such as agro-ecotourism. Coffee-dominated landscapes indeed have such value, with coffee tours offered by major coffee producers in Costa Rica fetching prices of up to US$20 per person (Café Britt's).

Figure 10.2 *A Rainforest Alliance-certified coffee farm on the Poas Volcano in Costa Rica*

Note: The scenic value of this landscape simultaneously promotes a sense of harmony with nature via the integration of trees among coffee plants (though the volcano in the background does not hurt), while contributing to both production and conservation values. The linear rows of trees on the farm serve as wind breaks, protecting the coffee, but also serve as important corridors for wildlife, including species that contribute pollination and pest control services on the farm.
Source: Thomas Husband, University of Rhode Island

Biological connectivity, which is recognized as a service by FONAFIFO and which can be considered a support service, is also tremendously spatially explicit at multiple scales. At regional scales, biological corridors have been designated because of their placement between protected areas. At local scales, farms that are surrounded by a matrix of sugar cane have less to contribute to connectivity than farms that are located between two forest patches (Estrada and DeClerck, 2011). If the objective of a payment is to obtain a specific service which can be measured in litres of water produced, megawatts of energy produced or dollar value of the crop not lost to pests (see Chapter 5), then the buyer should be willing to identify that portion of the landscape most capable of providing the service, or should be willing to pay more for those locations than for others since they guarantee more service provided per dollar invested. To illustrate this point, and to conclude this chapter, we provide four specific examples of studies that evaluate the relationship between services provided by biodiversity and their economic value.

Biodiversity and Coffee: Managing Biodiversity for Farm Value

Coffee is an ecologically, economically and culturally important crop in Central America, grown under a variety of shaded conditions ranging from full sun to what is locally called rustic coffee grown under the shade of a natural forest canopy. Intermediate to these extremes are commercial polycultures with differing degrees and diversities of shade (Moguel and Toledo, 1999). From an ecosystem services point of view, what is important to understand is that by altering the tree species richness and composition in these systems, we also alter their capacity to provide critical ecosystem services. We focus on two specific case studies below where biodiversity measures were tied to the provisioning of key ecosystem services, pest control (Kellermann et al, 2008) and pollination (Ricketts, 2004; Ricketts et al, 2004), and where those services were given an economic value.

Kellermann et al (2008) consider the economic value of pest control services provided by birds on Jamaican Blue Mountain coffee fields. In this study, they focus primarily on the coffee boring beetle (*Hypothenemus hampei*), which is coffee's primary pest. The female beetle bores into the coffee fruit and lays its eggs inside the seed. As the eggs hatch, the young larvae feed on the seed, essentially destroying the economic value of the fruit. Several studies have suggested that bird communities play an important role in controlling arthropods in coffee (Phillpott et al, 2009); however, few have focused on specific pests and even fewer have attempted to put an economic value on this service.

Kellermann et al (2008) used an exclosure experiment to prevent access to coffee plants by birds, and measured the degree of pest infestation on exclosed and exposed plants. They found that infested fruits on coffee plants that remained exposed were 1 to 14 per cent less than those plants where birds were excluded. Thus, the service provided by the avian community resulted in a greater quantity of saleable coffee beans that had an average economic benefit of US$75 per hectare at the time of the study, excluding additional savings, such as the value of reduced insecticide application. In this study, Kellermann et al (2008) estimated the economic value of bird predation to coffee farmers by first quantifying each farmer's average increase in saleable berries resulting from reductions in borer infestation. This value was translated into an economic contribution of birds to each farm by multiplying the increase in saleable berries by the mean production value per hectare per farm. In this example, the biodiversity that is being given this specific value is 17 bird species considered likely to contribute to pest control services (excluding hawks, hummingbirds and other non-insectivorous species). This valuation of biodiversity is based on the principle of avoidance costs (see Table 10.1), where the value of the biodiversity is the avoided cost of pesticide application due to the presence of birds.

In our second example, Ricketts et al (2004) considered the economic value of biodiversity associated with forest fragments adjacent to coffee fields in

Costa Rica. Increased quantity and quality of fruits (fewer peaberries) from increased pollinator activity was used as the primary ecosystem service of interest. By using pollination experiments, they found that forest-based pollinators increased coffee yields by 20 per cent within 1km of the forest and reduced the frequency of peaberries by up to 27 per cent. Ricketts et al (2004) combined the results of their experiment with data on farm yields and market prices to estimate the income contributed by neighbouring forest patches by assuming that pollination effects from the forest extend 1km into a coffee field. They calculated the income resulting from these patches as the area of the coffee farm within 1km from the coffee patch multiplied by the net increase in yield within this distance multiplied by the net income per unit of coffee. For the 1000ha farm where they conducted this study, the value of the two patches of forest totalling 157ha adjacent to the coffee was estimated at US$62,000 per year (7 per cent of the total farm income). Managers of PES have to be careful of hidden costs, however, as coffee physiologists suggest that if a plant sets more fruits with increased pollination, then it will likewise require greater fertilization in order to maintain this increased production. The compensation will probably not be total; but not taking this hidden cost into account would be misleading.

Ricketts (2004) provides a second piece of evidence not included in their economic analysis. In the Kellerman bird example, the biodiversity responsible for providing the pest control service is 17 species of birds. Pollination services, in contrast, were attributed to 11 species of bees, including the introduced European honey bee. In the study, the ten native species appear to be quite dependent on adjacent forest fragments, with the majority of their contribution to pollination targeting coffee plants located < 100m from the forest edges. In contrast, the generalist European honey bee is found throughout the coffee farm (> 800m from the forest edge). During the two years of the study, Ricketts (2004) observed that the majority of pollination service was provided by the European honey bee irrespective of location (edges and centre of the coffee), although pollination frequency at the centre of the farm is half that of locations on the forest edges. During the second year of this study, the honey bee exhibited an unexplained collapse, with a steep drop in the number of honey bees observed. Interestingly, the ten native species that provided secondary pollinator services increased their contribution to pollination near the forest edges in the absence of the honey bee with no noticeable drop in pollination services provided. In contrast, no native species were available to replace the European honey bee in the centre of the farms where more than 50 per cent of the already reduced pollination service was lost. In this case, the bee biodiversity not only contributed to increasing production through pollination, but increased the resilience of the services by having additional species on hand who were capable of maintaining continued pollination in the absence of the dominant pollinator. Thus, in addition to providing pollination, the bee community provided a sort of pollination insurance whose value has not been quantified.

Biodiversity and Energy Production: Economic Value

The Reventazón River in Costa Rica where the Costa Rican Institute of Electricity (ICE) managed several hydroelectric dams and is currently building several new ones (see Chapter 2), is another example. When completed, this single river in Costa Rica is expected to provide 35 per cent of the national energy need. Latin American rivers are silty by nature, a consequence of the frequent and often torrential rains. This is exacerbated by land-use conversion that can generate above average erosion events. Silt negatively affects electricity production in two ways. First, it fills some of the space in the reservoir that could be occupied by water, thereby reducing the water-holding capacity of the dam. Second, in order to remove the silt, dam operators are required to open the floodgates, releasing the silt along with the water stored behind the dam. In both cases, this represents a measurable loss in potential electricity generation. In a sense, ICE knows exactly how much it can invest in erosion control techniques, such as reforestation of steep slopes and agroforests with high tree densities, since this amount should be less than the value of the electricity generation lost to siltation. In this case, the value of biodiversity can be calculated as the amount of sediment retained by different land uses, including natural and agricultural ecosystems, since biologically diverse forest systems are among the best at regulating water flows and holding back sediments. Estrada and DeClerck (2011), using the same watershed, demonstrate that combining targeting for biodiversity and sediment control in this situation can provide a win–win of sorts. By mapping coffee farms that are critical for restoring biological connectivity, and mapping erosion hotspots, they were able to identify priority areas in the Reventazón watershed where joint payments by ICE and conservation groups could be bundled to increase the impact upon both conservation and erosion control.

A very similar and fascinating example of this is the Panama Canal, which operates on the same principle, and where large volumes of water are needed to operate the canal, which is, in essence, a series of dams. The canal has tremendous economic importance, with an average of US$65,000 paid per vessel and total revenues contributing up to 40 per cent of the national economy by some estimates (Dean, 2005). The passage of a single cargo vessel through the canal has a high ecosystem service cost, with 52 million gallons of water needed for each passage and up to 40 passages made per day. In this sense, sedimentation of the main reservoir, Gatun Lake, has very real economic cost in reducing the amount of water that can be stored in the reservoir, as well as in the cost of dredging the main canal to ensure sufficient depth for ship movement. Economic value can be assigned to the land uses around the lake both in terms of their capacity to hold onto and slowly release water, and in their capacity to control erosion, two distinct ecosystem services that are typically greater in natural forests and complex agroforests. In 1985, the 250,000ha Chagres National Park, which encompasses a large portion of the canal's watershed, was formed, an event which Stanley Heckadon of the Smithsonian Tropical Research Institute refers to as 'the day that Panama bought an insurance policy on the Canal'.

In these examples, rather than attempting to place an economic value on biodiversity itself, an impossible task, many would argue, the value is placed on the service received by distinct ecosystems. From a purely mechanistic and ecological point of view, the capacity of distinct ecosystems, including agro-ecosystems, to provide specific though usually multiple services affords the most direct means of valuing biodiversity. In our opinion, the future of PES depends on new and improved means of quantifying the relationship between biodiversity and ecosystem services, as well as in identifying stable markets for these services.

Conclusions

In this chapter we have argued that the real economic value of biodiversity is in its functional contribution to the provisioning of ecosystem services. We distinguish between these very mechanistic and functional values from biodiversity's intrinsic value, which can be defined as our moral obligation to protect global species biodiversity. In some circles, the notion of payment for ecosystem services emerged as a means of recognizing the rather concrete contributions that biodiversity makes to human well-being. Clean water flowing out of our taps 24 hours a day is a service that most people recognize as a valuable (if not a fundamental) human right, in contrast to the conservation of the three-toed sloth, which may mean little to many people, but which may be invaluable to others.

Although payment for ecosystem services in the Neotropics recognizes biodiversity as an ecosystem service, in reality the services paid for are carbon sequestration, clean water and energy, and scenic beauty. The biological diversity that is conserved with these interventions is more typically viewed as value added, or as the mechanism through which other services are provided. As PES continues to develop, several key gaps remain, the first of which is the development of better methodologies that tie ecosystems and their associated biodiversity to the provisioning of specific ecosystem services. Second, there must be continued discussion as to whether biodiversity piggybacking on ecosystem services is a reliable and effective method to protect biodiversity.

In terms of agricultural landscapes, which are the focus of this book, there is a further need to understand how on-farm interventions affect services that are provided at the farm level, compared to those provided at the local scale and higher. Economic values for services provided at the farm level, such as pest control and pollinator services described here, require different valuation tools than those provided at the local level, such as hydrological services.

Acknowledgements

We are grateful for the financial support provided by the European Union via CAFNET (Connecting, Enhancing and Sustaining Environmental Services and Market Values of Coffee Agroforestry in Central America, East Africa and India) and PolicyMix (Assessing the Role of Economic Instruments in Policy Mixes for Biodiversity Conservation and Ecosystem Services Provision) collaborative projects, and by the French Agence Nationale de la Recherche (ANR) through the SERENA (Environmental Services and Rural Land Uses) project. Support was also provided by Pôle de Compétences en Partenariat (PCP): Agroforestry Systems with Perennial Crops, CIRAD–CATIE–INCAE–Bioversity–CABI–Promecafé.

References

Adamson, M. (2001) 'Cuánto vale un Parque Nacional? Economía experimental y método de valoración contingente', in *Revista Ciencias Económicas*, vol XXI, no 1–2, Instituto de Investigaciones en Ciencias Económicas de la Universidad de Costa Rica, Costa Rica

Alpízar, F., Carlsson, F. et al (2008) 'Anonymity, reciprocity, and conformity: Evidence from voluntary contributions to a national park in Costa Rica', *Journal of Public Economics*, vol 92, no 5–6, pp1047–1060

Barbier, E. B., Baumgartner, S., Chopra, K., Costello, C., Duraiappah, A., Hassan, R., Kinzig, A., Lehman, M., Pascual, U., Polasky, S. and Perrings, C. (2009) 'The valuation of ecosystem services', in S. Naeem, D. E. Bunker, A. Hector, M. Loreau and C. Perrings (eds) *Biodiversity, Ecosystem Functioning, and Human Wellbeing: An Ecological and Economic Perspective*, Oxford University Press, Oxford, UK

Bunker, D. E., DeClerck, F., Bradford, J. C., Colwell, R. K., Perfecto, I., Phillips, O. L., Sankaran, M. and Naeem, S. (2005) 'Species loss and aboveground carbon storage in a tropical forest', *Science*, vol 310, no 5750, pp1029–1031

Carlsson, F. (2010) 'Design of stated preference surveys: Is there more to learn from behavioral economics?', *Environmental and Resource Economics*, vol 46, no 2, pp167–177

CBD (1992) Convention on Biological Diversity, text available at www.cbd.int, accessed 17 February 2011

Costanza, R., d'Arge, R., de Groot, R., Farber, S., Grasso, M., Hannon, B., Limburg, K., Naeem, S., O'Neill, R. V., Paruelo, J., Raskin, R. G., Sutton, P. and van den Belt, M. (1997) 'The value of the world's ecosystem services and natural capital', *Nature*, vol 387, no 6630, pp253–260

Daily, G. (ed) (1997) *Nature's Services: Societal Dependence on Natural Ecosystems*, Island Press, Washington, DC

Dean, C. (2005) 'To save its canal, Panama fights for its forests', *New York Times*, 24 May

Estrada, N. and DeClerck, F. (2011) 'Payment for ecosystem services for energy, biodiversity conservation and poverty alleviation', in J. C. Igram, F. A. J. DeClerck and C. Rumbaitis del Rio (eds) *Ecology and Poverty*, Springer, New York, NY

Faure, G. and Le Coq, J. F. (2009) *Estrategias de las cooperativas cafetaleras frente a los sellos ambientales en Costa Rica: informe en el marco del proyecto CAFNET*,

CIRAD, Marzo
Flynn, D. F. B., Gogol-Prokurat, M., Nogeire, T., Molinari, N., Richers, B. T., Lin, B. B., Simpson, N., Mayfield, M. M. and DeClerck, F. (2009) 'Loss of functional diversity under land use intensification across multiple taxa', *Ecology Letters*, vol 12, no 1, pp22–33
Freeman, A. M. (1993) *The Measurement of Environmental and Resource Values, Theory and Methods*, Resources for the Future, Washington, DC
Goulder, L. H. and Kennedy, D. (1997) 'Valuing ecosystem services: Philosophical bases and empirical methods', in G. C. Daily (ed) *Nature's Services: Societal Dependence on Natural Ecosystems*, Island Press, Washington, DC
Hearne, R. and Motte, E. (2001) *The Use of Choice Experiments to Investigate Public Preferences for Biodiversity Conservation within a Framework of Environmental Services Payments*, Post-doctoral research associate, CATIE, Costa Rica
Holdridge, L. R. (1967) *Life Zone Ecology*, Tropical Science Center, San José, Costa Rica
Kellermann, J. L., Johnson, M. D., Stercho, A. M. and Hackett, S. C. (2008) 'Ecological and economic services provided by birds on Jamaican Blue Mountain coffee farms', *Conservation Biology*, vol 22, pp1177–1185
Laliberte, E., Wells, J. A., DeClerck, F., Metcalfe, D. J., Catterall, C. P., Queiroz, C., Aubin, I., Bonser, S. P., Ding, Y., Fraterrigo, J. M., McNamara, S., Morgan, J. W., Merlos, D. S., Vesk, P. A. and Mayfield, M. M. (2010) 'Land-use intensification reduces functional redundancy and response diversity in plant communities', *Ecology Letters*, vol 13, no 1, pp76–86
Loreau, M., Naeem, S. and Inchausti, P. (eds) (2002) *Biodiversity and Ecosystem Functioning, Synthesis and Perspectives*, Oxford Biology, Oxford University Press, Oxford, UK
MEA (Millenium Ecosystem Assessment) (2005) *Our Human Planet: Summary for Decision Makers*, Island Press, Washington, DC
Mitchel, R. and Carson, R. (1989) *Using Surveys to Value Public Goods: The Contingent Valuation Method*, Resources for the Future (RFF), Washington, DC
Moreno, M. L. (2005) *La valoración económica de los servicios que brinda la biodiversidad: La experiencia de Costa Rica*, Inbio, Costa Rica, http://inbio.ac.cr/otus/pdf/valoracion-economica-biodiversidad-cr.pdf
Moreno Diaz, M. L., Salas Pinel, F., Otoya Chavarría, M., González Brenes, S., Cordero Rodríguez, D. and Mora Salas, C. E. (2010) *Análisis de las Contribuciones de los Parques Nacionales y Reservas Biológicas al desarrollo socioeconómico de Costa Rica*, UNA, CINPE, SINAC, Heredia, Costa Rica
Moguel, P. and Toledo, V. M. (1999) 'Biodiversity conservation in traditional coffee systems of Mexico', *Conservation Biology*, vol 13, no 1, pp11–21
Naeem, S. (2002) 'Ecosystem consequences of biodiversity loss: The evolution of a paradigm', *Ecology*, vol 83, no 6, pp1537–1552
Perrings, C., Baumgartner, S., Brock, W. A., Chopra, K., Conte, M., Costello, C., Duraiappah, A., Kinzig, A. P., Pascual, U., Polasky, J. T. and Xepapadeas, A. (2009) 'The economics of biodiversity and ecosystem services', in S. Naeem, D. E. Bunker, A. Hector, M. Loreau and C. Perrings (eds) *Biodiversity, Ecosystem Functioning, and Human Wellbeing: An Ecological and Economic Perspective*, Oxford University Press, Oxford
Philpott, S. M., Soong, O., Lowenstein, J. H., Pulido, A. L., Lopez, D. T., Flynn, D. F. B. and DeClerck, F. (2009) 'Functional richness and ecosystem services: Bird

predation on arthropods in tropical agroecosystems', *Ecological Applications*, vol 19, no 7, pp1858–1867

Ricketts, T. H. (2004) 'Tropical forest fragments enhance pollinator activity in nearby coffee crops', *Conservation Biology*, vol 18, no 5, pp1262–1271

Ricketts, T. H., Daily, G. C., Ehrlich, P. R. and Michener, C. D. (2004) 'Economic value of tropical forest to coffee production', *Proceedings of the National Academy of Sciences of the United States of America*, vol 101, no 34, pp12579–12582

TEEB (The Economics of Ecosystems and Biodiversity) (2010) *Ecological and Economic Foundations: An Output of TEEB*, The Economics of Ecosytems and Biodiversity (Pushpam Kumar, ed.), Earthscan, London

Venter, O., Laurance, W. F., Iwamura, T., Wilson, K. A., Fuller, R. A. and Possingham, H. P. (2009) 'Harnessing carbon payments to protect biodiversity', *Science*, vol 326, p1368

Villalobos, A. L. and Solano Valverde, V. (2007) *Tarifa Hídrica: Historia de un ejemplo pionero en América Latina*, Empresa de Servicios Públicos de Heredia S.A., Heredia, Costa Rica

11

PES and Eco-Label

A Comparative Analysis of Their Limits and Opportunities to Foster Environmental Services Provision

Jean-François Le Coq, Gabriela Soto and Cliserio González Hernández

Introduction

During the last decades, preoccupation with environmental issues has increased worldwide, leading to the development of instruments to promote environmentally friendly practices and environmental services (ES) provision. Two instruments are particularly innovative in the promotion of ES: eco-labelling of products and payment for environmental services (PES) schemes. During the last decade, both instruments have undergone rapid development. In the late 1990s, eco-labels in food and non-food products were negligible. By 2006, certified coffee alone represented approximately 4 per cent (more than US$434 million) of world green coffee exports (Giovannucci, 2008). In addition, since the late 1990s, implemented PES schemes grew to more than 300 by 2002 (Pagiola and Platais, 2002).

Compared to other instruments designed to foster ecosystems conservation, such as integrated development conservation programmes or command-and-control instruments, these instruments share two common characteristics: positive economic incentives and direct conservation orientation (Wunder, 2006). Nevertheless, there are few systematic and comprehensive analyses to compare these two instruments.

This chapter compares current eco-label and PES limits and opportunities in maintaining or promoting ES according to a sustainable development perspective. Based on a literature review of published articles in academic reviews and specific reports available on the internet or edited by local or international institutions, as well as interviews of stakeholders involved in the mechanisms development, the analysis concentrates on the main existing and documented PES schemes, as well as the most developed and documented regional case of eco-label certification: coffee.

Starting from their origin, definition and scope, these two instruments are compared in terms of their structure and functioning; their effectiveness, efficiency and socio-economic impacts; and their conditions of sustainability. A discussion of the main limits of both mechanisms and further research proposal ideas conclude the chapter.

Origin, Definitions and Scope of Development in Mesoamerica

Origin, definition and basic principles

Eco-label is one type of 'standard' that informs consumers of the intrinsic and extrinsic properties of a product. Eco-label copes with information asymmetries in the commodity chains between product producers and consumers regarding the environmentally friendly nature of the production process. As a type of private standard, eco-label has the following characteristics (Pattberg, 2005): it is *voluntary* in nature, since the choice to adopt it is free and depends on the willingness of the producer; and it is based on the *compliance* of some *product and process requisites*.

The PES mechanism has been developed as an alternative conservation promotional tool due to donor disappointment regarding the poor results of existing biodiversity-oriented supports (Wunder, 2006). For PES, we consider the most used definition, proposed by Wunder (2006), where it is defined as 'A *voluntary transaction* where *well defined ES* is bought by a service buyer from a service provider, if and only if a service provider secures service provision (*conditionality*).' Thus, although their origin and initial purposes are different, eco-label and PES mechanisms share common basic principles: both are voluntary and contractual market-based instruments.

Scope of development in Mesoamerica

In Mesoamerica, eco-label and PES mechanisms are well developed. The existing coffee eco-label is particularly well established in the form of Organics, Fair Trade, Utz Kapeh, Bird Friendly, Rainforest Alliance, Starbucks CAFE, Nespresso AAA, the Common Code for Coffee Community (4C), etc. (see Chapter 15 in this volume). To a lesser extent, cocoa, pineapple, banana and the forestry sector (Forest Stewardship Council) eco-labels are also developed.

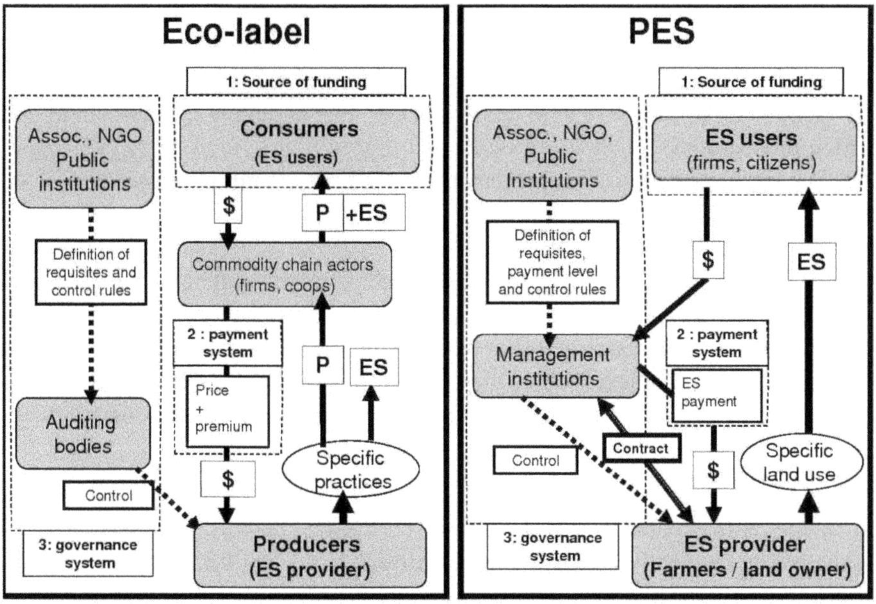

Figure 11.1 *Structure and functioning features of eco-label and PES mechanisms*

Notes: $ = financial flow; P = product flows; ES = environmental services flows.
Source: chapter authors

Mesoamerican PES mechanisms are well documented. Mayrand and Paquin (2004) inventoried eight Central American PES schemes. Vargas et al (2007) listed 12 local hydraulic PES schemes in Guatemala. Other local experiences include the Costa Rican Water and Power Utility Company in Heredia (ESPH); the various local PES schemes developed through the PASOLAC project in Nicaragua, El Salvador and Honduras; and the Honduran hydrological PES developed within the FOCUENCAS II project framework (see Chapter 16). Existing literature illustrates the multiplicity of experiences and the predominance of hydrological service-oriented schemes in the Mesoamerican region.

In the following section, we focus on two PES schemes with the most available information: PES-Costa Rica (see also Chapters 7, 12 and 14) and the payment for hydrological ecosystem services scheme in Mexico (PHES-Mexico) (see also Chapter 13), and highlight a few of the local experiences mentioned above.

Structure and Functioning

Since eco-label and PES mechanisms share basic principles, their structure and functioning have a comparable layout (see Figure 11.1). Both mechanisms tend to create an economic flow between two economic agent types: ES providers and ES users or consumers. With regard to eco-label, ES providers are producers who must comply with specific practices that lead to the production of so-called 'certified products' and associated ES. ES providers of PES schemes are farmers or landowners who have to develop or maintain a specific land use, which provides ES.

The basic structure of both mechanisms is similar and is composed of:

- a funding source;
- a payment system for ES providers (producers/farmers/landowners); and
- a governance system, composed of diverse actors (associations, NGOs and/or public institutions) that assume the mechanism regulating function by defining the rules, such as the specific practices or land-use requisites, as well as producer/ES provider compliance control and monitoring rules.

ES funding source

In this section we analyse and discuss the nature, level, characteristics and limits of the respective funding sources for eco-label and PES mechanisms (see Table 11.1).

The eco-label ES payment is embedded (internalized) in the final price of the eco-labelled products. Thus, the 'buyer' of ES is the *final product consumer*, who agrees to pay a premium for eco-labelled products that comply with specific requisites compared to 'conventional' products. This results in various consequences. First, the eco-label requisites address different services (water quality, soil erosion, biodiversity, etc.); but the additional payment is not specific to one ES. Second, funding is not specific to environmental contribution, with many certifications including social requisites such as Starbucks CAFE practices, Utz Kapeh or Rainforest Alliance certification (see Chapter 15). Third, as the ES contribution is embedded in the price of a product, the consumer willingness to pay a premium price for the eco-labelled products depends not only on ES provision, but on specific product attributes such as organoleptic quality. Consumers generally do not want to pay a high premium for a poor-quality coffee. Finally, the funding source level also depends on the coffee sellers' marketing strategies that define the final price of the eco-labelled products in comparison to conventional products and their market competitors.

The PES 'buyer' is theoretically the final ES user (Wunder, 2006). According to the ES considered, the final users may be different, as well as the nature of funding. For some ES that have global public goods characteristics, such as carbon sequestration to mitigate greenhouse effects, the ES beneficiaries include all societal classes worldwide. Practically all carbon funding sources are generally enterprises, from industrial countries, that compensate for their CO_2 emissions through the purchase of credits on regulatory or voluntary carbon markets (see Chapter 9). Hydrological services are mainly local and usually associated with a specific water use, such as human consumption, where the final users can be better targeted. The funding sources are generally

Table 11.1 *Principles and source of funding for eco-label and PES mechanisms*

	Eco-label	*PES*
Principles	ES provision associated with final products.	ES provision associated with specific land use.
Source of funding	Final consumer through commodity chain actors.	Final ES beneficiary, in theory, generally through intermediaries.
Nature of funding	Price of the eco-labelled product, including an additional price (premium) in comparison with conventional products.	Variable according to ES (carbon credit, water tariffs, park fees, earmarked tax, etc.).
Characteristics and limits	Funding not specific to one type of ES (bundle) and reflects various dimensions, such as social factors. Level of funding depends on consumer willingness to pay for higher prices for eco-labelled products (for some standards, no extra prices are paid by consumers). Source of funding included in already existing commodity markets. Level of funding is limited by the consumer willingness to pay (WTP). Level of funding is affected by global commodity market dynamics.	Funding is specific to ES type with the possibility of bundling. Level of funding depends on ES users' WTP for ES. Source of funding requires new markets to be developed. Identifying an effective funding source is difficult for some ES. Cost of access to funding source is difficult and expensive. Need for specific investment to identify funding source and to develop new markets (rules and institutions setting).

water supply or hydraulic power plant enterprises that can include an environmental services contribution in the water or electricity tariff paid by consumers (see Chapter 7). Few specific funding sources currently exist for biodiversity except conservation areas' entry fees or bio-prospecting payments (see Chapter 10). Funding sources can include tax earmarks, particularly in national PES programmes. PES-Costa Rica includes part of the oil tax paid by citizens to compensate for carbon emissions and PHES-Mexico has an obligatory water-use payment. Aside from final ES user funding sources, many PES mechanisms' specific projects are financed in their early development stage through multilateral or bilateral cooperation funds (see Chapter 16).

Due to the nature of the funding source, some specific and common limits of the funding mechanisms can be highlighted (see Table 11.1). The level of eco-label funding depends on:

- Market demand and the final consumers' willingness to pay a premium. The funding source is subject to general economic development in consuming countries, and consumers' awareness of environmental problems. Some certifications do not lead to premiums, such as Global Gap or 4C, which are requisites for supermarket brands to enter Europe.
- Global offer/demand for the product and the evolution of international market prices. The balance produces funding fluctuations. For example,

when the price of coffee is low due to global market overproduction, the possible funding to reward coffee producer practices is limited by the commodity price and the premium paid by the consumer.
- The marketing strategy and marketing power of the eco-label promoters. For example, the Smithsonian Institute's Bird Friendly eco-label did not develop a large demand, whereas Starbucks CAFE practices eco-label developed a higher market demand. Theoretically, the final consumer is the funding source of eco-label products; but the development of eco-labels also benefited from cooperation agencies' financial support. This is particularly true for organic or fair trade eco-labels where support programmes or projects are generally linked with poverty alleviation.

PES funding mechanisms are also subject to specific limits. First, the funding source can be difficult to define, as in the case of biodiversity, where final users are difficult to identify and the direct willingness to pay is still limited. Second, the access to the financial source can be difficult and costly, as in carbon credits in the regulated market (see Chapter 9). Third, no matter the ES, PES mechanisms rely on the identification of new funding sources and the creation of new markets, whereas the eco-label mechanism's funding source is derived from the existing commodity market. Thus, PES funding source development requires important initial financial and institutional investments in demonstrating ES provisions, identifying funders, measuring a willingness to pay, agreeing on a funding scheme, and designing rules and a specific governance system.

Finally, in both cases, information flow between funders and ES providers is generally limited to the consumers' certification requisites knowledge and the final ES users' PES schemes knowledge. Thus, information flows are still a crucial issue for both mechanisms. Lack of information may affect the trust of the final ES consumers/users upon the mechanism effectiveness and may affect their willingness to pay premium prices for ES. The key factors necessary to build trust between all actors involved are effectiveness and efficiency of the governance system, accuracy and credibility of the control and monitoring system, and scientific evidence of ES provision through land-use practices. Thus, the distance between the funding source and the service provider has led to the necessity to build a trustworthy institutional set-up (governance system), which is time consuming and costly.

Nature and level of payment

In this section, we compare the nature and level of the payment received by ES producers, which are the producers in the case of eco-label or the landowners in the case of PES (see Table 11.2).

Eco-label payment is based on a product unit (i.e. US$ per kilogram), which is defined in the transaction between producers and buyers. This has various consequences in terms of payment level received by the ES provider. First, the potential level of payment depends on production volume, but not

Table 11.2 *Nature and level of payment*

	Eco-label	*PES*
Nature of payment	Purchase of certified product though commercial transaction with product buyer of the commodity chain.	Contractual payment.
Level of payment	Payment based on product unit sold by the producer (US$/kg). Buying price of product including a 'premium'.	Payment generally per land unit area (US$/ha).
Characteristics and limits	Differential payment according to eco-label, but also other factors such as quality of product, market demand, etc. No differential payment according to environmental interests of the area and intensity of ES provision. No guarantee on the level of payments due to: • fluctuation of the production at farm level; • fluctuation of commodity market price; • no guarantee of the volumes of product sold effectively as certified products; • no insurance of premium level for all certified production; • no guarantee of full transmission of the premium due to asymmetries in the commodity chain.	Possible differentiated payment according to investment level, opportunity costs and intensity of ES provision. • Possible targeting and modulation of payment according ES provision. • Guarantee of multi-year payment through contract schemes. • Guarantee of contractual payment may be affected by institutional or political changes. • Levels of payment are based on negotiation among stakeholders and may better reflect a compromise between leading actors than opportunity costs or 'real' ES value.

directly on the ES agroecosystem provision. Second, the payment level depends on the market commodity price. For example, the international coffee market experienced large price fluctuations during the last decade, which ranged from less than US$40 per bag in 2002 to 2003 to more US$140 per bag in 2009, whereas the premium price paid to farmers for Fair Trade organic coffee had reached as much as US$100 per bag during the 2002 to 2003 coffee crisis period, and dropped to US$10 to $20 per bag or less in a context of global high price during the 2007 to 2008 international market (Le Coq et al, 2009). Third, the level of payment effectively received by farmers depends on the volume of products sold as certified products. Thus, it depends on market demand for certified products and the ability of producers or their organizations to find buyers that are interested in certified products. Producers commonly have to sell a part of their certified production as conventional production. For example, in 2006, the volume of certified coffee production was, depending on the certification, up to six times higher than the volume effectively sold as certified products (authors, based on Pratt and Kilian, 2008). Fourth, the level of payment received by the producers depends on the negotiation between producers and its buyer. Thus, it depends on the governance of the commodity chain, which affects the additional price (or premium) transmission efficiency paid by consumers to producers. The global governance of the commodity chain and the structure of power affect the distribution of the added value among the commodity chain actors. In most of the global commodity chain (and eco-labelled products, as well), the margin distribution between producers, processing and marketing agents (wholesalers, retailers) is generally unbalanced. The share of income received by the producers is generally lower than those received by processing, wholesale and retail. Moreover, except in the case of Fair Trade certification, there is no regulation and control of the premium paid to the producers. The additional price paid by the consumer may not be totally reflected at the producers' price level. At the farm level, the payment received by producers is based on the volume sold of one farm product, rewarding one part of the producers' activities, whereas the ES production depends on all the activities developed. For example, producers of certified coffee have to develop shade trees in their field, infrastructures to limit soil erosion on their farms (e.g. Rainforest Alliance) or not to use any chemical fertilizer throughout the entire farm (e.g. organic); but they only receive payment for their coffee production, although certification implies systemic investments or restrictions that affect all of their activities.

The payment in PES mechanisms is set by contractual agreement between the ES providers and the management unit of the PES mechanism. The payment is generally defined according to a land unit area (US$ per hectare) and is paid on a pluri-annual basis according to defined rules. These rules have different consequences. First, the PES mechanism payment level is guaranteed for the duration of the contract signed between the landowner and the agency in charge of implementing the mechanisms. It is thus theoretically secured for several years for the landowner. For example, PES-Costa Rica's contractual agreement secures payments ranging from US$64 ha^{-1} y^{-1} over five years for the conservation modality up to US$816 ha^{-1} over ten years for the reforestation modality (Pagiola, 2008). Moreover, as payment is set in a contract, the level of payment can be defined according to the ecosystems and practices and better reflects the differential investments and opportunity costs for the landowner. For example, PES mechanisms developed by the Water and Power Utility Company of Heredia (ESPH) consider different payment levels according to activity type: US$92 ha^{-1} y^{-1} over ten years are paid for conservation of existing forest or regeneration and US$854 ha^{-1} y^{-1} over five years for a ten-year reforestation contract (Villalobos and Solano Valverde, 2007). Third, PES mechanisms payment can be targeted toward zones of specific environmental interest and modulated to better reflect the intensity of ES provision (see also Chapter 13). Lastly, the level of payment reflects the power of negotiation between leading groups representing providers and beneficiaries of ES in the rule-setting mechanisms. As payment level is the result of a negotiation, it may differ from the ES provision opportunity costs or ES user willingness to pay.

In both mechanisms, the main issue is the attractiveness of the payment as an ES provision incentive. Whereas PES schemes enable differentiation and targeting of payment according to ES provision, eco-label payment depends more on market dynamics and commodity chain functioning. PES schemes offer more guarantees in terms of payment level and security to the ES provider according to short- and medium-term timeframes in comparison with eco-label, and engage the ES provider in a medium-term commitment. In both cases, the governance system is a key factor for the adjustment of the nature and level of payment.

System of governance

The eco-label and PES mechanism system of governance consists of three elements:

1 the rules-setting process;
2 the monitoring, control and evaluation process; and
3 the conditionality (see Table 11.3).

Table 11.3 *Rules-setting, monitoring, control and evaluation, and conditionality processes*

	Eco-label	*PES*
Rules-setting	Various forms of rules-setting processes according to eco-label (enterprise, NGO/association or public-led process). Diversity of process of consultation. → Trade-off between environmental and trading interests.	Various forms according to PES (state, NGO/association led or enterprise led). Diversity of form (formal setting to more informal). → Trade-off between environmental and social interests.
Monitoring, control and evaluation	Included in certification process. Private independent auditors. Independent from buyers. → Possible variability in the interpretation of norms according to auditing and certification bodies. → Cost for farmers (or producer organizations).	Included as part of PES mechanism. Implementation/management body. Generally link with payment function. → Costs included in the functioning of the implementation/management structure.
Conditionality	Defined in the normative of the certification. Producers' compliance to specific management practices. → Generic criteria (all over the world). → Mix of environmental but also social or quality criteria. → Debate change practices or valorizing existing states.	Defined in contractual agreement signed between ES provider and management body. Beneficiary commitment to maintain a specific land-use state (conservation) or change land use (reforestation, tree planting). → Local definition of criteria. → Possible targeting of area (priority). → Limitation of access due to fund availability.

For each system of governance element, we describe their respective features for both mechanisms and discuss some of their implications.

Rules-setting

The eco-label key rules-setting component is the certification rules which encompass:

- requisite rules of definition compliance at the farm and commodity chain level (norm definition process);
- the requisite content (the norm); and
- the rules that govern the certification process itself (auditing and certification process).

Norm definition processes can be led by private enterprise, associations or NGOs, or by public institutions. For example, eco-label coffee requisites can be defined by the buyers, such as Starbucks CAFE practices; NGOs such as the Red de Agricultura Sostenible for Rainforest Alliance certification or the Fair trade Labelling Organization (FLO) for Fair Trade certification; or by public institutions such as organic certification (Regulation UE 2092/91 for Europe, or Organic Foods Production Act – Farm Bill 1990 for the US). The norm definition process can be more or less open to producer participation. The requisites are generally developed with an international scope, but can be fine-tuned to reflect local conditions.

PES mechanism rules-setting concerns:

- the definition process of conditionality and payment characteristics (which are included in the contract); and
- the monitoring and control rules definition process.

Much like eco-label, the PES mechanism rules-setting can be led by a private enterprise (e.g. ESPH), by NGOs or associations (e.g. Defensores de la Naturaleza for the Sierra de las Minas water fund in Guatemala; see Vargas et al, 2007), or by the state (e.g. PES-Costa Rica or PHES-Mexico). The decision structure and formality of the rules-setting process differs according to PES scheme types. In the national PES mechanisms/government-financed PES, such as PES-Costa Rica or PHES-Mexico, the conditionality, scope of application and level of payment are fixed by national legislation (law, decrees), and the procedures manuals are drafted by a specific administration (the PES programme management unit) and sector representatives. In local PES mechanisms/user-financed PES (association/NGO- or enterprise-led process), the rules definitions are prone to be set by a direct negotiation between the representative of ES users and ES providers.

The certification requisite/norm (eco-label) or conditionality (PES) definition processes rely on scientific evidence, as well as negotiations between representative ES providers and user stakeholders. This process reflects a trade-off between different objectives (environmental, social and economic). Eco-label maintains a trade-off between the potential number of providers (and available product quantity to sell) and environmental constraints. For example, the organic norms environmental requisites are very restrictive due to public institutions, ecological movements and consumer representatives dominating the definition process and imposing their safety and environmental concerns. With Nestlé's Nespresso AAA and Starbucks CAFE practices, coffee traders control the process and impose their concerns about coffee quality and adequate production supply. The PES mechanism trade-off is between the number and nature of potential beneficiaries (and, thus, social/political orientations) and environmental objectives. For example, many local Guatemalan PES schemes promoted and supported by cooperation agencies focus on poverty alleviation and therefore target higher poverty index areas. Concern for poverty issues in the PES-Costa Rica is growing; however, the predominant mechanisms' orientation is conservation and reforestation (Pagiola, 2008). This reflects the dominant forestry and environmental sectors' interests, who promoted PES creation while still maintaining the rule definition (Le Coq et al, 2010). PHES-Mexico has a more balanced emphasis on both conservation and water service provision, and poverty alleviation and benefit distribution (Wunder et al, 2008), and is marked by an evolution that reflects opposition between social forces (McAfee and Shapiro, 2010).

Monitoring, control and evaluation process

The eco-label monitoring, control and evaluation process is embedded in the certification process and is independent of the payment mechanism. Requisite compliance control in terms of farmers' practices is systematic and generally done on an annual or biannual basis. Requisite compliance verification is performed during the audit and is critical in obtaining certification. One critical issue is the independent nature and accuracy of the control, which is generally done by independent (private) auditing agencies. Although each eco-label is different, the tendency is to develop a control system with three separate functions (third-party certification):

1. The norms-setting organization is in charge of defining the certification process's requisites and rules of the game and the certifying bodies' accreditation schemes.
2. The certification bodies are accredited to deliver producer certifications according to audit results.
3. The auditing agency in charge of the producers' compliance control sets the norm standard (see Chapter 15).

ISO 65 regulates auditing and certification functions because monitoring and control are done by various enterprises among producers and countries, which can result in various interpretations of the same norms. Thus, the control criteria and methods standardization is an important issue for eco-label.

The certification process is independent of the payment system; therefore, the cost is charged to the producers. Certification costs (including auditing costs and investment costs for requisite compliance) affect the potential producers' mechanism access. For example, in 2004, Fair Trade certification costs (not including investment costs) were more than US$1000 per year for organic certification and between US$2000 and $5200 per organization according to size and type of co-operatives (Villalobos, 2004). The role of farmers' organizations is crucial in enabling small farmers' access to certification by grouping small farmers' certification processes in order to reduce individual certification costs. For example, organic certification costs represented US$20 to $50 per year per producer, respectively, for members of associations with 20 to 100 producers (Villalobos, 2004). Producers' organizations also benefit from economy of scale with regard to research transaction costs, negotiations and monitoring contracts with auditing bodies (Faure and Le Coq, 2009). If norm compliance monitoring and control are embedded in the certification process and done on a regular basis, evaluation of the effects of ES provision or socio-economic impacts are conducted more punctually and generally completed by scholars disconnected from the monitoring and control process.

The ES provider's contractually agreed-upon PES mechanism's compliance monitoring and control is generally conducted by management bodies, and focuses on contractual land-use or practices compliance. For example, for PES-Costa Rica, FONAFIFO is the management unit in charge of payment, but also monitors forest cover through remote sensing. A centralized geographic information system (GIS) enables follow-up of the parcel under contractual agreement (localization, owners, etc.). This information system enables ex-ante control by avoiding multiple payments on the same parcel, and by combining GIS information and satellite imaging to monitor the current land-use state. Remote monitoring and control is combined with regular field visits. As the cost of monitoring and control by the management body is included in the implementation bodies' structural costs, ES providers in the PES mechanism do not have to pay directly for control and monitoring costs.

Scholars generally evaluate the effects of PES on ES provision or socio-economic situations more punctually. Socio-economic evaluations are prone to development in project-funded PES because it is generally included in the project design. However, national PES such as PHES-Mexico can also develop a multi-criteria evaluation system.

For both mechanisms, the accuracy, quality and fairness of monitoring and control are critical issues since they affect the trust of the founder and the legitimacy of these instruments. In both cases, systematic evaluation of the effects of the mechanism in terms of ES provision and socio-economic impact is critical.

Conditionality

Eco-label conditionality is the producers' requisite certification compliance included in the norm. Two features characterize the certification process's conditionality and affect the scope and accessibility to farmers:

1 the requisites (norm criteria); and
2 the norm form of application.

The norm criteria encompass various dimensions, according to specific eco-label objectives. While most include a variable requisite set of management practices to ensure that producers adopt eco-friendly practices (i.e. a ban or reduction of chemical use, level and type of shade cover, etc.), many also include social criteria (child labour-use ban, respect for legislation on a worker's minimum wage, etc.) or quality criteria. Theoretically there is no limitation on the eco-label application scope or farmers' access to certification if they comply with requisites. However, some eco-labels, especially those developed by coffee traders, emphasize quality and adopt practices that restrict farmers' access, in specific production areas, to their certification programmes. The norm form of application can also differ. Organic or Fair Trade certification audits capture the state of norm compliance, and in the case of requisite non-compliance, the producers are not certified. For other certifications, such as Rainforest Alliance, Starbucks CAFE practices or Nestlé's Nespresso AAA, producers can be certified even if they do not comply with all the requisites. Farmers can enter 'by grade' according to their level of compliance. Thus, the second certification form is gradual and pays more attention to the improvement process of the farmers' practices, and in this regard is more inclusive, whereas the first form of certification pays attention to the current state of the producers' practices and is more exclusive.

The PES conditionality is set in terms of land-use characteristics or changes in land use (such as reforestation or afforestation). Two specificities of the PES mechanism can be highlighted in comparison with the eco-label mechanism. First, although the PES mechanism beneficiary can prove the conditionality compliance in terms of land use, payments are usually granted according to the PES priorities zones defined by their importance in ES provision, whereas in most cases there is no targeting for eco-label. Second, although PES mechanism producers comply with requisites, access to the PES mechanism is generally limited by the amount of funding sources available. For example, the PES-Costa Rica budget is derived from various available funding sources. Many forest landowners apply for PES mechanism funding; but only half of them receive payment due to limitations in global funding sources. Third, the conditionality application can take on different forms. Conditionality compliance can be verified *ex-ante* upon entrance into the programme. For example, in PES-Costa Rica reforestation or plantation PES modalities, landowners received payment to reforest before planting trees as long as they complied with administrative requisites. The compliance can be verified *ex-post*, when the landowner has already made changes, such as the Regional Integrated Silvo-pastoral Ecosystem Management Project (RIMESP), where producers were paid according to the evaluation of the changes that occurred on their farms.

The nature and application of conditionality are key issues for the application scope, access conditions and legitimacy of both mechanisms. First, the nature of conditionality reflects ES provision proxies generally based on previous studies (ES provision of one set of practices or one ecosystem type) because the evidence of relationships between farmers' practices (eco-label) or land uses (PES) and ES provision is generally poorly demonstrated locally. Second, requisite adaptations to local conditions are critical and require further research. Finally, the conditionality verification form is still debatable in both mechanisms. When the conditionality is set on a desired state (after investments carried out by landowners), it limits the ES provider's capacity to upgrade the situation, but provides a guarantee to ES users/funders (i.e. organic certification or *ex-post* verification PES). When the conditionality is set towards a desired evolution, the mechanisms may better enable the ES providers to change their practices, but provides less of a guarantee to ES users/funders (i.e. gradual certification, such as Rainforest Alliance or *ex-ante* verification in PES). The dilemma still exists between paying for what is already provided or to support changes for higher future ES provision.

Effectiveness, Efficiency and Socio-Economic Impacts

Here we compare the results of eco-label and PES mechanisms according to two criteria:

1 the effectiveness and efficiency of the mechanism in ES provision; and
2 the effects of the mechanism on the poor.

Effectiveness and efficiency in ES provision

Effectiveness and efficiency in ES provision by the two mechanisms are subject to debate. The comparison is difficult since the literature does not adopt the same criteria or put the same emphasis on evaluating their effectiveness and efficiency. However, we propose considering the following criteria mainly derived from PES literature (Pagiola, 2005; Engel et al, 2008):

- the capacity of the mechanism to realistically maintain or improve ES provision;
- the capacity of the mechanism incentives to change practices that may not have occurred without payment (*additionality*);
- the inadvertent risk of displacement activities damaging the environment outside the PES intervention geographical zone (*leakage effects*);
- the long-term effects (*permanence*); and
- the efficiency of the mechanism (see Table 11.5).

ES provision incentive capacity

The mechanism's capacity to maintain or improve environmental practices is controversial in both cases. Two questions can be raised that address the effec-

Table 11.4 *Effectiveness and efficiency of eco-label and PES mechanisms*

Criteria	*Eco-label*	*PES*
ES provision incentive capacity	Changes in farmers' practices noticeable for some eco-label (mainly organic). Debate on the link between norms' requisites and effective ES provision.	Debate on the magnitude of the effects in reduction of deforestation rate. Debate on the biophysical link between land use and effective ES provision.
Additionality	Not clearly addressed.	Controversial or poorly demonstrated in national PES programmes. More prone to be demonstrated in local or project PES programmes.
Leakage effects	Not addressed.	Variable according to PES mechanism.
Permanence	Theoretically, no problem of permanence since eco-label mechanisms are not set on a limited timespan. Possibility of unilateral farmer withdrawal from the certification scheme at any time.	No evidence of cuts/need for more studies.
Efficiency	Need for more evidence on the premium transmission among the commodity chain from consumers to producers. Risk of premium capture by intermediaries of the commodity chains. Lack of studies, including all the transaction costs of the system (normative development, certification and auditing costs, etc.).	Variable transaction cost evaluations according to evaluation methods. Importance of start-up transaction costs in small-scale PES. Some evidence of superior efficiency of PES mechanism compared with traditional conservation policy instrument.

tiveness of the mechanisms on ES provision:

1 Do the eco-label certification and PES mechanisms effectively change farmers' practices and land uses?
2 Do these farmers' practices or land uses really provide better ES provision?

Capacity to effectively change practices or land uses. There is varying evidence that eco-label eco-certification affects farmers' practices. A case study of Costa Rican coffee farmers shows significant differences in practices between certified and conventional farmers, especially for those with organic certification (Quispe, 2007). Nevertheless, the amplitude of the changes may be prone to be limited as farmers may elect the certification that is closest to their current practices in order to limit additional investment. With regard to PES, literature pays more attention to the effectiveness of PES mechanisms to protect forest or mitigate deforestation, and shows controversial results.

Relationship between practices/land uses and ES provision. In both cases, this relationship is poorly documented and sometimes controversial. Most of the eco-label norms define requisites based on previous agronomic studies, which, in turn, define restrictions in terms of soil management practices, fertilization and pesticide management, etc. All of these requisites are proxies of ES provision. If coffee agroforestry systems provide ES (see Chapter 15), there is still a need to improve measurement of its ES provision (Rapidel, 2008). In PES programmes, the linkages between land use and ES provision are little measured or are subject to controversy (Wunder et al, 2008). Most ES programmes are based on the concept that a natural ecosystem (generally, forests) provides all desirable ES. In particular, links between land use and hydrologic services are poorly demonstrated (Chomitz and Kumari, 1998; Bruijnzeel, 2004; Chagoya Fuentes, 2008).

In both cases, the main issues are to further analyse the amplitude of the changes to practices due to the participation in the mechanism; and to further address the link between specific practices or land uses and real ES provision. Further development of baseline studies and monitoring to measure the evolution of the practices in their context, as well as biophysical evaluation of ES provided by ecosystems in local contexts, should be conducted.

Additionality

The lack of additionality is the risk of paying farmers or landowners for changes that they would have done without a payment ('money for nothing'). This criteria is poorly documented in eco-label literature; but is often discussed in PES programmes since it is an eligibility criterion for access to regulatory carbon sequestration markets. For PES projects, or specific carbon-oriented PES programmes, additionality has been generally demonstrated. For example, an analysis of projects in Quindio, Colombia, under the RISEMP framework shows that PES recipients significantly altered land use in comparison with a control group of non-participants (Pagiola and Rios, 2008, cited in Wunder et al, 2008). For national and multi-objective PES programmes, the lack of additionality has been highlighted, such as PES-Costa Rica (Wunder et al, 2008) and PHES-Mexico (Alix-Garcia et al, 2009).

Leakage effects

The leakage (or spillage) effect is the risk that intervention at one local site displaces the problem to other sites. Leakage is only relevant when the spatial scope of the intervention is lower than that of the desired effects in terms of service (Wunder et al, 2008). Since the eco-label mechanism functioning is not site specific but international, the criteria is poorly addressed in literature. Further assessment is needed of whether eco-label product development has an adverse affect in a community, or on a larger scale, leads to site production intensification or has environmental affects in other regions or countries.

Currently, PES programme leakage problems are more often cited; but, in practice, few quantitative estimates exist except for specific carbon sequestration PES. Given their small size, most local user-financed PES are unlikely to induce indirect leakage effects due to their size; but the potential increases in government-financed programmes that are on a much larger scale (Wunder et al, 2008). Authors argue that there is limited qualitative evidence of leakage and that the perception of widespread leakage is often exaggerated (Chomitz, 2006; Wunder et al, 2008). In both cases, assessment of leakage effects is difficult, and even more so in terms of reliable measurement of trans-country indirect leakage (especially for eco-label and national PES).

Permanence

Permanence can be addressed in two complementary ways: the permanence of the mechanism's effect, which we consider a sustainability factor; and permanence as the capacity to maintain a benefit after the programme (and payment) ends (Swart, 2003, cited in Wunder et al, 2008). Eco-label permanence is theoretically high because producers can still enjoy participation in eco-label schemes and receive additional payments over time, as long as they comply with the certification norms. It can also be argued that there is no guarantee of eco-label permanence since farmers can withdraw from certification schemes at any time because there is no long-term contractual agreement to link them to the certification scheme. Thus, de-certification and a return to 'business-as-usual' practices are likely if requisites are difficult to achieve or benefits derived from certification are limited in comparison with costs.

PES permanence is a more debated issue as PES mechanisms are generally conceived as a programme, and incentives are paid on a long-term contract basis. Theoretically, there is no reason to believe that a service will be provided after payment ends, as permanence of the effects can be interpreted as an indicator of low additionality (Wunder et al, 2008). Nevertheless, in the case of adopting a new specific land use that provides economic income (e.g. reforestation or tree planting in agroforestry systems), permanence effects are more likely to be effective. In those cases, cash flow from exploitation can be derived to further tree replanting; however, there is little empirical evidence and analysis of these practices.

In both cases, analyses of permanence are still limited. Further analyses are necessary to analyse the activities of actors after they leave certification schemes (eco-label) or after the end of payment (PES).

Mechanism efficiency

Mechanism efficiency can be defined as the results amount divided by financial investments used to obtain this result. The efficiency depends on the achieved results, which could include ES provision as well as other objectives, such as poverty alleviation (see the following sections). Here, we consider the efficiency to provide ES provision. Efficiency can be analysed from two perspectives:

1 the internal financial efficiency (i.e. the ability to transfer most of the funding source to the beneficiary, which is considered transaction costs); and
2 the mechanism efficiency compared to other mechanisms.

Internal financial efficiency of the mechanism. Eco-label criteria used for assessing financial efficiency of the mechanism comprise the differential between the premium price paid by the consumer and the premium price received by the producers. This criterion, which captures the price transmission along the global commodity chains, is affected by existing asymmetries of power between actors of the commodity chain. Few systematic studies have been done to capture this price transmission. Nevertheless, observations tend to show that the premium tends to be captured more by downstream actors than by producers. For example, in 2004, the price difference paid at the farmer level between conventional and eco-label coffee was less than US$0.50 per pound ($lb^{-1}$), and the price difference of roasted coffee at the consumer level in Europe between conventional and eco-label coffee was up to US$1.50 lb^{-1} (authors, based on Villalobos, 2004).

For PES, particular attention has been paid to the transaction costs of the mechanism, which captures the administrative costs to deliver funds to ES providers (Wunder et al, 2008). National PES schemes transaction costs (i.e. costs of administration of the implementation body) have been regulated by law. PES-Costa Rica and PHES-Mexico administrative costs were limited in 2008 to 7 and 4 per cent of the payment, respectively (Wunder et al, 2008). More thorough evaluations tend to yield different values when intermediary costs and taxes paid by programme beneficiaries are included. For example, PES-Costa Rica transactions costs have been evaluated at between 12 and 18 per cent at the FONAFIFO level (Miranda et al, 2003; Rojas and Aylward, 2003) and up to 22 to 25 per cent of the payment, including all other taxes paid by the beneficiary (Baltodano, 2000, cited in Locatelli et al, 2008). PES literature also highlights the importance of start-up transactions costs, which are the costs to design PES schemes that can be particularly important for small-scale PES projects (Wunder et al, 2008).

Compared efficiency of the mechanism. Eco-label literature poorly compared the efficiency of eco-label with other tools in terms of the effect on ES. However, as PES has been developed to overcome the limitations of traditional conservation instruments, comparison of PES efficiency with other traditional conservancy instruments, such as land purchased, is more current. For example, authors show that PES-Costa Rica was more efficient at reaching conservation purposes than national parks land purchases (Tattenbach et al, 2006). Moreover, Engel et al (2008) point out two cases of limited efficiency:

1 the failure to adopt practices whose social benefits exceed their costs, which occurs when the payments are insufficient to induce adoption of socially desirable land uses; or
2 the adoption of practices whose benefits are smaller than their costs.

To cope with this possible limitation, better targeting of PES schemes is often proposed (Wünscher et al, 2008).

Finally, in both cases, better measurements are required to enable comparison. In particular, eco-label analysis of norm development costs (which can be assimilated in PES start-up transaction costs) should be developed, as well as integral analysis of transaction costs, including cost of auditing, etc.

Conditions of access and impacts upon the poor

Although it is not explicitly the primary goal of these mechanisms, the effects of eco-label and PES on poverty alleviation is a criteria of evaluation of interest for many scholars, actors and development agencies. In order to analyse the mechanism effect on poverty alleviation, we consider two criteria:

1 the condition of access and participation of the poor to the mechanisms; and
2 the direct and indirect effects generated by mechanisms' participation in livelihood conditions (see Table 11.5).

Access to the mechanism and participation of the poor

This is controversial for both eco-label and PES mechanisms. Access to certification could be considered an additional barrier to entry for export markets since poor farmers often lack human and/or financial capacities to comply with

Table 11.5 *Conditions of access and impacts of eco-label and PES upon the poor*

	Eco-label	*PES*
Access and participation	Controversial level of participation. Main limitation factors for access to the poor are the lăck of inversion and administrative capacity. Importance of co-ops for smallholders and poor farmers in terms of access.	Controversial level of poor participation depends on targeting of the programme. Main limitation factors for access to poor are lack of land title and administrative capacity. Importance of the role of intermediary organizations in access to small landowners.
Direct and indirect effects on livelihood	Effects on smallholder incomes are limited due to low production levels (a result of restricted land access and productivity). Evidence of some indirect effects.	Effects on incomes are limited due to limited land tenure. Indirect effects need to be further analysed.

requisites. Small producers generally have access to certification through producers' organizations that:

- finance the necessary investments to comply with requisites;
- strengthen the capacity of small farmers through specific training;
- assist farmers in preparing for the auditing process (organizing records, pre-certification evaluations, etc.); and
- generally assume certification costs for their members, such as auditing costs (Faure and Le Coq, 2009).

PES access and participation of the poor is also controversial. According to Grieg-Gran and Bishop (2004), markets for ecosystem services can affect the poor in the following ways:

- as buyers/users of ecosystem services;
- as employees of operations producing ecosystem services;
- as users of natural resources affected by a market initiative; or
- through more indirect effects (e.g. the impacts of changing land use upon food prices, rural wages or the multiplier effects of local purchases).

Whereas studies showed limited access to poor people in the PES scheme in Costa Rica (Zbinden and Lee, 2005), studies of PHES-Mexico demonstrated an important participation of marginalized population centres (Muñoz-Piña et al, 2005; see also Chapter 13 in this volume). The main problems that currently limit small farmers in accessing PES are the lack of information, land title and administrative capacity to comply with requisites (Pagiola et al, 2005). As for eco-label, the role of intermediary organizations is crucial in smallholders' access to PES. In Costa Rica, 50 per cent of PES contracts pass through local organizations that assume the technical and administrative role and pool PES contract demand.

Direct and indirect effects on the poor

The direct impacts of eco-label and PES on poverty are also controversial. Eco-label is considered an option to increase small farmers' incomes in the short term (Kilian et al, 2006). Nevertheless, the impacts on poverty are generally negligible due to low production intensity, which limits net additional incomes so that farmers are unable to escape from the poverty trap (Valkila, 2009).

For PES, evidence of direct incomes is also still controversial. In Costa Rica, Locatelli et al (2008) showed that the impact of PES reforestation upon the poorest landowners (small farmers and working-class landowners) was notably positive in most dimensions (cultural, institutional), except income, and that this impact was generally higher for small landowners than for upper-class landowners. Moreover, the strongest positive impacts were for landowners applying to the PES through a local NGO.

The comparison highlights the limitations of both mechanisms in terms of effects on poverty alleviation. As the poor are characterized by limited endowments and access to key factors, which are calculated on the payment level of both mechanisms, the level of payment is restricted: additional incomes from eco-label are limited by the low level of coffee production due to restricted land access and/or productivity, and payments received by smallholders in PES schemes are limited due to restricted land tenure. In both mechanisms, intermediary organizations (forestry or producer organizations, NGOs, co-operatives or associations) are key actors in facilitating small farmers' access to the mechanisms. Moreover, institutional frameworks, such as land tenure policy (especially for PES programmes) are crucial, as well as market and labour protection regulations. Better targeting is most commonly proposed to specifically improve both access to and impacts upon the poor, especially for PES mechanisms (Locatelli et al 2008). Nevertheless, a trade-off should be found between poverty and conservation objectives (Kosoy et al, 2007). Finally, the literature review revealed limited studies that made integral evaluations of the effects on poor farmers, including indirect (labour, etc.) and comprehensive effects (including all types of capital that affect livelihood conditions of the local population). Thus, there is a need for further evaluation of the economic effect of both mechanisms on poor livelihoods, including direct and indirect outcomes.

Conditions of Sustainability

As already mentioned, both eco-label and PES mechanisms are market-oriented tools that attempt to orient new funding towards conservation purposes. In this section, we compare and discuss the mechanisms' sustainability condition, taking into account two main features: their financial sustainability and their institutional sustainability (see Table 11.6).

Financial sustainability

Financial sustainability can be defined as the capacity of the system to maintain sources of funding over time. The eco-label mechanism's financial sustainabil-

Table 11.6 *Conditions of financial and institutional sustainability of eco-label and PES mechanisms*

	Eco-label	*PES*
Financial	Depends on coffee market demand and overall production. Equilibrium between increased market demand and certified offer.	Depends on capacity to connect various financing sources. Depends on international negotiation.
Institutional	Importance of confidence, trust and legitimacy of the mechanism. Political stability and legislative framework. Continuous learning and innovation processes.	

ity depends on the balance between offer and demand for specific segments of eco-labelled product markets, and thus the ability to maintain attractive premiums. The demand side depends on the final consumer willingness to pay for eco-label products, which reflects:

- the overall sensibility of the consumer towards environmental issues;
- the marketing investments that sensitize consumers to specific eco-labels; and
- the system confidence that depends on the credibility and proven effects of the eco-label.

The offer side depends on:

- the interest of producers to participate in certification schemes, which depends on certification costs/benefits as well as their environmental consciousness;
- the interest of traders to buy certified products; and
- the setting of a conducive institutional environment that facilitates producers' adoption of certification schemes.

If demand-side perspectives are good in terms of the general rise of consumer concerns regarding environment issue, the offer-side perspective should be evaluated. First, the production level can be affected by a low cost–benefit ratio of eco-label products compared with conventional products due to:

- poor productivity as a result of the compliance of norms in comparison with conventional production (e.g. organic production productivity in Central America is 20 to 50 per cent lower than conventional production productivity; authors, based on Haggar and Soto, 2010);
- difficulties in complying with even stricter norms;
- a rise in certification process costs;
- a rise in prices for conventional products.

Second, potential production can be affected by a rise in competition within the same eco-label commodity market segment. This competition among producing countries with different comparatives advantages (e.g. cost of labour, access to inputs, quality of the products, etc.) and among producers with different production efficiency (due to technology access) may lead to the exclusion of less efficient farmers. Finally, some authors argue that as the market for eco-labelled products matures and concurrence rises, the premium will probably decrease over the long term (Villalobos, 2004; Killian et al, 2006).

PES mechanisms' financial sustainability is a critical issue since most PES mechanisms are quite recent. Conditions of financial sustainability differ according to the types of PES mechanisms and ES. For many local PES experiences that developed with cooperation funds from pilot projects, with the assumption that sustainable funding sources would be found during the project's duration, the financial sustainability depends on the capacity to identify and maintain the appropriate funding source or to negotiate new support projects or institutional mechanisms. For National PES programmes, the funding mechanisms are set by laws and benefit from certain sustainability. However, the financial sustainability of these programmes depends not only on political decisions, but also on the capacity to bundle various sources of funding (see Chapter 12). PES financial sustainability also depends on the type of ES. For carbon sequestration ES mechanisms, sustainability depends on international mechanisms of compensation and the development of voluntary carbon markets. The former is subject to complex negotiation among countries and the final decision on an agreement is still uncertain. Moreover, these mechanisms are often complex and access requires high financial and technical capacities (see Chapter 9). The latter is still limited but is developing rapidly with the rise of the social and environmental responsibility of enterprises. For hydrological PES, especially for local PES mechanisms, the financial sustainability depends on the willingness to pay and the management capacity of the local managing enterprises (such as the water delivery enterprises and/or energy production enterprises). The development of a legislative framework to secure principal payment in contribution to ES provision in water regulation can facilitate the sustainability of funding these mechanisms. Finally, whatever PES mechanisms are considered, key factors for financial sustainability are the capacity of the mechanism to demonstrate an effective provision of ES (and to measure it economically) and to exhibit a transparent and cost-effective mechanism.

Institutional sustainability

The second element of the sustainability of both mechanisms comes from institutional and political features. In both cases, confidence and trust in the mechanisms by the final user or consumer is the key factor. This confidence and trust relies on:

- the evidence of mechanism effects (and, thus, the quality of the monitoring and control process);
- the transparency of fund management in the mechanism (especially in the PES mechanisms);
- proper communication to final consumers/ES users.

The diffusion of information is crucial in both mechanisms in order to create the necessary trust, since producers/ES providers and consumers/ES users are generally geographically located far from each other.

In both cases, sustainability depends on the national institutional framework. Institutional frameworks affect eco-label sustainability in various dimensions. First, the national commercial regulations framework (e.g. tariffs, regulations, etc.) affects the competitiveness of the national producers. Second, the social and environmental framework (e.g. water, labour regulations) affects the possibility of compliance with the eco-label norm. For example, the environmental regulation of the coffee sector in Costa Rica fosters investment of coffee enterprises through an eco-friendly approach, which can be considered a facilitating factor for eco-label compliance norms in comparison with countries with loose regulations. Third, agricultural policy and the existence of active producers' organizations can also contribute to eco-label sustainability as it may provide technical or financial assistance to farmers that facilitates and secures their necessary investments to comply with eco-label requisites. The institutional framework is also a key factor for sustainability of the PES mechanism. First, the recognition of ES provision in the legal framework may facilitate development of ES mechanisms. Second, environmental regulations may also facilitate the efficiency of ES mechanisms. Third, the existing legal framework regarding land tenure regulation and contractual enforcement also facilitates the sustainability of PES. Finally, the existence of sound, trustworthy and legitimate institutions and organizations is also important in the development of ES, since they facilitate access to ES mechanism producers by providing necessary assistance and defence of the PES programmes.

In both cases, political stability is an important factor for sustainability. Political stability has a diffuse impact upon eco-label mechanisms since it affects the confidence of the market agent in the whole economic system of the country. The importance of political stability is more straightforward in terms of PES. It is especially important for national PES or government-financed PES where funding comes from governmental sources. But it also affects the sustainability of local PES/or private (buyer) PES, even if they are based on private arrangement, since it contributes to the reliance of the legal framework that enables enforcement of PES contractual agreements.

The continuous revision of rules and the setting of a learning and innovation process are two key factors for the sustainability of the mechanisms. In terms of eco-label, the revision of requisites and certification process are regularly done to integrate consumers concerns and to strengthen the confidence and accuracy of monitoring and control processes. In the case of PES, examples of long-lasting PES programmes such as PES-Costa Rica and PHES-Mexico show a continuous evolution of the modalities, conditionalities, levels of payment, and monitoring and control methodologies in order to adapt PES mechanisms to specific situations and to improve their effectiveness, efficiency and legitimacy.

Finally, investment in national education and public awareness that improves the public conscience regarding environmental issues is bound to facilitate the effective implementation of PES mechanisms.

Conclusions

Although eco-label and PES mechanisms are both positive economic incentives, their differences in terms of principle, structure and functioning have consequences regarding their respective advantages, limits and scopes. Eco-label is subject to commodity chain governance, whereas PES mechanisms are based on territorial governance. Thus, eco-label development and scope depend on the existence of a market demand and can cover international markets and producing countries. On the other hand, PES mechanisms are a new institutional construction that requires local investment in an institutional setting to be put in practice. Their scope is local or national and affects local territories.

Eco-label and PES can both be considered interesting approaches in promoting ES provision, but they face common limitations in showing the evidence of their impact upon ES provision and poverty alleviation. Thus, additional research and studies are necessary to better evaluate their effectiveness and efficiency. Moreover, in order to more accurately compare these two mechanisms, homogenization of methods to evaluate efficiency and effectiveness still need to be developed.

Although they are both economic and market-oriented mechanisms, they depend on the setting of local, national or international institutions and public policy frameworks. The participation of beneficiaries in rule design and/or the institutions in charge of managing the schemes still need to be promoted.

Due to their distinct characteristics and scopes, eco-label and PES mechanisms have strong potential complementarities: the first is more connected to existing dynamic commodity markets and the second allows better targeting of incentives. Moreover, eco-label is more likely to be efficient in capturing funding from distant areas to support difficult-to-measure ES provision, whereas PES mechanisms may be more efficient at capturing local funding for well-identifiable and measurable ES provision. Nevertheless, the possible synergy between both mechanisms needs to be further documented and explored at various levels. Further research is needed:

- at the producer level to analyse how producers combine these different mechanisms to develop environmentally friendly practices at the farm level;
- at the territorial level to analyse how the different local stakeholders mobilize and combine these instruments and thus affect ES provision at the landscape level; and
- at the institutional setting and promoter level to design mechanisms that take advantage of both schemes.

This may lead to a new generation of innovative mechanisms, such as 'landscape labels', which could be an interesting perspective to couple the territorial and market approaches; or the coupling of PES payment and product premiums, where the commodity chain acts as brokers to access international ES markets, such as carbon.

Acknowledgements

We are grateful for the financial support provided by the European Union via CAFNET (Connecting, Enhancing and Sustaining Environmental Services and Market Values of Coffee Agroforestry in Central America, East Africa and India) collaborative project and by the French Agence Nationale de la Recherche (ANR) through the SERENA (Environmental Services and Rural Land Uses) project. Support was also provided by Pôle de Compétences en Partenariat (PCP): Agroforestry Systems with Perennial Crops, CIRAD–CATIE–INCAE–Bioversity–CABI–Promecafé.

References

Alix-Garcia, J., de Janvry, A., Sadoulet, E. and Torres, J. M. (2009) 'Lessons learned from Mexico's Payment for Environmental Services Program', in L. Lipper et al (eds) *Payment for Environmental Services in Agricultural Landscapes: Economic Policies and Poverty Reduction in Developing Countries*, FAO, Springer Science and Business Media, LLC, Rome, Italy, pp163–188

Bruijnzeel, L. A. (2004) 'Hydrological functions of moist tropical forests: Not seeing the soil for the trees?', *Agriculture, Ecosystems & Environments*, vol 104, pp185–228

Chagoya Fuentes, J. L. (2008) *Multidisciplinary Approach to Support the Design of a Local Policy of Payment for Hydrological Ecosystem Services, in a Micro Watershed Located in Northern Veracruz, Mexico*, CATIE, University of Bangor, Turrialba, Costa Rica

Chomitz, K. M. (2006) *At Loggerheads? Agricultural Expansion, Poverty Reduction, and Environment in the Tropical Forests*, World Bank, Washington, DC

Chomitz, K. M. and Kumari, K. (1998) 'The domestic benefits of tropical forests: A critical review', *World Bank Research Observer*, vol 13, pp13–35

Engel, S., Pagiola, S. et al (2008) 'Designing payments for environmental services in theory and practice: An overview of the issues', *Ecological Economics*, vol 65, no 4, pp663–674

Faure, G. and Le Coq, J. F. (2009) *Estrategias de las cooperativas cafetaleras frente a los sellos ambientales en Costa Rica: informe en el marco del proyecto CAFNET*, CIRAD, Marzo

Giovannucci, D. (2008) 'Trends toward differentiation and sustainability', in B. Bagley (ed) *El futuro del café en Colombia: Essays on the Political Economy of Coffee*, Federación Nacional de Cafeteros and Planeta, Bogotá, Colombia

Grieg-Gran, M. and Bishop, J. (2004) 'How can markets for ecosystem services benefit the poor?', in Dilys, R. (ed) *The Millenium Development Goals and Conservation – Managing Nature's Wealth for Society's Health*, International Institute for Environment and Development, London

Haggar, J. and Soto, G. (2010) *Análisis del Estado de la Caficultura Orgánica*, Consultoría para la Coordinadora de Comercio Justo en América Latina, Turrialba, Costa Rica

Kilian, B., Jones, C., Pratt, L. and Villalobos, A. (2006) 'Is sustainable agriculture a viable strategy to improve farm income in Central America? A case study on coffee', *Journal of Business Research*, vol 59, pp322–330

Kosoy, N., Martinez-Tuna, M., Muradian, R. and Martinez-Alier, J. (2007) 'Payments for environmental services in watersheds: Insights from a comparative study of three cases in Central America', *Ecological Economics*, vol 61, no 2–3, pp446–455

Le Coq, J. F., Soto, G. and Gonzalez, C. (2009) 'Voluntary standards in coffee sector and PES mechanisms, incentives to ES provision by AFS: The Costa Rican experience', Communication presented to the 2nd World Congress of Agroforestry, Nairobi, 23–28 August

Le Coq, J. F., Froger, G., Legrand, T., Pesche, D. and Saenz, F. (2010) 'Payment for environmental services program in Costa Rica: A policy process analysis perspective', Communication presented at the 19th annual meeting of the Southwestern Social Science Association, Houston, TX, 31 March–3 April

Locatelli, B., Rojas, V. and Salinas, Z. (2008) 'Impacts of payments for environmental services on local development in northern Costa Rica: A fuzzy multi-criteria analysis', *Forest Policy and Economics*, vol 10, pp275–285

Mayrand, K. and Paquin, M. (2004) *Payments for Environmental Services: A Survey and Assessment of Current Schemes Unisfera International Centre (for the Commission for Environmental Cooperation of North America)*, Montreal, Canada

McAfee, K. and Shapiro, E. N. (2010) 'Payments for ecosystem services in Mexico: Nature, neoliberalism, social movements and the state', *Annals of the Association of American Geographers*, vol 100, no 3, pp579–599

Miranda, M., Porras, I. T. and Moreno, M. L. (2003) *The Social Impacts of Payments for Environmental Services in Costa Rica: A Quantitative Field Survey and Analysis of the Virilla Watershed*, International Institute for Environment and Development, London

Muñoz-Piña, C., Guevara, A., Torres, J. M. and Braña, J. (2005) *Paying for the Hydrological Services of Mexico's Forests*, Instituto Nacional de Ecologia, Mexico

Pagiola, S. (2005) *Assessing the Efficiency of Payment for Environmental Services Programs: A Framework for Analysis*, World Bank, Washington, DC

Pagiola, S. (2008) 'Payments for environmental services in Costa Rica', *Ecological Economics*, vol 65, pp712–724

Pagiola, S. and Platais, G. (2002) *Market-Based Mechanisms for Conservation and Development: The Simple Logic of Payments for Environmental Services*, Environmental Matters – Annual Review, July 2001–June 2002 (FY 2002), World Bank's Environment Department, Washington, DC

Pagiola, S., Arcenas, A. and Platais, G. (2005) 'Can payments for environmental services help reduce poverty? An exploration of the issues and the evidence to date from Latin America', *World Development*, vol 33, pp237–253

Pattberg, P. (2005) 'What role for private rule-making in global environmental governance? Analysing the Forest Stewardship Council (FSC)', *International Environmental Agreements*, vol 5, pp175–189

Pratt, L. and Killian, B. (2008) 'How much farther can coffee markets take us?', Presentation, November 2008, La Laguna, Tres Ríos Costa Rica, http://www.sintercafe.com/images/downloads/presentations2008/How_much_farther_can_coffee_markets_take_us.pdf

Quispe, J. L. (2007) *Caracterización del impacto ambiental y productivo de las diferentes normas de certificación de café en Costa Rica*, CATIE, Turrialba, Costa Rica

Rapidel, B. (2008) 'Bienes y servicios ambientales de la caficultora', Communication to the 19th Congress on Coffee, Anacafe, 28–29 July 2008, Guatemala

Rojas, M. and Aylward, B. (2003) *What Are We Learning from Markets for Environmental Services? A Review and Critique of the Literature*, Markets for Environmental Services Series, IIED, London

Tattenbach, F., Obando, G. and Rodríguez, J. (2006) *Mejora del excedente nacional del pago de servicios ambientales*, FONAFIFO, San José

Valkila, J. (2009) 'Fair Trade organic coffee production in Nicaragua: Sustainable development or a poverty trap?', *Ecological Economics*, vol 68, no 12, pp3018–3025

Vargas, F., Rojas, O. and Hernandez-Vela, O. (2007) *Sistematizacion de experiencias de compensacion de servivios ambientales hidricos en Guatemala*, REGSA, CONAP-PARPA, MAGA, Guatemala

Villalobos, A. (2004) *Precios y premios del cafe sostenible en America Latina, EEUU y Europa*, Centro de Inteligencia sobre Mercados Sostenibles (CIMS), Alajuela, Costa Rica

Villalobos, A. L. and Solano Valverde, V. (2007) *Tarifa Hídrica: Historia de un ejemplo pionero en América Latina. Empresa de Servicios Públicos de Heredia S.A.*, Heredia, Costa Rica

Wunder, S. (2006) 'Are direct payments for environmental services spelling doom for sustainable forest management in the tropics?', *Ecology and Society*, vol 11, no 2, p23

Wunder, S., Engel, S. and Pagiola, S. (2008) 'Taking stock: A comparative analysis of payments for environmental services programs in developed and developing countries', *Ecological Economics*, vol 65, pp834–852

Wünscher, T., Engel, S. and Wunder, S. (2008) 'Spatial targeting of payments for environmental services: A tool for boosting conservation benefits', *Ecological Economics*, vol 65, pp822–833

Zbinden, S. and Lee, D. R. (2005) 'Paying for environmental services: An analysis of participation in Costa Rica's PSA program', *World Development*, vol 33, pp255–272

Part III

From Theory to Practice: Tales of Success and Lessons Learned

12

Leveraging and Sustainability of PES

Lessons Learned in Costa Rica

Rodrigo Murillo, Bernard Kilian and René Castro

Introduction

Forestry, agroforestry and agricultural systems provide a wide range of ecosystem services, among which biodiversity, carbon sequestering and soil erosion prevention are extremely valuable. However, landowners, both private and governmental, do not receive financial retribution from society for providing these services. The simple concept of applying fair trade concepts to the supply and use of ecosystem services has led to the creation of payment for environmental services (PES) schemes, established in several areas throughout the world. At present, many of those programmes are heavily supported by donations from international organizations and companies in search of high impact and visible initiatives regarding social responsibility and the enhancement of corporate reputation. Some programmes receive considerable governmental funding, resources which, in some cases, are obtained through loans from institutions such as the World Bank, which in the long run must be paid back to the granting agency by the taxpayers of the recipient country, as is the case in Costa Rica. This chapter questions the sustainability of PES programmes in Costa Rica under the current funding model and explores the possibility of leveraging such initiatives through contributions of common citizens, whether they are locally based or members of the international community. Common citizens are considered a new and robust source of funding for PES schemes.

This chapter starts by exploring the background, evolution and funding sources of PES schemes in Costa Rica, throughout their 30 years of existence, and then describes in detail the forces that influenced the creation of FONAFIFO, the National Forestry Financing Fund, an institutionalized governmental entity in charge of managing the country's PES programmes.

Perspectives regarding new funding sources, such as the ones provided by the carbon market and its Clean Development Mechanisms (CDMs), as well as the potential contributions of international corporations and global non-governmental organizations (NGOs), are explored. Furthermore, the chapter focuses on the driving forces behind PES awareness in international companies in order to better understand the constraints that must be overcome so that donations become a viable option. The chapter proposes the common citizen as an alternative new funding source and explores in depth a real case of a self-sustaining water PES scheme created by a water/power utility company in the city of Heredia, Costa Rica, known as ESPH (Empresa de Servicios Públicos de Heredia), which is fully supported by consumers. Considerable attention is given to the characteristics of both this water PES and also to the city of Heredia, in addition to the supportive legal framework that allowed the creation of a water tariff. This analysis also considers the common citizen's involvement and describes the tools used by the ESPH to create local citizen awareness regarding the water PES scheme, as well as citizens' willingness to pay for this service.

The last section of the chapter summarizes the lessons learned by FONAFIFO and the ESPH regarding PES schemes, in general. Finally, there is an analysis of the requirements to sustainably replicate ESPH's water PES scheme in a national or an international environment.

Costa Rica's Environmental Commitment and the Evolution of PES

A PES may provide a mechanism for balancing responsibility: those who create negative externalities must internalize their actions by compensating those who deliver positive environmental services to society. It can also be used to ensure that providers (e.g. owners of forestry and agroforestry systems with watershed protection) receive economic retribution from users (e.g. a water bottling company or a water utility company and their consumers). The value of environmental services has suffered distortions as such services are commonly considered free of charge. Customers have believed that they should only pay for manufactured goods and traditional services, and not for environmental services such as water or a beautiful landscape.

Costa Rica's environmental commitment is the result of 50 years of efforts. In the first half of the last century, the Costa Rican government (and international financing agencies), through its agro-export economic development

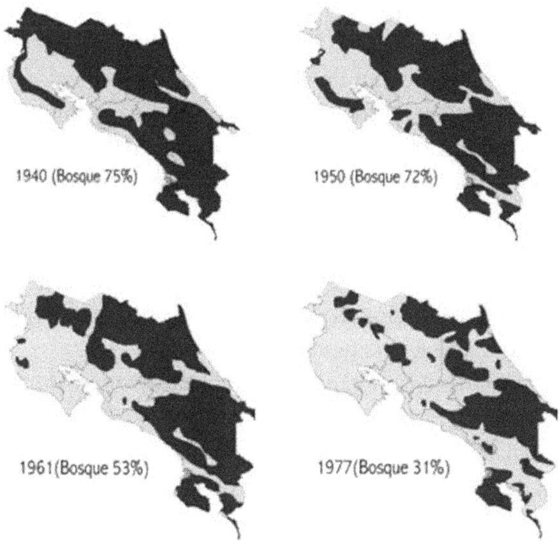

Figure 12.1 *Forest coverage in Costa Rica from 1940 to 1977*

Note: 'Bosque' is the Spanish word for 'forest'.
Source: FONAFIFO

policy, actually encouraged deforestation under the assumption that cleared land could be used in more productive ways, such as for livestock and agriculture. Figure 12.1 shows how the rate of deforestation in Costa Rica increased between 1940 and 1977.

It was not until the late 1960s that the Costa Rican government began to be concerned about the depletion of the nation's forests, thus setting the stage for the approval of the Forest Bill in 1969.This bill allowed companies tax deductions on any expenses related to reforestation programmes. Although it was not put into practice until 1979, the Forest Bill institutionalized the forest management activity by creating the General Bureau of Forestry (Dirección General Forestal, or DFG), as a dependency of the Ministry of Agriculture and Livestock (Ministerio de Agricultura y Ganadería, or MAG). However, subsequently, major flaws were acknowledged: tax credits were offered exclusively to taxpayers managing reforestation programmes, while existing forest owners were excluded from receiving benefits. In 1984, a study showed that only 26.1 per cent of the original dense forests remained, a fact which compelled the DGF to create, in 1986, a new mechanism called Forestry Payment Certificates (FPCs), which targeted the provision of financial incentives for reforestation and afforestation initiatives – namely, tax exemptions for both national and international firms importing capital goods, and also for landowners.

In 1992, the first known PES local transaction occurred when a group of farmers called *Los Jilgueros*, located near the town of Guápiles, in the Caribbean region of Costa Rica, agreed with the Central Volcanic Mountain Range Development Foundation (Fundación de Desarrollo de la Cordillera

Volcánica Central, or FUNDECOR) to exploit 1600ha in a tourism-related project. FUNDECOR, a local NGO founded in 1989 with the goal of protecting and developing the large and dense forests of the central volcanic mountain range where several national parks are located, granted the farmers access to a primary and secondary tropical humid forest territory so that they could develop an ecotourism project called *Ecocampo*. Economic incentives were provided for clearing trails in these forests so that tourists could take guided tours, enjoying the scenic beauty provided by the mountains and forests of this region. This successful pioneer PES programme inspired the Costa Rican government to continue supporting environmental initiatives as a means of promoting social and economic development.

In early 1997, Costa Rica became the first country in the world to trade in a voluntary carbon market, selling US$2 million to Norway: 200,000 tonnes CO_2 equivalent (t CO_2e) at a price of US$10 per tonne. The transaction set the foundation for Certificates of Emissions Reductions (CERs), credits issued for emission reductions under the rules of the Kyoto Protocol as one of the CDMs implemented in 2005 within the framework of the regulated international carbon market (see Chapter 8 in this volume).

Table 12.1 shows a breakdown of the more 200,000ha included in the forest conservation and reforestation governmental incentives that were first offered in 1979 and ended by 1996. The areas registered did not get two different incentives at the same time (i.e. these economic resources were spread as widely as possible).

During a span of ten years (1987 to 1997), the forest coverage doubled. Figure 12.2 shows that national forest coverage climbed to 42 per cent by 1997. However, these maps have very different grain sizes limiting true comparisons. An explanation of all the factors involved that leveraged this massive forestry growth is beyond the scope of this chapter; however, it is very clear that the prior increasing deforestation tendency was reversed.

Table 12.1 *Impact of preliminary forest incentives in Costa Rica (1979–1996): Number of hectares under protection and management per incentive type or funding source*

Incentive type	*Total area registered (ha)*	*Period(years)*
Income tax deduction	35,597	1979–1992
Soft credits (loans from financial institutions allocated at interest rates below the average commercial interest rates)	2802	1985–1995
Forestry Payment Certificate (FPC)	45,482	1986–2000
Advance Paid Forest Certificate (APFC)	40,747	1988–2000
Forest Development Fund (FDF)	12,789	1989–1995
Forest Payment Certificate for Management Purposes (FACMP)	45,222	1992–1999
Forest Protection Fund (CPF)	22,200	1995–1996
Total	204,839	1979–1996

Source: FONAFIFO

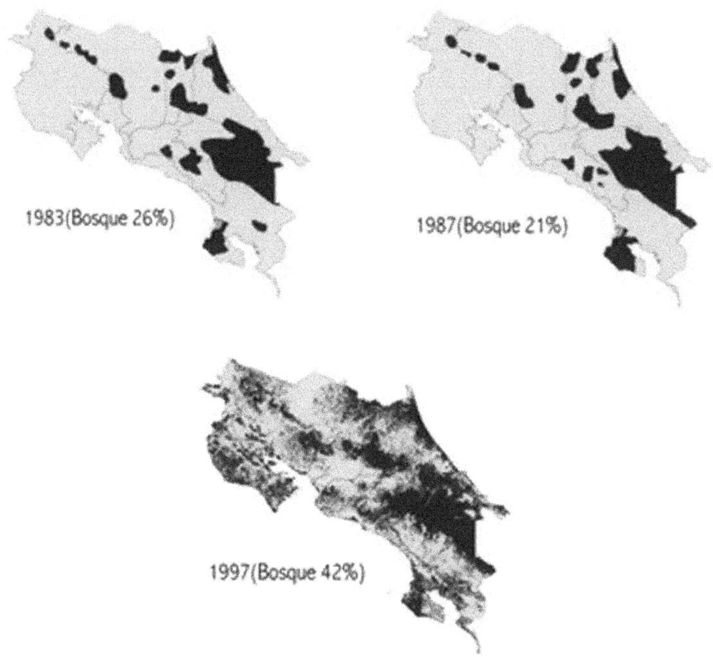

Figure 12.2 *Forest coverage in Costa Rica from 1983 to 1997*

Note: 'Bosque' is the Spanish word for 'forest'. These maps were provided by the same source; however, the tracking systems used in 1997 were more sophisticated than the previous ones, so grain size is much smaller and the information is more accurate.
Source: FONAFIFO

FONAFIFO Creation and Its Funding Sources

FONAFIFO's history starts in 1990 with the enactment of the Forestry Bill, which entitled the DGF to establish a trust fund for forestry activities. A seed capital of US$425,000 was fully allocated from Costa Rica's national budget in 1991.

Both previous Costa Rican PES self-initiatives and the ongoing impetus ignited by the Earth Summit in Rio de Janeiro in 1992 created the ideal environment for the Costa Rican government to concentrate all of its forest conservation efforts in one specialized office in charge of financing the forestry sub-sector and managing all national PES programmes. The FPCs, created in 1986, paved the road, both legally and institutionally, for the implementation of a formal nationwide governmental PES programme run by FONAFIFO, which was institutionalized in April 1996. Table 12.2 shows the official and legally acknowledged environmental services included in FONAFIFO's programmes.

FONAFIFO's primary objective was to promote forest management through different mechanisms, including granting credits to small and medium-sized landowners for forest plantation processes and reforestation; forest

Table 12.2 *Environmental services included in FONAFIFO's PES scheme in 1996*

Service	*Application*
Greenhouse gas (GHG) mitigation	Fix, reduce, sequester, store and absorb greenhouse gas.
River basin protection	Urban, rural or hydropower use.
Protection of biodiversity	Conservation, sustainable scientific and pharmaceutical use, research and genetic improvement. Protection of ecosystems and different life forms.
Protection of scenic beauty	For scientific and tourism purposes.

Source: FONAFIFO

Table 12.3 *Resource allocation of FONAFIFO's PES programmes from 1997 to 2005*

Source	*Area covered (ha)*	*Proportion of total (%)*
Forest protection (primary and secondary forest)	451,420	88.9
Reforestation	27,096	5.3
Established tree plantations	1248	0.2
Forest and agroforestry management	28,066	5.5
Total	507,830	100

Source: FONAFIFO

nurseries; agroforestry systems; recovery of depleted areas; and technological changes for the exploitation and industrialization of forest resources. The level of economic incentives depended on the type of activity. PES projects for private forests and agroforestry systems have been the main programmes executed by FONAFIFO since its creation and currently consume around 70 per cent of its annual budget. Table 12.3 shows how funds were distributed, between 1997 and 2005, for a total of 507,830ha under FONAFIFO's PES scheme.

From the demand side, in most cases the average US$816 per hectare for five years reforestation incentive covers the opportunity cost of many landowners. However, as in the case study from Mexico (see Chapter 13), there are cases in which the opportunity cost is not covered, putting a lot of pressure on land, especially when a forestry use has to compete with alternative usages such as for livestock and agriculture. Additionally, FONAFIFO deducts 10 to 15 per cent out of the paid lump sum; thus, the effective payment per hectare is lower than the US$816.

Figure 12.3 shows that by 2005, forest coverage of Costa Rica had reached 51 per cent. A project run by FONAFIFO, called Ecomarkets I, which is described below, is considered to have contributed to this positive effect.

The initial funding sources

Since its foundation, FONAFIFO was entitled to invest in financial instruments to accomplish national objectives and policies. The Forestry Bill stipulated that

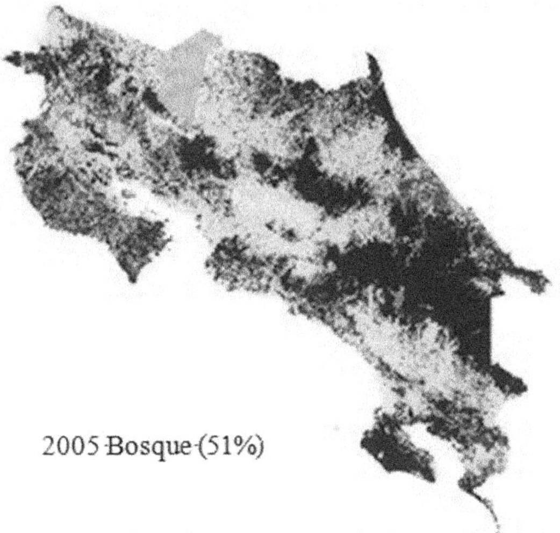

Figure 12.3 *Total forest mass coverage in Costa Rica in 2005*

Note: 'Bosque' is the Spanish word for 'forest'.
Source: FONAFIFO

FONAFIFO's funding would be mainly constituted by government contributions, donations, credits from international organizations, fundraising and the economic benefits of carbon credit investments, 40 per cent of revenues derived from taxes on wood usage, and the issue of forestry bonds or certificates. Conversion of foreign debt and PES provided by private or public, national or international organizations were also considered part of FONAFIFO's funding.

FONAFIFO's first funding source originated from the carbon certificates sold to Norway in 1997, in addition to one third of the sales tax on fossil fuels collected by the Costa Rican central government in accordance with the 1996 national bill. One of its first tasks was to fund an existing PES project called *Plama Virilla*, which targeted watershed protection and water quality for the Virilla River Basin where five hydropower plants of the National Power and Illumination Company (Compañía National de Fuerza y Luz, or CNFL), a major national utility company, are located.

In 2000, a project called Ecomarkets I was proposed to expand FONAFIFO's budget so that PES programmes could be increased and reinforced. By means of agreements, loans and donations, US$49 million was raised between 2000 and 2005. Table 12.4 shows the breakdown of Ecomarkets I funding sources.

The Ecomarkets I project contributed to the protection of almost 50 per cent of all land registered or PES programmes in Costa Rica between 1996 and 2005 (more than 0.5 million hectares). Ecomarkets I was the first documented initiative where a tropical country acquired a debt with an international financial organization in order to invest in forest conservation and biodiversity. It was also the first time that the World Bank and the Global Environment

Table 12.4 *Funding sources for the Ecomarkets I project*

Source	*Type*	*Amount (US$ million)*	*Proportion of funding (%)*
World Bank	Loan	32	65
Global Environment Facility (GEF)	Donation	8	16
Costa Rican government	Donation	9	19
Total		49	100

Source: Sills et al (2005)

Facility (GEF) had allocated funds to a PES programme. Since the major component of Ecomarkets I was a loan from the World Bank, the scheme included an agreement that the loan would be repaid through the collection of a fixed 3.5 per cent over the total annual fossil fuel taxes charged to Costa Rican consumers.

In addition, agreements with local companies, such as Energía Global S.A., Platanar S.A. and Hydroplant, helped to procure revenues to provide protection for water resources. Additionally, PES contracts were sold to CNFL and Florida Ice & Farm, a brewery and water bottling company.

Between 2000 and 2005, other funding sources were obtained for various PES programmes in Costa Rica:

- In 2002, the German Development Bank (Kreditanstalt für Wiederaufbau, or KfW) allocated a donation of US$11.2 million for the protection and reforestation of northern Costa Rica.
- Several local agreements led to the procurement of US$0.5 million in annual funding. A water tariff was placed on those who used water as a 'raw material' in their production activities, either in the private or public sector, so that watershed protection projects could be financed.

The current funding scenario

This chapter has described the initial 30 years of successful achievements with Costa Rican PES projects (i.e. since 1979) and it can be argued that these experiences were the precursors of some of the current worldwide mechanisms. Between 1997 and 2005, PES benefits were provided for around 10 per cent of Costa Rica's territory (i.e. 500,000ha of forests and plantations were registered in these schemes). However, by the end of 2006, one year after Ecomarkets I ended, only 250,000ha (5 per cent of Costa Rica's territory) were still covered by a PES scheme, representing only half of the area previously considered (i.e. local and national efforts were insufficient to achieve FONAFIFO's goal of maintaining 500,000ha in PES registered schemes).

FONAFIFO's dependence on the fossil fuel tax, along with short-term donations, was not sufficient to support an expansion of the PES programmes throughout Costa Rica. This conclusion led to a search for new mechanisms to establish sustainable (long-term) PES funding. Options considered included:

- private purchasing agreements for ecosystem services;
- sale of carbon certificates in international carbon markets; and
- issue of Environmental Services Certificates (ESCs) as a financial instrument targeting the preservation of existing forests, generating new ecosystems and guaranteeing PES.

Expecting to collect US$1.35 million per year, they basically focused on companies interested in watershed/forest protection and carbon footprint offsetting. The mechanism was designed so that FONAFIFO would receive funding from companies and institutions benefiting from PES, who wanted to compensate forest owners for preserving their forests. The amount invested depends on the number of hectares an individual or a company desires to protect. The smallest investment area is 1ha, with an average value of US$57 ha^{-1} y^{-1} for contracts that are usually for five years.

Consolidation of a nationwide tariff on water exploitation permits (e.g. wells, springs and hydroelectric use), approved through a national bill in 2006, was the only new mechanism in FONAFIFO's portfolio. The origin of this new mechanism can be traced to former voluntary agreements between FONAFIFO and large-scale water consumption companies whose 'spirit of contribution' had led to their willingness to pay (WTP). Until the 'inelasticity' of water prices became apparent, yearly charges to companies with high levels of water consumption was very low, ranging for a particular company from US$25 per year in the case of irrigation purposes, either pumped or by gravity, to US$400 per year for brewing beer. The new water tariff aimed at increasing the 'price of water' for all consumers by nearly threefold over a seven-year period, charging in proportion to the amount of water consumed and according to the type of consumer. The original bill faced strong opposition so the tariff was lowered from the originally proposed figure to a charge rate below what most water users were willing to pay. The water tariff was conceptualized to provide FONAFIFO with a cash flow of US$390,000 over a five-year period.

Within the Ecomarkets II project, starting in 2007 and running until 2011, it was proposed to include the 250,000ha of the unsatisfied demand that remained from Ecomarkets I, plus an additional 38,000ha, all under 20-year contracts instead of the previous 5-year terms. Table 12.5 breaks down the Ecomarket II budget allocation for this additional area according to the new PES modalities, which by 2006 had changed in comparison to Ecomarkets I modalities, partially because some technical disputes over forest management had been resolved (e.g. natural regeneration was separated from forest protection).

Ecomarkets II includes natural regeneration and reforestation as potential income sources to generate carbon sequestering credits in the global carbon market, under the Kyoto Protocol Clean Development Mechanisms. It is expected that 64 per cent of the project's present value will be generated by CO_2 trading.

Ecomarkets II has not only been designed to target the development and implementation of sustainable mechanisms, but also to provide positive social

Table 12.5 *Resource allocation for PES programmes under the Ecomarkets II scheme*

Modality	*Area covered (ha)*	*Proportion of total (%)*
Forest protection (primary forest or consolidated secondary forest)	31,770	85.0
Natural regeneration (land use switch to consolidated secondary forest)	3530	9.4
Reforestation (forestry plantations)	1995	5.3
Agroforestry systems (payment by the square footage and not by hectares)	92	0.2
Total	37,387	100.0

Source: Castro and Vega (2006)

Table 12.6 *Funding sources for the Ecomarkets II project*

Source	*Type*	*Amount (US$ million)*	*Proportion of funding (%)*
Global Environment Facility (GEF)	Donation	10	11.07
World Bank	Loan	30	33.22
Costa Rican government	Donation	47.5	52.66
Bio-carbon	Donation	2.5	2.83
Other	Donation	0.2	0.21
Total		90.3	100

Source: Castro and Vega (2006)

impacts regarding poverty reduction (again, see similarity to the Mexican case study in Chapter 13) and a transfer of technical know-how as key elements in sustainable rural development. It aims to increase both local and global PES services. In order to enhance the programme's attractiveness to donors and financing institutions, FONAFIFO focused its PES programmes on Costa Rican sections of the Mesoamerican Biological Corridor. Table 12.6 shows the proposed funding sources for Ecomarkets II.

Compared to Ecomarkets I, only 33 per cent of the funds (not 65 per cent) were provided by loans. Costa Rica's government increased its direct financial support by providing more than 50 per cent of funding, compared to 19 per cent provided for Ecomarkets I. This demonstrates the deeply rooted commitment to preserve the environmental services provided by forests and agroforestry systems.

Perspectives for Future Funding Sources

If Ecomarkets II repeats the success of Ecomarkets I, an Ecomarkets III project is very likely to be proposed in 2011, probably with even higher financial commitment from the government. The negotiating efforts for this kind of project require that short-term funding be collected in order to support long-term initiatives, a task which is quite cumbersome and adds a great deal of uncertainty to future projections (e.g. 20- to 30-year schemes instead of the original 5-year contracts). New sources of funding are needed to provide sustainability for FONAFIFO's efforts.

Funding constraints in the carbon market

Resources collected from carbon trading have been seen as a way to pay Costa Rican landowners' opportunity costs and support PES programmes. After the creation of the carbon markets, several proposals have been made in Costa Rica to include carbon market-based instruments as emerging funding mechanisms. In 1997, an initiative called the Protected Areas Project (PAP) was devised, under the Kyoto Protocol Joint Implementation concept, to finance the purchase of 10 per cent of Costa Rica's forests from their owners and put them directly under a governmental protection perpetual PES scheme as Wildland Conservation Areas (WCAs). Thus, 25 per cent of the country's territory could be consolidated under government protection. However, those efforts have been unsuccessful, even though recently there has been intensive international negotiation.

REDD (reduce emissions from deforestation and degradation) credits are another carbon market-based mechanism, delivering 'co-benefits' such as biodiversity conservation and poverty alleviation, and offering the opportunity to utilize funding from developed countries to reduce deforestation in developing countries. Unfortunately, REDD credits cannot be considered as sustainable income sources because of the uncertainty of international carbon market trading even when there is real demand for them in the voluntary schemes. Legal and technical disagreements regarding their inclusion as climate change mitigation mechanisms are still to be resolved. FONAFIFO hopes to sell REDD credits through a project called Carfix, proposed since 1997. However, these REDD credits are not recognized under the Kyoto Protocol baseline for the country, so Costa Rica has been proposing that the baseline be moved back until at least 1990 in order to be able to market its 'old REDD credits'. The success of this initiative is not only contingent upon the carbon market, but also upon the approval of Costa Rica's National Controlling Bureau.

So far, the only successful REDD experience in the country has been an initiative with the World Bank which has supported Costa Rica's efforts to reduce GHG emissions through the Forest Carbon Partnership Facility (FCPC), which donated US$3.6 million to FONAFIFO to specifically develop REDD credits in order to sell them in the future.

Constraints of Clean Development Mechanisms

The envisioned long-term revenue sources, based on the approved mechanisms of the Kyoto Protocol regarding afforestation and reforestation, are not providing as much funding as expected. The majority of CDM projects worldwide are related to energy instead of forestry or agroforestry.

To mitigate this preference for energy-related projects, FONAFIFO tries to enhance the perceived quality of CDM land use-related projects developed in Costa Rica by offering 'gourmet carbon'. This quality image relies on 'old REDD credits' produced by a government committed to national and international environmental policies; a robust background implementation of PES projects; a pool of successful experiences in PES mechanisms used as models for initiatives around the world; the social impact of the countries' projects that take into account small and medium-sized companies, as well as native landowners; high levels of involvement of women; and protection of biodiversity-rich areas. The combination of these factors has given FONAFIFO a high reputation as an international carbon credit producer, leveraging the value of its projects on the voluntary market to an average of US$8 per t CO_2e compared to projects in other tropical countries whose prices are as low as US$2 per t CO_2e. However, a project to offset hydropower of the same scale could trade carbon at 30 Euros per t CO_2e, evidence of the price elasticity of CDM projects.

The process to implement a CDM or agroforestry project is much more cumbersome when compared to an energy-related project, as was presented in Chapter 8 of this volume. From a pragmatic perspective, it is more efficient for FONAFIFO to produce credits by growing forests for 10 to 15 years and then to sell the credits in the voluntary carbon market. However, this is not feasible in many cases since there is always a huge pressure on cash flow.

International corporations, international citizens and worldwide NGOs

Costa Rica has had a commitment to nature conservation since the 1970s, when it began creating the national parks systems throughout the country, currently used for recreational and research purposes. The fact that Costa Rica has been a front-runner in environmental policies related to PES and CDM under the Kyoto Protocol framework has supported the international positioning of the country generated through 'country brand' in natural tourism.

Costa Rica has successfully used its 'green image' and fame for 'environmental conscience' to leverage some international donations in the past. In the context of this country, it is necessary to identify and develop the market variables that could drive donations as a source of sustainable funding. Unfortunately, in terms of total size and areas that could be protected, Costa Rica is not significant in the context of worldwide forest cover. Countries such as India and especially Brazil (because of the Amazon) are more appealing to international donors since their total impact upon the world's carbon emission offsetting is higher. Additionally, those countries still suffer high deforestation rates, so financial support is perceived as more necessary compared to investing in a country such as Costa Rica, where deforestation has basically been stopped. From this perspective, attracting large amounts of money to Costa Rica for these purposes is an extremely difficult task, especially if an international NGO or a private firm seeks impact projects to justify 'investment' decisions that are short-term public relation instruments in social and environmental responsibility schemes. An exception is Nature Air, a low-budget airline based in Costa Rica offering domestic flights, 60 per cent of them to Wildland Conservation Area destinations, who entered the carbon neutral network worldwide by offsetting its carbon emission through the purchase of PES from FONAFIFO. Unfortunately, cases like this are not common. For a company with a 'green image' such as Nature Air, it is strategic to support PES schemes in its geographical area of operation. However, for an international company such as Lufthansa, it is probably more appealing to support PES programmes in Brazil; thus, the practice of redeeming mileage to support PES schemes in the Amazon has become common.

New mechanisms to attract international funding and donations could be based upon the same historical tourist elements that have been used to build the 'country brand's' positioning and awareness. A Costa Rican PES brand-awareness drive could eventually be implemented through agreements with international organizations and government environmental agencies in developed countries. However, persuading these international PES supporters to participate requires that large economic disbursements be allocated to marketing campaigns, resources which unfortunately are not at FONAFIFO's disposal. Additionally, FONAFIFO lacks a specialized department exclusively devoted to marketing activities: its payroll of only 50 people is barely enough to oversee managing, monitoring and controlling functions of existing PES schemes.

Due to all the previous reasons and in order to support the growing and sustainable PES in Costa Rica, local citizens should be seen as a potential fresh funding source. From a financial point of view, and if compared to an international supporter, it is easier and less expensive to influence a local citizen's willingness to contribute to a particular PES programme, transforming him or her into a potential sustainable PES source, even when their marginal contributions might be less than that which an international supporter could provide.

The funding potential of local citizens: ESPH's water PES project

The water market around the world is extremely distorted. A study presented at the World's Water Forum in 2003 in Japan reported that customers were willing to pay as much as 4000-fold more for 1 litre of bottled water, when compared to 1 litre of water provided by a public utility company. This price difference was due to perceived attributes of quality and transportability.

The basic idea behind a water PES project is to compensate landowners upstream for their contributions to the protection of water quality and quantity in rivers and springs. This retribution is made by the water end users downstream because they directly benefit from the protection of the watershed. If end users are to continue to enjoy good-quality water and a sustainable future supply, they should pay an incentive to landowners to avoid damaging land-use changes in the recharge areas of aquifers and where water is collected. Methods to measure these services are presented in Chapter 2 and the techniques for marketing those services are detailed in Chapter 7.

In the past, Costa Rica has offered environmental services mostly related to forest protection; however, the concern for maintaining water resources has increased. Costa Rica has made significant progress, providing safe drinking water throughout the country (the index of coverage for human consumption is 82.2 per cent); nevertheless, pollution rates have risen.

The ESPH and the city of Heredia

The ESPH is a public service water and power utility company under an incorporated legal framework, in charge of supplying drinking water and electricity to the four counties that make up the city of Heredia. Currently, the company has around 55,000 registered customers and complies with all health requirements for safe water provision.

A substantial amount of the water consumed in the city comes from local springs in the forests, which represent only 6 per cent of the total land use over the aquifer of the entire area of Heredia. Several factors threaten the supply from these springs: absence of an urban growth plan and irresponsibility and/or ignorance related to livestock practices. The goal of the ESPH is to protect or restore the 1900ha surrounding the upstream water springs that it manages (i.e. an area that ranges from 0.5km to 1km above them). These protected areas provide 'inexpensive water' from springs since they simply collect the water flow from the mountains before it is piped to households by gravity after a disinfection and treatment process. The cost of extraction of groundwater in the lowlands, by means of pumps, is almost four times higher because of the energy costs.

Some important characteristics of the city of Heredia must be highlighted for this discussion. The city represents around 16 per cent of the total Greater Metropolitan Area of Costa Rica (over 2 million inhabitants), with a population density of around 642 inhabitants per square kilometre. 75 per cent of the inhabitants are either urban or semi-urban, with high and increasing social standards. When compared to the rest of the metropolitan area, the average citizen in Heredia has an above average educational level, an above average purchasing power and a clean record of public service payments.

Supportive legal framework for the ESPH's water tariff

Besides the Forest Bill that created FONAFIFO (Bill 7575), Costa Rica's legislation had identified water as an economically valuable asset, establishing the legal framework for the ESPH to develop and charge a 'water fee' based on the consumption of each customer. Among the legislation, the following bills could be cited:

- Public Services Regulating Authority Bureau (Autoridad Reguladora de Servicios Públicos, or ARESEP) Bill 7593;
- Biodiversity Bill 7788;
- Environmental Bill 7554.

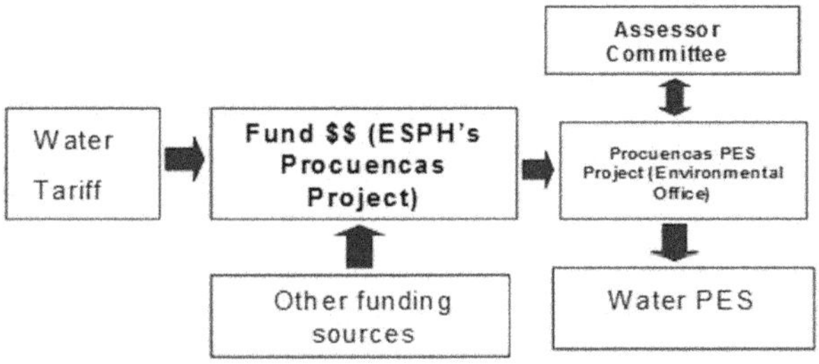

Figure 12.4 *ESPH's water tariff operational scheme*

Source: from an ESPH electronic document provided by Luis Gámez-Hernández

The 'water fee' was justified as representing an 'environmentally adjusted' tariff for a freshwater supply, which placed a value on the country's water resources and also on the environmental cost of forest restoration/protection of upstream springs.

ESPH collects funds from every user by charging a proportional consumption amount on every monthly bill. The funds are deposited in a special account and are allocated to watershed PES. In specific cases the ESPH has used the funds to directly purchase the land to be protected, which additionally has required soft bank loans to cover the total cost of the purchase. However, the total amount of the funds collected is not enough to fully cover a land purchasing scheme since real estate pressure reaches up to US$100 per square metre in some cases, meaning that the most beneficial long-term use of funds is to distribute them through agreements with existing landowners. 'Renting' low/medium vulnerability lands from their owners, through a PES scheme, is more cost effective than the allocation of resources to a more focused 'one-time buy' of high vulnerability land. Currently, the total of the ESPH annual funding is around US$150,000, with around 25 landowners benefiting from payments, and thus protecting 1200ha out of the 1900ha targeted in priority areas. Figure 12.4 describes how this PES scheme operates.

ARESEP, as the Costa Rican national regulatory entity in charge of fixing prices for public products and services, required the ESPH to provide justification for the cubic metre tariff to be charged to consumers. Several components were integrated within the environmentally adjusted tariff structure:

- water supply forecasting potential for the micro-river basins;
- economic–ecological valuation of the water;
- environmental factors calculated for fixing the tariff;
- demand study or payment willingness from the local water consumers.

In order to implement the initiative, a cooperation agreement was made with the Ministry of Environment and Energy (Ministerio de Ambiente y Energía, or MINAE), in charge of the development of environmental guidelines, and FUNDECOR, in charge of providing third-party technical, managerial and legal knowledge to assess and monitor the PES projects. Two approaches were combined when calculating the final water tariff value:

1 value of water sequestering potential from forest, defined as quantity and quality of water: economical value of forests' productivity from a water service point of view (US$0.008 per cubic metre);
2 value of river basin restoration: first year costs for creating forestry plantations on the upper watershed areas (US$0.014 per cubic metre).

Even though the two component values of the tariff add up US$0.022 per cubic metre of water, ARESEP only approved a compounded fixed amount of US$0.005 per cubic metre in 1999. When the water tariff was first presented, landowners were compensated according to the following guidelines:

- Conservation/natural regeneration projects: US$68.86 ha^{-1} y^{-1} compensation (close to the amount determined by landowners' water PES provision willingness study). In 2004, the incentive was increased to US$110 ha^{-1} y^{-1} for ten-year contracts due to an increment from US$0.005 per cubic metre to US$0.009 per cubic metre approved by ARESEP.
- Reforestation projects: US$571 ha^{-1}, distributed over five years. Additionally, an incentive of US$326 ha^{-1} was provided to landowners if the project required planting new trees. During the first five years, total PES service was US$898; however, when tariffs increased to US$0.009 per cubic metre in 2004, the paid incentive also increased to US$1058 ha^{-1} for each 5 years during a 20-year contract.

Based on the data collected in a 1999 survey, a linear regression model was constructed showing that 92 per cent of Heredia's citizens were willing to pay as much as US$0.046 per cubic metre to ensure a future high-quality water supply. This amount represented around 1 per cent of the average household monthly income in Heredia of around US$585 (i.e. about US$4.63 per month average household water bill). The original 1999 customer payment willingness exceeded almost ten times the real US$0.005 per cubic metre water tariff, so the logical step would be to present new data to ARESEP in order to request a further tariff increase approval.

An ESPH monthly bill is composed of basically three items: regular water consumption, wastewater management fee and a water tariff. Every customer is uniquely identified in every monthly bill through the use of GIS. Figure 12.5 shows an ESPH water bill where the water tariff is outlined by an ellipse.

It is desirable that ESPH makes a request to ARESEP for a further increase of the water tariff, so that more resources can be collected to consolidate the programme, purchase high vulnerability lands, keep up present payment value and compensate for the increasing costs of opportunity. However, tariff increase constraint is not really the payment willingness factor, but rather the lack of demand for PES from the highland forest owners.

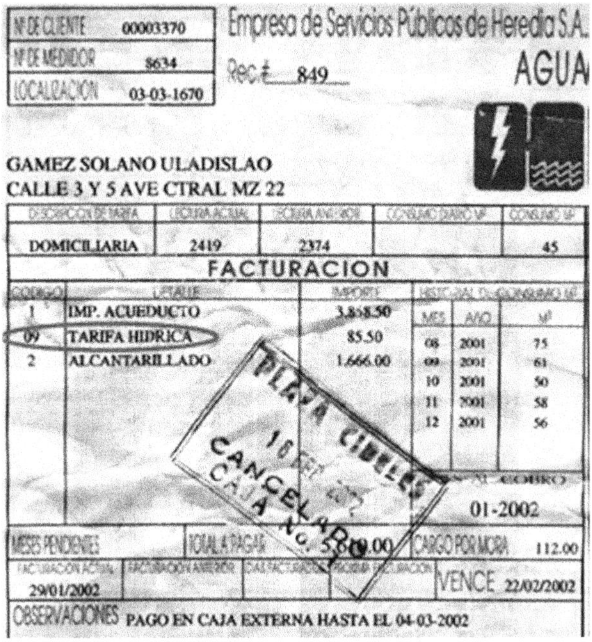
Nº DE CLIENTE 00003370
Nº DE MEDIDOR 5634
LOCALIZACION 03-03-1670
Empresa de Servicios Públicos de Heredia S.A.
Rec ₡ 849
AGUA
GAMEZ SOLANO ULADISLAO
CALLE 3 Y 5 AVE CTRAL MZ 22
DOMICILIARIA 2419 2374 45
FACTURACION
1 IMP. ACUEDUCTO 3.868.50
09 TARIFA HIDRICA 85.50
2 ALCANTARILLADO 1.666.00
08 2001 75
09 2001 61
10 2001 50
11 2001 58
12 2001 56
01-2002
MESES PENDIENTES
TOTAL A PAGAR 5.6[illegible]9.00
CARGO POR MORA 112.00
29/01/2002
VENCE 22/02/2002
OBSERVACIONES PAGO EN CAJA EXTERNA HASTA EL 04-03-2002
PLAZA CIBELES
CANCELADO
CAJA No. 5

Figure 12.5 *An example of an ESPH water bill*

Source: from an ESPH electronic document provided by Luis Gámez-Hernández

The concept of land usage as an economical issue enters the scenario again. The forest highland areas of Heredia are under pressure as real estate, extensive livestock raising, strawberries/flowers and ornamental crops (ferns) or simple family ownership traditions, which in some cases justifies having 'idle' land and resisting reforestation and natural regeneration programmes. Unless there is data to prove to ARESEP that there is a higher PES demand than can be met with current resources, there is no way to push for a new tariff increase.

From a supply perspective, ESPH's scheme could be considered more attractive to landowners than FONAFIFO's; for a reforestation project, ESPH grants are currently US$818 ha^{-1} for five years, while FONAFIFO delivers US$816 ha^{-1} for five years minus a 15 per cent deduction for management expenses, providing the landowner with only US$693 ha^{-1} for five years. In the case of protection schemes, ESPH grants are US$86 ha^{-1} y^{-1} compared to the average US$64 ha^{-1} y^{-1} (US$55 ha^{-1} y^{-1} when the 15 per cent administrative fee is discounted for the first year) paid by FONAFIFO.

Local citizens' involvement

Among the major benefits of this ESPH programme, environmental citizen responsibility and public programme recognition are the most important since they are the foundation of its sustainability. ESPH has deployed strong awareness campaigns in the city of Heredia to promote watershed protection. For instance, an annual 'march for water' is held in the city every year, where informative advertisements are distributed to citizens. Additionally, there is a specific budget allocated to educational programmes in primary schools in order to create environmental awareness among children. The monthly water bill is actually used to deliver awareness messages regarding watershed PES importance.

In general, 'water marketing' campaigns are not only targeted to create environmental awareness, but also to unveil to the public the distortions in the water market. To illustrate this, the case of a water bottling company located in this city is frequently mentioned: the owner simply employed rudimentary equipment to fill plastic containers with ESPH's faucet water to later sell it on 'nicely branded' in supermarkets (i.e. the same water, in a different packing, created a different customer perception of the same commodity). This profitable business was being supported by all watershed protectors in mountains above the city of Heredia. The most interesting fact was that the bottled water was actually of lower quality when compared to the ESPH's faucet water, due to the bottling process.

If, in the future, water pollution, wastewater disposal and air pollution, among other environmental concerns, are to be included in a PES scheme besides the watershed PES; significantly more efforts are required to increase environmental awareness; but law reinforcement mechanisms also need to be developed.

Key success factors

The model developed by the ESPH is considered a good example of a water PES in Costa Rica. It was the first national effort to create a sustainable fund fully independent from donations or loans, in which citizens have taken an active role in an environmental initiative, rewarding positive externalities provided by landowners of the watershed areas. The legal, institutional, organizational and environmental awareness experiences of ESPH established a milestone for Costa Rica's environmental protection initiatives. Several key success factors have been identified regarding the sustainability of the ESPH programme:

- An initiative develops within the ESPH, without any external funding intervention, making it a self-supported company which collects its own sustainable income.
- An economic model is developed to internalize environmental variables.
- Local citizen/user/customer identifies the relationship between quality of water/quality of living and upstream forest watershed.
- ESPH transparency and legitimacy result in a historical trust link between the company and Heredia's citizen.
- ESPH owns facilities for water/power provision by means of a consortium of the four municipalities that compound the city of Heredia so that it can freely make decisions (municipalities play only an auditor role).
- A formal accounting system which provides a means of water usage track-

ing legitimizes the water tariff by creating a sense of trust in the contributor. This allows ESPH to fairly charge users in proportion to his/her consumption and avoids the subsidizing effect that occurs in other municipalities throughout Costa Rica and Latin America, where there is an absence of such tools.
- ESPH uses a sophisticated global positioning system (GPS) and a geographical information system (GIS), which allows both property monitoring and customer pinpointing.
- Internal political willingness in ESPH's management staff/board of directors enable decisions to be made at the very top of the organizational structure, leveraging the implementation phase.
- Medium population density and users' purchasing power are above the average for Costa Rican citizens.

A medium population density is one determinant from a capital investment/customer benefits perspective because the water distribution operation achieves economies of scale. The higher the number of customers, the higher the amount of money collected, so the more beneficiaries spread over a wider area and the higher the sustainability of PES schemes.

ESPH has attempted to convince the largest water utility company in Costa Rica, the Water Piping and Sewage Company (Acueductos y Acantarillados, or AyA), and other municipal water utility companies to launch a similar scheme in San José, Costa Rica's capital. Efforts have included directly outsourcing experiences from ESPH; but so far they have been unsuccessful, essentially because of the absence of internal political willingness.

Lessons Learned

Even when Costa Rica's government has previously invested in marketing campaigns to increase local consciousness and greater environmental awareness, improving land use continues to be a complex issue. Even if conservation and reforestation are promoted through PES programmes, if a landowner's opportunity cost is not covered, that land will be used in other more productive investment opportunities.

Since Ecomarkets I was a solid success in FONAFIFO's funding scheme, the Ecomarkets II project was launched as a reasonable next step. Unfortunately, these projects require major governmental involvement, putting a lot of pressure on the national budget of a developing country such as Costa Rica.

ESPH's scheme is local and applicable to lands where the opportunity cost is very high when compared to FONAFIFO's scheme, where the opportunity cost is not usually high. The natural effect has been that some landowners previously participating in FONAFIFO's programmes decided to resign and change to the ESPH scheme, confirming the hypothesis that land use is still first and foremost an economic matter. In this line of thought, to a certain extent, the water fee that

FONAFIFO counts on as an additional funding source works essentially like the ESPH scheme, but on a nationwide basis.

The interest of local citizens can be capitalized in a PES scheme as the ESPH experience has proven: small-scale self-sustainable PES projects can be carried out under the right conditions in a developing country, totally supported by local citizens/users, though administrative costs of such programmes may need to be 'subsidized' by a company.

Even if local law does not expressly state that PES must be charged to local users, with institutional willingness, interpretation of existing laws could help to materialize projects that are totally supported by local service users. On the order hand, a well-designed campaign regarding environmental services is necessary to positively influence and change the existing distortions in ordinary citizens' minds regarding PES schemes. This certainly requires the heavy involvement of local governments in order to allocate resources to achieve such an effect.

Currently, Costa Rica is enjoying benefits provided by some international environmentally conscious players due to its international 'green image', even though the area under PES schemes in this country is not significant in a worldwide context. The international donor community tends to support PES schemes in other latitudes where there is a perception that contributions are more necessary and will have a higher impact. Mechanisms to direct this funding towards FONAFIFO need to be created, using the leverage of all the positive attributes that have given Costa Rica a 'country brand' in natural tourism worldwide.

Further studies must be done to define the driving factors that create PES awareness and involvement in national and global scale schemes, so that the concepts of self-sustainability can be extended to take advantage of support from the general public. For local schemes, the characteristics of the citizens have to be considered: demographic factors, purchasing power, awareness and population density can be the key to the success of PES.

References

Asquith, N. and Wunder, S. (2008) *The Bellagio Conversations: Payment for Watershed Services*, Fundación Natura Bolivia, Santa Cruz de la Sierra, Bolivia

Castro, R. (1999) *Valuing the Environmental Service of the Permanent Forest Stands to the Global Climate: The Case of Costa Rica*, UNDP, Mexico

Castro, R. and Vega, E. (2006) *Análisis del Proyecto Ecomercados II*, CLACDS-INCAE Research, INCAE Business School, Alajuela, Costa Rica

Cordero, D. (2002) *Lineamientos para la formulación de una estrategia para la sostenibilidad financiera del Programa PROCUENCAS de la ESPH S.A. bajo un Modelo de Inversión Ambiental Compartida (MIAC),* Universidad Nacional, Heredia, Costa Rica

ESPH – SEED (1999) *Estructura tarifaria hídrica ambiental ajustada: Internalización del valor de variables ambientales*, Empresa de Servicios Públicos de Heredia, Heredia, Costa Rica

Gámez-Hernández, L (2004) *Los recursos hídricos como servicio ambiental y aplicaciones prácticas de su valoración: El Caso de la Empresa de Servicios Públicos de Heredia*, Empresa de Servicios Públicos de Heredia, Heredia, Costa Rica

Hope, R., Porras I. and Miranda M. (2005) *Can Payments for Environmental Services Contribute to Poverty Reduction? A Livelihoods Analysis from Arenal, Costa Rica: Negotiating Watershed Services*, Unpublished report, CLUWRR Centre, University of Newcastle, IIED, Winrock International and Universidad Nacional de Costa Rica, Costa Rica

Pagiola, S. (2002) 'Paying for water services in Central America: Learning from Costa Rica', in S. Pagiola, J. Bishop and N. Landell-Mills (eds) *Selling Forest Environmental Services: Market-Based Mechanisms for Conservation and Development*, Earthscan, London

Redondo-Brenes, A. and Welsh, K. (2006) *Payment for Hydrological Environmental Services in Costa Rica: The Procuencas Case Study*, Tropical Resources Bulletin, Yale University, vol 25, spring

Rodríguez, J. (2005) *The Environmental Services Program: A Success Story of Sustainable Development Implementation in Costa Rica: FONAFIFO, Over a Decade of Action,* FONAFIFO, San José

Sills, E., Hartshorn, G., Ferraro, P. and Spergel, B. (2005) *Evaluation of the World Bank–GEF Ecomarkets Project in Costa Rica*, Panel Evaluation Report, North Carolina University, United States and the World Bank, Washington, DC

World Bank (2006) *Project Appraisal Document: Mainstreaming Market Based Instruments for Environmental Management (ECOMERCADOS 2)*, Report No 36084-CR, Washington, DC

Websites

FONAFIFO:
www.fonafifo.com/paginas_english/environmental_services/sa_requisitos.htm;
www.fonafifo.com/paginas_espanol/servicios_ambientales/sa_estadisticas.htm

GEF Project Executive Summary GEF Council Submission:
http://207.190.239.143/Documents/Work_Programs/documents/Costa_Rica_Mainstreaming_Instruments_ExecSumm.pdf

Personal interviews

Interview with Alexandra Saenz and Maria Elena Herrera held on 12 November 2008 at FONAFIFO's office, San José, Costa Rica

Interview with Luis Gámez held on 20 November and 28 November 2008 at ESPH offices, Heredia, Costa Rica

13

The Mexican PES Programme

Targeting for Higher Efficiency in Environmental Protection and Poverty Alleviation

José Eduardo Rolón Sánchez, Ina Salas Boucher and Iván Islas Cortés

Introduction

Around 500,000ha to 800,000ha of forest are lost each year in Mexico, placing the country in the second worst position regarding deforestation in Latin America. Worldwide, between 2000 and 2005, the country had the fourth highest rate of deforestation in terms of loss of primary forest (FAO, 2006). Public policy until the end of the 1990s, regarding the dynamics of land-use change, followed a pattern that favoured an increase of field cropping areas, as well as induced and cultivated grasslands in the country's forest areas. The forest sector had a weak capacity to compete economically with these alternative uses, including a low level of financial support. In an attempt to decrease the deforestation rate, the Mexican federal government implemented several strategies that included actions to develop the commercial forest sector; promote forestry plantations; reforest areas that have suffered deforestation; and protect and conserve native forest that is important in terms of environmental services.

The national payment for hydrological environmental services (PHES) is a federal programme that was established to support the strategy of conserving native forest. Created in 2003, the Mexican PHES is part of a general strategy focused on promoting a market in environmental services related to forests. The overall forest area covered through this strategy is considered to be the largest in Latin America (e.g. 350,000ha were incorporated in 2008 with a budget of 500 million Mexican pesos, around US$39 million) (CONEVAL, 2009).

This chapter describes and analyses the programme's initial implementation, evolution and results up to 2010. It explains how the programme was initially created, and the later transformation based on new information about the effectiveness of PHES. The chapter is divided into five sections. The following (second) section describes the initial setting of the PHES, considering its objective, scope, governance structure and source of funding. The initial working of the programme, through a process to select providers and purchasers, payable amounts, the duration of their participation and the monitoring system, is presented in the third section. This programme has evolved since its initial conception, following a process of constant evaluation. The fourth section analyses this evolution, including the political pressures to include poverty targets and other social and environmental concerns. The concluding section analyses how PHES reached the objectives that were originally established and the current process to improve the programme.

The Mexican Payment for Hydrological Environmental Services Programme

In 2003, Mexico's National Forestry Commission (Comisión Nacional Forestal, or CONAFOR) implemented its PHES programme, based on the premise of a positive relationship between forests and water quality and availability. As such, its main objective is to improve or maintain forest hydrological services by payment to forest owners for the provision of hydrological environmental services from their forest (DOF, 2003). The importance of this environmental service is justified: 101 aquifers (about 15 per cent of the total) from 653 of the country's aquifers are overexploited (SEMARNAT, 2008, p42). Therefore, the programme addresses a clear environmental and social need.

The programme was designed to cover forest areas in a good state of conservation. Other federal forest programmes addressed sustainable forest management or reforestation activities – for example, the Forestry Development Programme (Programa de Desarrollo Forestal, or PRODEFOR) and the Commercial Plantation Programme (Programa de Plantaciones Comerciales, or PRODEPLAN) that were intended to promote commercial forest plantations. The National Programme of Reforestation (Programa Nacional de Reforestación, or PRONARE) covered reforestation and restoration activities. PHES has its niche in forest conservation as part of a strategy to include economic incentives linked to the environmental services that well-conserved forests can provide.

The programme structure

The PHES is a federal programme of the Ministry of Environment and Natural Resources (SEMARNAT); it is directly managed and implemented within this ministry by CONAFOR. Given the characteristics of this environmental service, the programme was initiated and managed at a federal level, with the aim of looking for local markets during a later phase. The PHES was subsequently included by CONAFOR within ProArbol, an umbrella programme created in 2006 that encompassed all initiatives that provide economic incentives for forest management, restoration and conservation from this agency.

Different actors are involved in the operative structure of this programme. The financial body, the Mexican Forest Fund, is the financial entity that channels funding to the programme for payment to forest owners that provide environmental services. Decisions on how much to charge, how much to pay, whom to pay and the activities to be financed are all decided by the National Technical Committee. The Technical Advisory Council is the advisory body for the programme, including representatives from government, civil society, academia and forestry organizations representing providers of environmental services. This council has no decision-making capacity; it can only make recommendations. In order to ensure greater transparency, the Graduate School of the University of Chapingo and the Mexican Civil Council for Sustainable Forestry, a civic organization, act as external evaluators of this payment for environmental services (PES) programme. Every year these

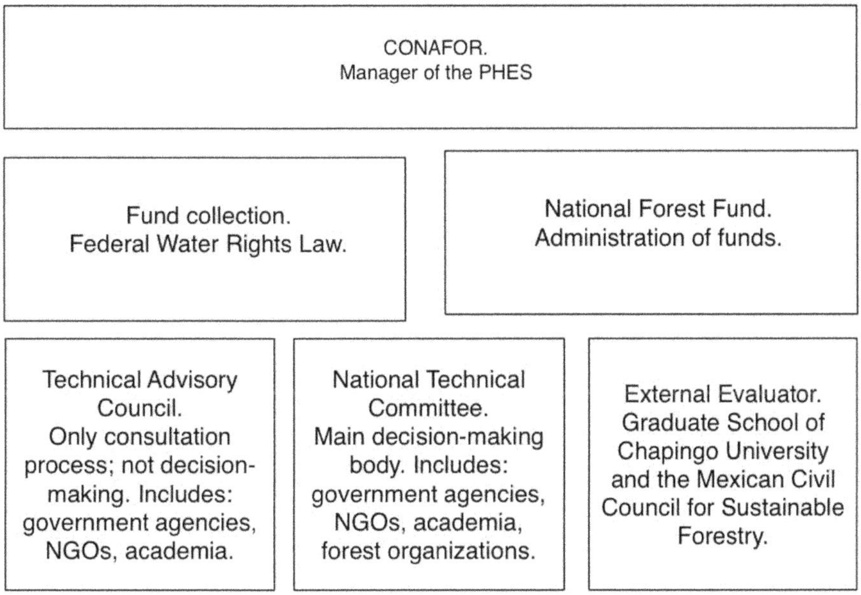

Figure 13.1 *Governance structure of the PHES programme in Mexico*

Source: chapter authors

organizations evaluate the programme performance in environmental, economic and social terms.

The government established a mandatory payment for the use of national waters into federal law. The taxpayer does not have a direct relationship with the provider of the hydrological environmental service and is forced to contribute a payment without being able to know if the service is actually delivered. Originally, the programme was assigned an amount of 200 million Mexican pesos (around US$15.5 million) from these charges; this has to be approved each year by the National Congress (Muñoz Piña et al, 2008).

Initial Settings

The process that leads to decisions on how to obtain funding for PHES (or the purchaser of environmental services), the selection of forest owners who provide these services and the amount that they were initially paid are presented in this section. Some other key decisions, related to the duration of the payment (over how many years) and the monitoring system, in order to verify that commitments are honoured, are also presented.

In the Mexican PHES, the federal government is the only purchaser of hydrological services, acting as a third party or intermediary on behalf of resource users (i.e. Mexican society as a whole), which provides the funding through the charge for water use previously mentioned. Thus, the federal government has a monopoly and, in practice, can select the best strategy to obtain the greatest environmental benefit from the economic resources allocated to the programme.

The fact that the Mexican government is the sole purchaser provides the possibility of choosing among offers from different service providers, considering limited programme funds (i.e. choose providers who maximize the environmental benefit at the lowest possible cost). It represents an opportunity to seek 'additionality' and 'conditionality' in the selection of the participating areas. However, the federal government also has pressure from interest groups that criticize the PES for not increasing payments or not accomplishing general development goals included in their government programmes, such as poverty alleviation (Alix-García and Javry, 2008).

The PHES programme has focused on the search for areas where there is a greater provision of hydrological environmental services and where the condition of additionality is addressed. For this purpose, in order to access PHES during the early years, the forest areas had to be located within 'eligible zones' annually modified and published by CONAFOR in agreement with established operation rules. The eligible zones were based on percentage of forest coverage (greater than or equal to 80 per cent of its area), critical areas of overexploited aquifer recharge, areas of superficial water shortage, and location of priority mountains or natural protected areas.

Definition of payable amounts

Payment for hydrological environmental services seeks to benefit conservation by recognizing the positive benefit generated by the existence of forest areas. However, since the quantification of the water service provided by forest coverage has not been determined, the PHES uses an alternative measure to calculate the amount to be paid, considering the potential profits generated by alternative land uses under the opportunity cost approach. Other services, such as the relationship between forest coverage and precipitation, groundwater infiltration, erosion and degradation, and the prevention of disasters caused by floods and landslides, are not considered.

Under this approach, the amount of the payment is the difference between the opportunity cost of maintaining forest land and the income generated by alternative activities. A payment of approximately 200 Mexican pesos per hectare (about US$16 per hectare) results from a comparison of the productivity of maize and beans, crops providing the lowest income, and was chosen to be able to start the programme with a low budget that could still target a large number of potential suppliers.

The PHES also introduced a payment difference depending on the type of forest, under the assumption that some kinds of vegetation (cloud forest) provides a better service than others, using a forest classification (importance for watersheds) provided by a group of environmental experts. They agreed that cloud forest is the most important for water capture during the dry season; they did not arrive at a consensus regarding other forest types and thus only recommended a greater payment for cloud forest. This decision was also the result of a political negotiation with producers and farmers who argued for higher amounts. As a consequence, the amounts were increased to 400 Mexican pesos per hectare for cloud forest (about US$31) and 300 pesos (about US$24) for other types of forest (Muñoz Piña et al, 2008).

Duration

The programme provides yearly payments to forest owners, renewed automatically during a period of five years if the forest providers comply with the conservation agreement. During the initial period (implementation of the programme), different organizations and actors involved discussed the possibility of renewing an agreement after the fifth year. The possibility of contract renewal was agreed, but stipulated that the programme should contribute to the creation of a market structure with private-sector participation. Since this is not a developed market, it is not easy to determine an appropriate timeframe; there is no formula to calculate how many years will be needed to create a PHES national market. It is also necessary to take into consideration that the PHES programme seeks to internalize positive externalities that, by definition, do not have a market because of their characteristics, and probably the creation of a market would not be possible without government intervention.

Monitoring

The verification of the provision of services (conditionality) is an essential part of the national PHES programme. Operation of the monitoring mechanisms is the responsibility of the federal government through CONAFOR, which performs regular evaluations of national forest coverage. This is done through analysis of high-resolution satellite images. If the image interpretation yields negative or dubious data, technical staff from CONAFOR visit these areas to verify the state of forest vegetation.

Evolution of the Programme

We can describe the evolution of the PHES programme as a process of learning by doing. From the initial settings presented in the previous section, the programme has been modified as new information about its effectiveness to conserve native forest was obtained. Of particular interest is the use of a targeting system using different criteria to select eligible forest owners. Within the eligible zones, the government could not prioritize areas when it had an excess of demands in terms of budget availability; the targeting system intended to correct this situation.

One criterion, in particular, the deforestation risk index, was used to focus on those forest owners where payments really make a difference in land-use decisions (i.e. an area that would be deforested without a PHES payment). Criteria designed to favour areas that could provide higher benefits, in terms of hydrological services or in environmental terms (e.g. part of a natural protected area), were also included.

Other criteria responded more to a general concern of policy-makers to address the social situation of forest owners. In Mexico, according to the poverty index of the National Population Council (CONAPO), more than 70 per cent of forest areas coincide with populations with high levels of marginalization or who are considered poor (Muñoz Piña, 2008). This led to the inclusion of the alleviation of poverty as a criterion in the PHES programme, one of the general goals of the federal government. Another change in the programme was the level of payment; offering the same amount throughout the country was not appropriate. Other changes responded to concerns about the need to transform the incentives for forest conservation in a more permanent manner, after the payment by the federal government was no longer available. The evolution of the programme is visualized in Figure 13.2.

Targeting criteria

PHES has been modifying its selection criteria of service providers to meet its objectives in a gradual manner as more information is gathered about the status of forest lands in Mexico and as a result of the implementation of the programme, seeking to achieve more efficiency in terms of service provision and covered area.

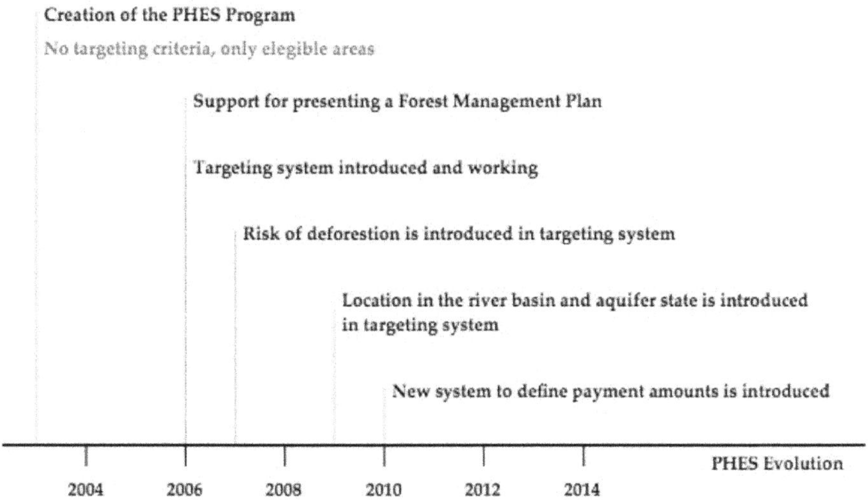

Figure 13.2 *Historic evolution of the PHES programme*

Source: adapted from DOF (2004, 2006, 2007, 2008, 2009, 2010)

One initial modification was the introduction of targeting criteria. During the first two years of the programme, targeting was only based on a map of eligible zones that could apply to the programme. These maps did not discriminate among different forest owners that applied to the programme, and all were within the eligible zones. As a result, areas with a high or low provision of hydrological services could be included, decreasing the efficiency of the programme. Overall, the previous system gave a single value to each criterion and, therefore, was more discretionary.

In 2005, the PHES included new criteria for targeting the financial beneficiaries of the programme in an attempt to improve performance by being more selective regarding the areas to be accepted for receiving a payment. The system was modified, giving points to applicants derived from several criteria, such as the percentage of forest cover in the area or if the property was inside a natural protected area. This system was not implemented that year as the federal budget could cover the demand from forest owners to participate in the programme.

In 2006 the targeting system was first applied, introducing as additional criteria the degree of marginalization of the population in the area and if the property was in an overexploited aquifer. The inclusion of marginalization as a criteria reflected a concern of policy-makers that almost all forest owners are in areas classified as high or very high marginalization. Addressing poverty and marginalization was also considered an objective of the ProArbol umbrella programme, where the PHES was located that year. On the other hand, the scoring system also gave higher scores for forest areas located within overexploited aquifers than for those with a medium or balanced overexploitation, trying to focus payment on those areas where water scarcity is of more concern.

Additionally, since 2007, a deforestation risk index was included in the criteria. The index was developed following a concern regarding additionality of payments. Many of the participants had a lower opportunity cost than the payment or could conserve forest without a payment. The deforestation risk index provided a measure to identify areas under high or very high deforestation pressure where payments would have more impact in conserving native forest. The index is based on a model that includes as variables primary vegetation, secondary vegetation, altitude, slope, temperature, accessibility to cities, marginalization, migration, corn yield, type of property, type of farmer, and protected natural areas, among others. The final model is represented geographically on a map entitled *Risk of Deforestation*, where there are five risk levels according to the priority for payment for environmental services: very low, low, middle, high and very high. This index, developed by the National Institute of Ecology (INE), gives higher scores to forest areas with higher deforestation risk where payments are more likely to make the difference between conserving and not conserving the forest.

Criteria that reflected the agenda of several of the participants in governing the PHES programme were also added. For example, the programme included higher scores for applicants in natural protected areas, and in federal, state or municipal protected areas, which are part of the agenda of the National Commission of Natural Protected Areas (CONANP) and of environmental organizations that participate in the technical committees.

Other criteria were designed to take into account environmental attributes that relate the presence of forests with water quality and quantity, such as biomass density, level of soil degradation or prevalence of evergreen forest ecosystems. Additional criteria reflected a preoccupation about the permanence of conservation action by participants, considering the PHES as a transition to develop PES markets in the forest sector. For example, a higher score is given to 'promising areas to develop PES markets', or 'if the applicant can show the support from a third party that is willing to pay for the hydrological service'. Another type of criteria was designed to support those applicants with more organizational capabilities to honour contract commitments, such as if the communal landowners (or *ejido*) have a monitoring committee, or if there is an association of forest owners in the area.

As a result of this evolution, the assignation system considers the criteria presented in Table 13.1.

Value of payments

The value of the initial fixed payment per hectare was adequate in some regions; but in others this payment was less than the opportunity cost of alternative activities. In some areas, with a high risk of deforestation and important in terms of environmental services, 400 Mexican pesos per hectare was not

Table 13.1 *Targeting criteria in PES*

General criteria for PES in ProArbol	*Score*
Property location regarding:	
• A natural protected area (federal)	Yes: 5 No: 1
• A natural protected area: state, municipal or private	Yes: 3 No: 1
• Limited within the 60 priority mountains	Yes: 3 No: 1
• A watershed where there are other PES	Yes: 5 No: 1
• The forest area has approved land management	Yes: 5 No: 3
• Risk of deforestation:	
– very high	Yes: 5
– high	Yes: 3
– average	Yes: 1
• Promising area for providing environmental services	Yes: 5 No: 3
• Communal landowners (*ejido*) have a monitoring committee	Yes: 3 No: 1
• Association of forest owners in the area	Yes: 5 No: 3
• Willingness of another actor to contribute to PES	Yes: 5 No: 1
• Biomass density	
– high	Yes: 5
– average	Yes: 3
– low	Yes: 1
Specific criteria for PHES	*Score*
Percentage of forest coverage:	
• Higher than 70%	Yes: 5
• Between 61% and 70%	Yes: 3
• Between 50% and 60%	Yes: 1
The property is located in:	
• Overexploited aquifer (greater than or equal to 100%)	Yes: 5
• Overexploited aquifer (but less than 100%)	Yes: 3
• Strategic areas for restoration	Yes: 5 No: 3
• Areas with low timber production	Yes: 3 No: 1
The property is in a watershed with:	
• Availability less than 4 (corresponds to high water scarcity zones in the classification of the National Water Commission of Mexico):	
– upper watershed	Yes: 5
– middle of the watershed	Yes: 4
– lower watershed	Yes: 3
• Availability between 4 and 7 (water scarcity zones):	
– upper watershed	Yes: 5
– middle of the wathershed	Yes: 3
– lower watershed	Yes: 2
• Availability more than 7 (water availability zones):	
– upper watershed	Yes: 4
– middle of the wathershed	Yes: 2
– lower watershed	Yes: 1
Soil degradation:	
• Low	Yes: 5
• Average	Yes: 3
• High	Yes: 1
Evergreen forest ecosystem prevalent:	
• Over 50%	Yes: 5
• Between 20% and 50%	Yes: 3
• Less than 20%	Yes: 1

Source: adapted from DOF (2009)

attractive. Differentiating payments can be extremely difficult to manage and subject to political contestation by groups that received a lower payment. In 2010, CONAFOR decided that payment amounts would be defined by deforestation risk and the type of ecosystem. The payments per hectare can vary from 280 Mexican pesos (US$21) to 1100 pesos (US$86) (Rolon and Reyes, 2009).

The programme budget has also increased year by year, reaching over 965 million Mexican pesos in 2009 (approximately US$80 million). This increased funding was achieved through the inclusion of additional financial sources, such as private and international contributors, and due to the approval by the National Congress for additional federal funds, which demonstrates the intention to cover the demand created by this programme (Rolon and Reyes, 2009).

Duration of payments

The initial one-year contract period, renewed automatically for a maximum of five years, has been retained with some minor modifications. One modification has been that the provider has to propose a plan for forest management when seeking to renew the contract. It is the aim that after five years of PHES payments, sustainable forest activities will be competitive and, thus, the forest will be conserved, or that alternative sources of payment, at local, regional and even international levels (e.g. carbon markets), will have been identified. Implementation of the forest management plan should have commenced by the end of this five-year period, permitting this transition.

Monitoring

The initial methodology of monitoring service provision by considering forest conservation has prevailed as it can cover a wide area of the national programme. Additional information that can be considered is from the National Forest Inventory, which seeks to evaluate forest quality and expansion. This system is like a forest census and is actualized every five years. At the regional level, CONAFOR also initiates technical visits to evaluate forest management and to approve the annual payment.

This methodology to monitor provision of hydrological services has some problems: there is no direct relationship between forest cover and water quantity and quality. This may lead to direct payments in eligible zones that do not coincide with underwater charge zones, or without knowing the actual direction of the infiltrated water and the corresponding discharging zones (Manson, 2007). This is very important in terms of demonstrating to local or regional purchasers that they are actually paying for a service that is being provided.

One of the recommendations made in a forum in Mexico during November 2009 on the use of PES for hydrological services was that the national programme should seek to advance more detailed studies to improve monitoring of service provision. However, the cost and the time needed to collect and analyse information is an important barrier for implementing these more detailed studies. This problem is exasperated because the use of the federal budget and the programme's operation rules are subject to concrete deadlines (Rolon and Reyes, 2009).

The possibility of using other types of methodology to evaluate provision of hydrological services and not only forest cover has been explored in projects at the regional and local scale. This includes some experience that uses community monitoring to obtain a more precise and less costly analysis of service provision (Manson, 2008). This type of experience could be adapted to the PHES programme, but at a more regional scale.

Results of the Programme

The objective of PHES is to obtain a hydrological environmental service from a provider; the payment makes a difference in the decision regarding the provision of the service. There is a lack of information to evaluate the effectiveness of targeting these payments in zones that provide hydrological services because the evaluation is based only on conservation of forest, using satellite images and, in some cases, *in-situ* verification, but not on measuring the hydrological services that this conservation may deliver. This is one weakness in the design of the programme. Only a few isolated studies by other research institutions exist on the performance of the programme (Peñuela-Arévalo and Carrillo-Rivera, 2009).

However, based on an assessment of the selection and targeting system, the effectiveness of the programme can be discussed. During the period between 2003 and 2008, the programme focused on forest land located in:

- extremely overexploited aquifers;
- areas with water shortage;
- highly marginalized areas;
- areas at high risk of deforestation; and
- protected natural areas.

Figure 13.3 illustrates the percentage of fund allocation each year, classified according to the characteristics of the forest areas. If the payments were equally allocated between these different criteria, a symmetrical diamond figure would result. This is clearly not the case in any year and the relative priority given to each variable has changed from year to year. For example, the priority given to overexploited aquifers decreased from 2005/2006 to 2007/2008; but the reverse trend occurred regarding the priority given to highly marginalized areas. The priority given to areas with a high and very high risk of deforestation also increased during the same period. However, a further increase in the protection of such lands is needed.

Progress in protecting areas with a high/very high risk of deforestation has been made (see Table 13.2). According to INE data, one indicator of the

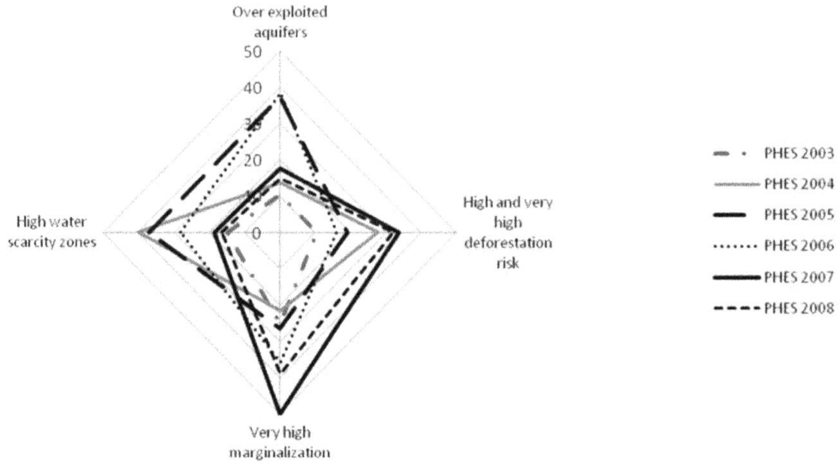

Figure 13.3 *Use of four basic criteria to select PHES participants (2003–2008)*

Source: Dirección General de Investigación en Economía y Política Ambiental, INE (2009)

'additionality' of the programme is the progress made in assigning resources to zones with a high and very high risk of deforestation. In 2003, the first year of the programme, 11 per cent of the PHES were assigned to these zones; in 2007 the figure was 34 per cent. However, the trends are erratic, with 50 per cent of the payments, and sometimes more than 70 per cent (in 2003 and 2006), going to zones with low and very low risk of deforestation.

As has been described above, forest owners can receive payments for a maximum of five years. After this time they have to reapply to the general selection process. One reason for limiting payments to five years is to encourage the creation of PES markets outside of the programme. Alternatively, this period gives forest owners the opportunity to develop economic activities that are compatible with forest conservation. In this regard, a survey was conducted by the Graduate College of the University of Chapingo to evaluate CONAFOR PES from 2003 to 2007. It was found that 33 per cent of people interviewed would like to dedicate their forest area to forest-related activities, which could

Table 13.2 *Distribution of PHES according to deforestation risk (2003–2008)*

Deforestation risk	*PHES 2003 (%)*	*PHES 2004 (%)*	*PHES 2005 (%)*	*PHES 2006 (%)*	*PHES 2007 (%)*	*PHES 2008 (%)*
Very high	4	11	7	6	14	11
High	7	17	13	10	20	16
Average	17	20	21	16	18	20
Low	30	30	27	25	22	26
Very low	42	22	33	43	27	27
Total	100	100	100	100	100	100

Source: Dirección General de Investigación en Economía y Política Ambiental, INE (2009)

Table 13.3 *Distribution of PHES according to the poverty marginalization criterion (2003–2008)*

Marginalization (poverty)	*PHES 2003 (%)*	*PHES 2004 (%)*	*PHES 2005 (%)*	*PHES 2006 (%)*	*PHES 2007 (%)*	*PHES 2008 (%)*
Very high	25	22	26	36	50	37
High	47	61	53	47	42	45
Average	18	8	14	12	5	13
Low	8	6	6	4	3	5
Very low	2	3	1	1	0	1
Total	100	100	100	100	100	100

Source: Dirección General de Investigación en Economía y Política Ambiental, INE (2009)

provide environmental services under a good forest management regime after the end of the PHES payment. However, 36 per cent reported that without a payment they would use the land for crop production or livestock; these landowners did not perceive economic alternatives to maintain the forest (Colegio de Postgraduados, 2008, p47).

As a result of the inclusion of PHES in the ProArbol programme, another objective was added. One of the objectives of the ProArbol programme is to decrease poverty and marginalization in forest areas. In the PHES programme, between 72 and 92 per cent of the payments were assigned to zones classified with a high and very high marginalization index (see Table 13.3). This occurred even before the marginalization criterion was included in the rules of operation of the PHES programme in 2006 because more than 70 per cent of forest areas are located in poverty zones.

In order to improve the efficiency of the programme, it is important to analyse how the payments have helped these populations to improve their situation. Although a systematic analysis of this topic has not yet been carried out, the annual evaluations of the programme, made by the Graduate College of the University of Chapingo, report that, for the period of 2003 to 2007, 79 per cent of the people interviewed considered the payment of PES as very important or important in terms of income. However, only 38 per cent considered that this payment increased their income compared to alternative activities before applying to PES; two-thirds of those interviewed mentioned an improvement of their standard of life, while one third said that this had not changed (Colegio de Postgraduados, 2008, p46).

The way ahead

Two central discussions regarding PHES are how to ensure that forest conservation continues after the five-year period and how to include a greater number of forest owners in the programme (the number has been increasing from year to year).

One way to respond to these concerns is to develop a scheme of shared funds (fondos concurrentes) that intends to involve more third parties in the financing of PHES – for example, CONAFOR would contribute 50 per cent to the shared fund and a third party (a state or municipal government, civic organization or private investor) the other 50 per cent (CONAFOR, 2010).

Through this scheme it is intended to expand the area that can be covered by the PHES programme, while promoting the idea that other sources of funding, which might be more directly connected to forest areas of interest to local and regional actors, are accessed. For example, the state government of Mexico (*Estado de Mexico*) has expressed an interest in this scheme to provide support in forest areas that are not necessarily within the 'eligible zones' or that achieved low scores within the general targeting system of the PHES. Although, the shared fund scheme gives more flexibility, it is necessary to assess if it will lead to forest conservation in areas with a high risk of deforestation or that provide hydrological services.

Conclusions

The goal of the PES described in this chapter is to obtain a hydrological environmental service from a forest owner by means of an annual payment, and that this payment actually makes a difference regarding the service provision decisions of the forest owner. However, in the case of this Mexican PHES programme, alleviation of poverty and marginalization has been added as an additional objective. Dealing with poverty alleviation within the PHES programme was a challenge. As a national programme, it should cover all forest areas; but different population groups with different economic levels live in these forest areas. These social groups have different capabilities to access the programme or influence the terms in which they reach agreements within it (e.g. regarding payment amounts or the type of areas that will be considered). These factors have led to changes in the PHES (e.g. of the eligibility criteria to receive payments).

Nonetheless, it is worth mentioning the importance of using instruments that contribute to strengthening the performance of the programme, such as the estimated opportunity cost of land and the deforestation risk, while including marginalization criteria for targeting payments within the programme. However, these modifications result in a cost, in terms of efficiency, since the inclusion of other objectives, such as poverty alleviation, can lead to payments being directed at areas containing highly marginalized communities where there is not a high deforestation risk. In other words, there is a trade-off in terms of efficiency when a PES scheme also seeks to alleviate poverty.

After nearly seven years of offering this PHES programme, it is necessary to evaluate the impact that it has had on avoided deforestation, including how this avoided deforestation may have been transferred to somewhere else in the country. This phenomenon, known as 'leakage', could have led to deforestation outside of the PHES areas. It is also timely to study areas where the payment did not succeed in avoiding deforestation in order to analyse what deforestation drivers remained in force and then either modify the PHES programme or use alternative policy instruments.

Finally, to evaluate its efficiency for long-term forest conservation, the ability of the PHES programme to create a capacity of forest owners to conserve their forests, even after the termination of programme payments, should be analysed. In some cases, the programme may have facilitated the capacity of forest owners to develop complementary conservation activities. In other cases, in the absence of a payment, economic activities derived from forest conservation will not compensate for the opportunity cost of alternative land uses, such as agriculture (including livestock), leading to a strong economic pressure to eliminate the forest. Such reflections are necessary to reinforce environmental policy designed to halt deforestation.

References

Alex-García, J. and Javry, E. S. (2008) *The Role of Deforestation Risk and Calibrated Compensation in Designing Payments for Environmental Services*, www.ine.gob.mx/areas/dgipea, accessed 21 September 2010

Colegio de Postgraduados (2008) *Evaluación Externa de los Apoyos de los Servicios Ambientales Ejercicio Fiscal 2007*, COLPOS-CONAFOR, Mexico

CONAFOR (2010) 'Lineamientos para promover mecanismos locales de pago por servicios ambientales a través de fondos concurrentes', CONAFOR, www.conafor.gob.mx/portal/index.php/tramites-y-servicios/apoyos, accessed 21 September 2010

CONEVAL (2009) 'Informe de la Evaluación Específica de Desempeño 2008', PROÁRBOL – Programa de Pago por Servicios Ambientales (PSA), Comisión Nacional Forestal, Mexico, www.coneval.gob.mx/contenido/eval_mon/3522.pdf, accessed 20 July 2010

Dirección General de Investigación en Economía y Política Ambiental, INE (2009) 'Pago por servicios ambientales en México: situación actual y objetivos de futuro', presentation at *Tercer Seminario de Divulgación, Servicios Ambientales: Sustento de la Vida*, Instituto Nacional de Ecología, Mexico, 7 August, available at www.ine.gob.mx/seminarios/827-seminarios3

DOF (2003) 'Acuerdo que establece las Reglas de Operación para el otorgamiento de pagos del Programa de Servicios Ambientales Hidrológicos', *Diario Oficial de la Federación*, 3 October, pp6–22

DOF (2004) 'ACUERDO que establece las reglas de operación para el otorgamiento de pagos del programa para desarrollar el mercado de servicios ambientales por captura de carbono y los derivados de la biodiversidad y para fomentar el establecimiento y mejoramiento de sistemas agroforestales (PSA-CABSA)', *Diario Oficial de la Federación, Segunda Sección*, 24 November, pp1–21

DOF (2006) 'ACUERDO por el que se expiden las Reglas de Operación de los Programas de Desarrollo Forestal de la Comisión Nacional Forestal', *Diario Oficial de la Federación, Tercera Sección*, 16 February, pp1–130

DOF (2007) 'ACUERDO por el que se expiden las Reglas de Operación del Programa Pro-Arbol de la Comisión Nacional Forestal', *Diario Oficial de la Federación, Segunda Sección*, 20 February, pp1–44

DOF (2008) 'ACUERDO por el que se expiden las Reglas de Operación del Programa Pro-Árbol de la Comisión Nacional Forestal', *Diario Oficial de la Federación, Cuarta Sección*, 28 December, pp1–129

DOF (2009) 'Reglas de Operación del Programa ProArbol 2010', *Diario Oficial de la Federación*, 31 December, pp1–61

DOF (2010) 'Reglas de Operación del Programa ProArbol 2010', *Diario Oficial de la Federación, Sexta Sección*, 31 December

FAO (United Nations Food and Agriculture Organization) (2006) *Evaluación de los Recursos Forestales Mundiales 2005: Hacia la ordenación forestal sostenible*, Rome, Italy, www.fao.org/docrep/009/a0400s/a0400s00.htm, accessed 10 June 2010

Manson, R. (2007) *Efectos del uso del suelo sobre la provisión de servicios ambientales hidrológicos: Monitoreo del impacto del PSAH*, INE, Mexico, www.ine.gob.mx/areas/dgipea, accessed 28 September 2008

Manson, R. (2008) *Efectos del uso del suelo sobre la provisión de servicios ambientales hidrológicos: Monitoreo del impacto del PSAH*, INE, Mexico, www.ine.gob.mx/dgipea, accessed 20 November 2009

Muñoz Piña, C., Guevara Sanginés, A., Torres, J. M. and Braña, J. (2008) 'Paying for the hydrological services of Mexico's forests: Analysis, negotiations and results', *Ecological Economics*, vol 65, no 4, pp725–736

Peñuela-Arévalo, L. A. and Carrillo-Rivera, J. J. (2009) 'La teoría de los Sistemas de Flujo de agua subterránea como herramienta para la definición de zonas de recarga en Programas de Pago por Servicio Ambiental Hidrológico', Congreso Mundial de Tierras Silvestres Wild 9, Mexico

Rolon, J. E. and Reyes, J. A. (2009) 'Los pagos por servicios ambientales hidrológicos en México: Documento síntesis de lecciones aprendidas de su instrumentación', INE, Dirección General de Economía y Pólitica Ambiental, Congreso Mundial de Tierras Silvestres Wild 9, Mexico

SEMARNAT (2008) *Estadísticas del Agua en México*, SEMARNAT, Mexico, DF

14

Assessing the Impact of Institutional Design of Payments for Environmental Services

The Costa Rican Experience

Juan Robalino, Alexander Pfaff and Laura Villalobos

Introduction

During the last few decades, the role that land use and land-use change have played in protecting biodiversity and in carbon sequestration has been widely recognized. This has led to a significant increase in the implementation of conservation policies around the world. Among these, policies that can simultaneously improve environmental outcomes and reduce poverty levels have gained special attention. It is under this context that researchers and policy-makers have focused their efforts on programmes of payments for ecosystem services.

By now, these programmes have been implemented in many countries of the Latin American region (e.g. Mexico, Ecuador and Colombia). But Costa Rica was one of the first developing countries to implement this policy nationwide, recognizing legally that forests generate services that need to be compensated. This pioneering effort was called the Payment for Environmental Services (PES) programme. It officially started in 1997 and is still under way.

This programme has been successful in different ways. However, there is an important measure of success that requires special attention. How much deforestation was the programme able to avoid? This is an issue that has continuously appeared as a potential problem in the design of conservation policies (Andam et al, 2008; Sims, 2010) and specifically for this programme (Pfaff et al, 2007; Robalino et al, 2008; Sierra and Russman, 2006).

If the programme has not been effective, policy-makers should consider strategies to address this problem. One strategy that has been discussed in the literature is improved targeting of high-threat areas. This consists of paying only those parcels with higher likelihood of deforestation. The amount of ecosystem services will increase if deforestation is actually reduced.

In this chapter, we describe the evolution of the PES programme payments in Costa Rica. The first years of implementation set the basis for what the programme has become. Important changes have been made since the beginning, such as the institution in charge of implementing the programme (2003), parcels selection criteria, and new offices that were opened in different areas of the country with the objective of reducing application costs.

Using 2003 as the starting point of when these changes took place, we discuss if they had a programme efficiency effect on reducing deforestation. We focus on forest conservation contracts because it is the most important category of the programme in terms of budget and amount of land enrolled. We use matching techniques, geographic information systems (GIS), characterize the areas where payments were implemented in each of the time periods using a long list of variables, and look for similar areas that did not receive payments. We find that, as other studies have found for this period (Robalino et al, 2008; Arriagada, 2008), the impacts are low but significant. While it seems that, overall, institutional changes have not had a significant effect on impact, we also look at the impacts of forest conservation contracts per office. We find that those offices located in areas with high deforestation tend to have higher impacts.

Efforts towards improving targeting based on the likelihood of deforestation will easily improve the programme's effectiveness. Evidence of this is that office contracts in areas where deforestation is high are significantly more effective. Shifts in budget distribution to these offices could lead to further impacts. Additionally, information such as distance to roads and markets can also be used to estimate the likelihood of a parcel to be deforested, which in turn can lead to an increase in the amount of avoided deforestation of future contracts.

The chapter is organized as follows: first we describe the evolution of the programme and then present our data. We discuss the methods used to estimate the impacts, before presenting the results. We then discuss our findings and finally present our conclusions.

Evolution of the Programme

The concept of PES as it is currently conceived is the result of a long policy process that Costa Rica has gone through for many decades. The country started to design policies to prevent deforestation with the inception of the first forestry law in 1969 (see Rojas et al, 2003, for evolution of forest incentives); but at the same time, agriculture and cattle activities were favoured as a strategy for rural development, which hindered the forest policies' success (Moreno-Díaz, 2005).

By 1996, two important events had occurred:

1 The country had to reduce subsidies to productive sectors, including the forestry sector, as a part of the structural adjustment programme negotiated with the World Bank.
2 The forestry sector had developed an influential institutional framework and exerted pressure against the elimination of their privileges (Rojas et al, 2003).

In 1996, Forestry Law 7575 introduced two important policies. First, the law prohibited land-use change. This clearly was going to hurt the forestry sector; therefore, the development of compensation mechanisms for retaining forest cover seemed a fair and reasonable next step. The Forestry Law therefore introduced the current PES system.

This law applies the user pays principle. The objective targeted small and medium-sized farmers who had a sustainable forest management plan certified by a licensed forester (Sierra and Russman, 2006) and compensated them in order to provide an incentive for retaining forest. The four environmental services recognized by the new forest law included:

1 mitigation of greenhouse gas emissions;
2 hydrological services, including provision of water for human consumption, irrigation and energy production;
3 biodiversity conservation; and
4 provision of scenic beauty for recreation and ecotourism.

The institution initially in charge of payments management was the National Conservation Areas System (SINAC). SINAC was formally created by the Biodiversity Law of 1998[1] as the institution responsible for forestry, wildlife and protected areas management. In 1996 the Forestry Law also formally established the National Forestry Financing Fund (FONAFIFO) as a fund aimed at financing the forestry sector for reforestation, protection and management activities that were included in the PES programme. When the PES became operational in 1997 and until 2002, SINAC was in charge of the PES programme management using FONAFIFO as the financing fund.

In 2002, Decree No 30762 MINAE of the Forestry Law was reformed so that FONAFIFO assumed administration of the PES contracts. According to the decree, SINAC would concentrate on its habitual responsibilities (conservation policies through protected areas management) and use the experience acquired by FONAFIFO during the past five years to expand the PES programme by improving the quality of the service to landowners.

This decree consigns FONAFIFO the regulation and determination of the administrative and technical procedures for PES, including the procedure manual, beneficiary selection, documents review and contracts formalization, the definition and extent (in hectares) of priority areas, and the terms of payment. According to this law, SINAC will help FONAFIFO in supervising the approved contracts through its regional offices.

Relevant changes in the programme took place in FONAFIFO. At the beginning of the programme, the PES programme reimbursed three types of actions by landholders:

1 forest protection;
2 sustainable forest management; and
3 reforestation activities.

In 2003, a forest management category was eliminated and agroforestry systems were introduced; in 2004, a natural regeneration category was also included (see www.fonafifo.com).

During the 2000 to 2002 period, SINAC also operated through ten regional offices, which are located according to the 11 conservation areas that divide the country. There are no PES contracts in the Isla del Coco Conservation Area.

In 2003, under FONAFIFO's administration, seven new offices were opened in strategic locations around the country (the main office remains in San José, the capital city). This reduces the costs of application for people in remote rural areas. Both the conservation areas of the SINAC administration and FONAFIFO's offices are illustrated in Figure 14.1.

Another important change in the programme was the amount and currency of the payments. Since the payments are distributed over a certain number of years (forest protection and reforestation are five-year contracts and agroforestry spans three years), landowners would receive a lower total payment amount because of inflation. Therefore, in 2005, higher payment amounts were approved and they were fixed in dollars currency instead of colones, the local currency, in order to compensate for inflation.

In order to finance the forestry sector, FONAFIFO receives funds through different financing sources: public funds in the national budget, donations, credits conceded by international organisms, private funds, own-generated funds, and timber and fuel taxes (see Chapter 12 in this volume and www.fonafifo.com). Also, in 2001 FONAFIFO created the Environmental Services Certificate (ESC), which is a financial instrument where FONAFIFO receives funds from companies and institutions interested in compensating forest owners for preserving them.

In 1997, US$21 million in payments were allocated to 88,830ha of forest protection, 9325ha of forest management and 4629ha of reforestation. By 1998, there was a substantial excess demand for participation in the programmes; the formal waiting list may be in excess of 70,000ha (Chomitz et

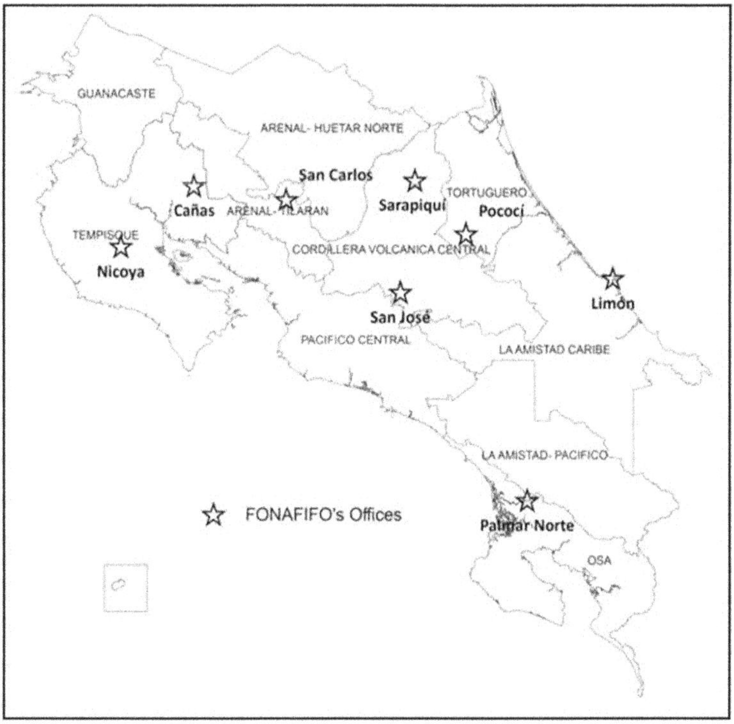

Figure 14.1 *Map of Costa Rica's conservation areas and FONAFIFO's offices*

Source: information from FONAFIFO

al, 1998). By 2007, US$24.8 million was allocated with 1180 contracts, 66,000ha (91 per cent in forest protection) and 541,531 trees.

When the PES programme started in 1997, targeting was ambiguous and the local offices were responsible for contract assignments according to their

Table 14.1 *Assigned amounts per hectare and/or trees for payment for environmental services, 2000–2008 (US$)*

Year	*Forest protection*	*Reforestation*	*Agroforestry systems (trees)**	*Exchange rate****
2000	214	548	-	308
2001	221	565	-	329
2002	220	563	-	360
2003	218	559	0.80	399
2004	219	559	0.80	438
2005–2008**	320	816	1.3	

Notes: * The amount of payment is per tree.
** Amounts fixed in US$.
*** Ministerio de Planificación de Costa Rica (MIDEPLAN) based on Banco Central de Costa Rica (BCCR).
Source: www.fonafifo.com

Table 14.2 *Distribution of hectares for payments for environmental services, 2000–2007*

Year	*Forest protection (ha)*	*Reforestation (ha)*	*Agroforestry systems (trees)*	*Total number of contracts**
2000	26,583	2457	–	271
2001	20,629	3281	–	287
2002	21,819	1086	–	297
2003	65,405	3155	97,381	672
2004	71,081	1557	412,558	760
2005	53,493	3602	513,684	755
2006	19,972	4866	380,398	619
2007	60,567	5826	541,531	1180

Note: * Includes contracts in forest management and established plantations.
Source: www.fonafifo.com

own land priorities. For instance, in 1997 the priority area for the assignment of payments was the entire country, although there were some general priority criteria that the offices could take into account when targeting. As the learning process developed, these criteria became simpler and clearer. From 2003 to the present, priority areas were confined to those lands that fit five specific criteria:

1 areas inside biological corridors;
2 projects which have expired contracts from prior years;
3 forest areas that function as watershed protection;
4 private areas inside protected areas;
5 within the above criteria, priority is given to those districts with a Social Development Index below 40 per cent.

The objective of introducing the last criteria was to reach the poorest landowners in rural areas, so that the programme could achieve both conservation and social outcomes. This is a noteworthy effort to improve the programmes' impacts. Nevertheless, it can be argued that targeting the poorest districts does not necessarily guarantee that the poorest are enrolled, since the spatial scale of reference and the district might, in some cases, be too general.

Methods of Evaluation

Identifying the overall net effect would be a simple task if payments for environmental services were randomly distributed across Costa Rica. Then, deforestation rates in areas that were not enrolled in the programme would be good estimates of the impacts. All other observable and unobservable factors that affect deforestation would, in all expectation, be identical in the two groups.

However, PES is not randomly distributed. Governments and policymakers have specific objectives and restrictions when choosing these sites. As discussed in the previous section, there are some prioritization criteria and,

Table 14.3 *Comparison of parcels enrolled and not enrolled in the PES programme (selected characteristics)*

Variables	*Parcels not enrolled in PES*	*Parcels enrolled in forest protection contracts*				*Parcels not enrolled matched with enrolled parcels*			
	(1)	*(2) SINAC*	*Difference in means (1) versus (2)*	*(3) FONAFIFO*	*Difference in means (1) versus (3)*	*(4) SINAC*	*Difference in means (2) versus (4)*	*(5) FONAFIFO*	*Difference in means (3) versus (5)*
Deforestation (fraction)	0.0251	0.0000	–0.0251*	0.0000	–0.0251***	0.0306	0.0306***	0.0338	0.0338***
Implied annual deforestation rate+	0.51%					0.62%		0.69%	
Land characteristics:									
Distance to local roads (logarithm) (m)	7.32	7.46	0.14	7.82	0.50***	7.41	–0.05	7.80	–0.02
Distance to national roads (logarithm) (m)	7.79	8.09	0.30***	8.35	0.56	8.10	0.01	8.34	–0.01
Distance to rivers (logarithm) (m)	6.82	6.94	0.12	6.90	0.08	6.97	0.03	6.77	–0.13*
Distance to San José (logarithm) (m)	11.52	11.48	–0.04	11.37	–0.15***	11.47	–0.01	11.37	0.00
Distance to Pacific (logarithm) (m)	10.01	10.32	0.31**	10.75	0.74***	10.32	0.00	10.73	–0.02
Distance to Atlantic (logarithm) (m)	11.35	11.14	–0.21**	11.00	–0.36***	11.18	0.04	10.99	–0.01
Distance to towns (logarithm) (m)	7.77	7.99	0.22***	8.16	0.39***	7.98	–0.01	8.16	0.00
Distance to sawmills (logarithm) (m)	9.70	9.73	0.03	9.76	0.06*	9.70	–0.03	9.76	0.00
Distance to schools (logarithm) (m)	9.37	9.47	0.10*	9.65	0.28***	9.42	–0.05	9.63	–0.02
Precipitation (mm^{-1} y^{-1})	3366.27	3668.70	303***	3601.73	235.45***	3659.0	–10.00	3619.0	18.00
Elevation (metres above sea level)	430.84	435.38	4.54	529.44	98.60***	472.0	36.70	531.0	2.00
Slope (%)++	59.43	60.80	1.37	44.63	–14.80***	66.69	5.89	46.15	1.52
Stock of forest (%)	50.57	57.31	6.74***	61.05	10.48***	57.08	–0.23	61.64	0.59

Notes: Statistical significance: *** $p < 0.01$; ** $p < 0.05$; * $p < 0.1$. No asterisk means no significant difference.

+ We calculated the implied annual deforestation rate using the following formula: total deforestation in 5 years = $(1 - \text{annual deforestation rate})^5$

++ 100% slope = 45°.

within these criteria, payments are assigned on a first come, first served basis. Therefore, it could be argued that not every landowner has the same probability of being chosen; rather, only those with better access to information, lower transaction costs and certain geographic conditions are enrolled.

Therefore, if the impacts of the programme were to be estimated by just comparing deforestation in the parcels enrolled in the programme with the deforestation payments outside the programme, a selection bias would be included in the effect (see Lee, 2005, and Caliendo and Kopeining, 2005, for how to estimate policies' effects).

This selection bias can be observed in our data. In order to determine if parcels enrolled in the programme are similar to those parcels outside the programme, we compared the mean characteristics of each of the groups (compare column 1 versus columns 2 and 3 in Table 14.3). Parcels not enrolled in the programme and parcels enrolled by SINAC have similar slopes and elevations. However, parcels enrolled by FONAFIFO have lower slopes and elevations. Parcels not enrolled in the programme are closer to local and national roads, towns, sawmills and schools than parcels enrolled by both FONAFIFO and SINAC. This suggests that the parcels enrolled in the programme are located in more remote areas, where the opportunity costs of land are lower.

This analysis illustrates that systematic differences exist in the characteristics of the groups. It is thus not easy to infer if the programme caused the differences in deforestation or if these differences are related to these characteristics. We used matching techniques to address the bias originated by the non-random allocation of PES contracts across Costa Rica.

Economists have applied matching techniques to overcome these problems (for reviews see Dehejia and Wahba, 2001; Caliendo and Kopeining, 2005). Within environmental economics, matching strategies have been used to evaluate the effect of air quality regulations on environmental outcomes (Greenstone, 2004) and on economic activity (List et al, 2003). However, just recently, matching techniques have been used to evaluate the direct effects of land restriction policies (Andam et al, 2008; Pfaff et al, 2007, 2009; Robalino et al, 2008; Sims, 2010).

The principle of this technique is to find an adequate control group by matching each treated observation to the most similar untreated observations. For example, parcels enrolled in PES are located far away from roads. Therefore, we will compare the deforestation rates of areas far away from roads enrolled in PES contracts with deforestation rates of areas far away from roads not enrolled in PES contracts. This eliminates the bias caused by the accumulation of PES contracts in areas far away from markets. Matching applies this principle to a multidimensional space of characteristics.

There are many matching strategies. We use propensity score matching (PSM) developed by Rosenbaum and Rubin (1983). One key advantage of using propensity score matching estimates is that results are less sensitive to the choice of functional form in the model (Rosenbaum and Rubin, 1983; Dehejia and Wahba, 2001; Ho et al, 2007). In other words, when using parametric methods, how other independent variables are included in the model (linearly, with squares or cubes) can affect the estimates of the effect of the policy. But given that before estimating the effect we make sure that treated and untreated observations are balanced in relation to those variables, the inclusion of different functional forms does not affect the result.

However, as with all approaches, matching requires certain conditions for the identification of the effect. All relevant characteristics that might affect both the likelihood of being treated and the pre-treatment outcome must be included when selecting the control group.

We first estimated a probability of being enrolled in the PES programme for all treated and untreated observations, based on a set of characteristics (see following section). Using this probability, we determined how similar treated (enrolled parcels) and untreated (parcels not enrolled) observations were. For each treated parcel, we looked for four untreated parcels with the most similar probability of being enrolled in the programme. In order to avoid choosing observations that were too different, of those four observations chosen we only used those that were within a 0.1 per cent probability distance.

Using this approach, we generated adequate control groups for the parcels that were enrolled by SINAC and for the parcels that were enrolled by FONAFIFO. When we compare the characteristics of the parcels enrolled in conservation contracts by SINAC and the characteristics of our chosen control group, we can see that they are, in general, more similar than when we compare SINAC's parcels with the rest of the parcels outside payments.

After the control group was properly chosen, we obtained two groups that were similar except for the fact that one received payments and the other did not. Therefore, we could compare the average deforestation among the groups and conclude that if there was any significant difference, the programme had an effect.

However, note that even after matching there might still be significant differences among the samples. For example, the distance to rivers for the parcels enrolled by FONAFIFO is statistically different compared to the distance to rivers of the matched control group for FONAFIFO. Slopes of parcels enrolled by SINAC are also different from the slopes of SINAC's matched control group. This indicates that in spite of the improvement, there are some characteristics that systematically differ between the samples, and additional corrections are needed to take out the effect these characteristics might have on the outcome. To account for this, we ran an additional regression, again using the control variables (e.g. distance to rivers) to reduce their role in the estimation of the impact (Ho et al, 2007).

Data

Using GIS, we randomly drew 50,000 locations across Costa Rica. Each of these locations is our unit of analysis and represents a parcel. On average, we sampled one parcel for each square kilometre across Costa Rica. We used forest cover maps for 2000 and 2005 (see Sanchez et al, 2007; Pfaff et al, 2007) that were based on aerial and satellite pictures. This information allowed us to determine the presence of forest in each of the 50,000 randomly drawn parcels for each year and, therefore, the dynamics of deforestation for 2000 to 2005. We dropped the parcels that were flagged as problematic due to the uncertainty about the presence of forest. We were left with 47,241 observations. Of those observations, we focused on deforestation decisions and therefore we only considered points that were covered by forest in 2000 (20,760 observations).

Maps of the PES contracts across Costa Rica are also available. Given that we studied only areas that potentially received payments for environmental services, we only considered locations outside protected areas or government land. Therefore, we were left with 9107 observations.

As previously discussed, there are different types of PES contracts. This analysis focused only on forest protection contracts. We eliminated sample parcels that were enrolled in other types of contracts. Contracts that were implemented before 2000 were also eliminated from the control group.

We were finally left with 604 observations that were enrolled in PES forest protection contracts between 2000 and 2005. Out of those locations with contracts, 72 were implemented in 2000, 61 in 2001, 19 in 2002, 166 in 2003, 190 in 2004 and 113 in 2005. This means that we have 152 locations selected by SINAC (2000 to 2002) and 469 selected by FONAFIFO (2003 to 2005).[2] We were left with 7523 observations that were untreated – that is, observations that did not receive payments.

GIS was also used to obtain parcel characteristics. Parcel characteristics allowed us to find an adequate control group. We obtained accurate measures of slopes of the terrain, precipitation, elevation, and distance to rivers and oceans that we classified as natural characteristics. We also computed distances to San José, population centres, sawmills, schools, national roads and local roads. Finally, we obtained the forest stock for each grid. All of these variables comprise the characteristics we used to find an adequate control group. The natural characteristics are related to the productivity of land, and the distances to relevant points indicate access to markets and the availability of infrastructure.

Impacts

As discussed in the section on 'Methods of evaluation', if payments were allocated randomly, we could use the deforestation in parcels not enrolled as an estimate of the counterfactual of deforestation with payments. In other words, the effect of the programme would be the difference between the deforestation of parcels with payments and the deforestation of parcels without payments. In Costa Rica, between 2000 and 2005, this difference is 2.51 per cent – that is, we would conclude that the programme prevented 2.51 per cent of the land enrolled from being deforested in a five-year period, or that the programme

Table 14.4 *Mean comparison and matching estimates of the impact of forest protection contracts during PES (by institution)*

Approach	*SINAC (2000–2002)*	*FONAFIFO (2003–2005)*
Mean comparison	−0.0251*	−0.0251***
Standard error	[0.0129]	[0.0073]
Annual impact (%)	0.50%	0.50%
Mean comparison after matching	−0.0306***	−0.0338***
Standard error	[0.0077]	[0.0051]
Annual impact (%)	0.61%	0.67%
Lineal regression after matching	−0.0301***	−0.0340***
Standard error	[0.0075]	[0.0053]
Annual impact (%)	0.61%	0.69%

Note: Statistical significance: *** $p < 0.01$; ** $p < 0.05$; * $p < 0.1$.

prevented 0.5 per cent of the land enrolled from being deforested every year (see Tables 14.3 and 14.4).[3]

As discussed in previous sections, since contracts are not randomly assigned, the deforestation of parcels not enrolled is not a good estimate of what would have happened with the enrolled parcels if there were no programme, and the estimated impact presented in the previous paragraph is therefore biased. After using matching to find an adequate control group, we again compared the mean deforestation rate between treated and control groups. We found that during SINAC's administration from 2000 to 2002, deforestation decreased by 0.61 per cent annually, and FONAFIFO stopped deforestation in 0.69 per cent of the land enrolled annually from 2003 to 2005. This suggests that the institutional and operative changes introduced by FONAFIFO's administration improved the programme's impact according to a small but significant magnitude.

Our conclusions did not change even after running a regression analysis using the control variables for any remaining imbalance (see Table 14.4). After comparing the treated observations with the matched untreated observations using other variables, the results were very similar. The parcels chosen by SINAC between 2000 and 2002 avoided 0.61 per cent of deforestation per year of the land enrolled, while the parcels chosen by FONAFIFO between 2003 and 2005 avoided 0.67 per cent of deforestation per year.

When we looked in detail at FONAFIFO's work by evaluating the impact that each office has had (see Table 14.5), we found that the annual effect of the parcels enrolled in Cañas, Limón and Nicoya is virtually negligible. This implies that virtually all of the parcels enrolled in the programme through these offices would not have been deforested anyway. In other words, these parcels have characteristics that result in no deforestation threat, so they would have remained in forest independent of the payment. This is an important result, since it suggests that for the additionality criteria to be met, better targeting has to be achieved so that payments are used to block actual deforestation.

Table 14.5 *Reduced deforestation of forest protection contracts by FONAFIFO's offices during 2000–2005*

Office	*Effect*
Cañas	0.0000
Standard error	0.0000
Limón	0.0000
Standard error	0.0000
Nicoya	0.0000
Standard error	0.0000
Palmar Norte	–0.0103
Standard error	–0.0078
Pococí	–0.0028
Standard error	–0.0132
Sarapiquí	–0.0238*
Standard error	–0.0129
San Carlos –0.0460***	
Standard error	–0.012
San José	–0.0205*
Standard error	–0.0106

Note: Statistical significance: *** $p < 0.01$; ** $p < 0.05$; * $p < 0.1$.

Palmar Norte and Pococí follow similar levels of impacts. In contrast, the offices located in Sarapiquí, San Carlos and San José were able to choose parcels with characteristics that our evidence shows would have been deforested if they had not enrolled in the programme. There is an important issue that we should point out. If deforestation around Limón, for example, is insignificant, even with large efforts of targeting, the officers from Limón would not have been able to improve their impact. However, in San Carlos, for example, where deforestation rates are extremely high, it is easier for officers to choose those that are highly likely to be deforested.

Conclusions

We find that PES for forest conservation in Costa Rica between 2000 and 2005 has had a significant effect in reducing deforestation. When comparing the results from previous periods to these new results, we can certainly conclude that these policy efforts have become more effective over time. However, there is still room for improvement. The levels of impact are still low. We find that the reduction in yearly deforestation ranges between 0.6 and 0.7 per cent of the land enrolled in the programme. In other words, between the start and the end of a typical forest conservation contract (five years), deforestation would have been avoided in 3 to 3.5 per cent of the land enrolled.

Improving these numbers can be a difficult task given the low levels of deforestation across the country and the lack of accurate opportunity costs data. Efforts towards improving targeting based on the likelihood of deforestation will improve the programme's effectiveness. An example of this is that the contract offices in areas where deforestation is high are significantly more effective.

Moreover, information such as was used in the analysis (e.g. distance to roads, rivers and markets, and costs) can be used to estimate the likelihood of a parcel to be deforested and therefore increase the amount of avoided deforestation in future contracts.

Finally, we emphasize the importance of estimating the impacts of a policy by finding an adequate comparison group. This method has been particularly useful when evaluating the impacts of PES since it allows us to determine what would have happened if the parcels were not enrolled in the programme. According to our results, the programme's impact was higher once we controlled for other factors that could have an effect on the deforestation outcome. Therefore, policy-makers can use more precise information when learning and making decisions on how to improve the programme's impacts.

Notes

1 Even though SINAC was formally established in 1998, PES operations started in 1997.

2 There are 17 observations that had payments both with SINAC and FONAFIFO. Therefore, the sum of SINAC and FONAFIFO contracts is more than the observations enrolled in PES forest protection contracts.

3 We calculated the implied annual deforestation rate using the following formula: total deforestation in 5 years = $(1 - \text{annual deforestation rate})^5$.

References

Andam, K., Ferraro, P. J., Pfaff, A., Sánchez-Azofeifa, G. A. and Robalino, J.A (2008) 'Measuring the effectiveness of protected area networks in reducing deforestation', *Proceedings of the National Academy of Sciences of the United States of America*, vol 105, no 42, pp16,089–16,094

Arriagada, R. (2008) *Private Provision of Public Goods: Applying Matching Methods to Evaluate Payments for Ecosystem Services in Costa Rica*, LACEEP Policy Brief No 8

Caliendo, M. and Kopeinig, S. (2005) *Some Practical Guidance for the Implementation of Propensity Score Matching*, IZA Discussion Papers 1588, Institute for the Study of Labor (IZA), Germany

Chomitz, K. M., Brenes, E. and Constantino, L. (1998) *Financing Environmental Services: The Costa Rican Experience*, Central America Country Management Unit, Latin American and the Caribbean Region, World Bank, Washington, DC

Dehejia, R. H. and Wahba, S. (2001) 'Propensity score-matching methods for nonexperimental causal studies', *Review of Economics and Statistics*, vol 84, no 1, pp151–161

Greenstone, M. (2004) 'Did the Clean Air Act cause the remarkable decline in sulfur dioxide concentrations?', *Journal of Environmental Economics and Management*, vol 47, no 3, pp585–611

Ho, D. E., Imai, K., King, G. and Stuart, E. A. (2007) 'Matching as nonparametric preprocessing for reducing model dependence in parametric causal inference', *Political Analysis*, vol 15, pp199–236

Lee, M. J. (2005) *Micro-Econometrics for Policy, Programme, and Treatment Effects*, Oxford University Press, Oxford, UK

List, J. A., Millimet, D. L., McHone, W. W. and Fredriksson, P. G. (2003), 'Effects of environmental regulations on manufacturing plant births: Evidence from a propensity score matching estimator', *Review of Economics and Statistics*, vol 85, no 4, pp944–952

Moreno-Díaz, M. L. (2005) *Pago por Servicios Ambientales, la experiencia de Costa Rica* [*Payments for Environmental Services: The Experience of Costa Rica*], Instituto Nacional de Biodiversidad INBIO, Heredia, Costa Rica

Pfaff, A., Robalino, J. A. and Sánchez-Azofeifa, G. A. (2007) 'Evaluating the impacts of payments for environmental services', Presentation at NBER Summer Institute, July, www.pubpol.duke.edu/research/papers/SAN08-05.pdf

Pfaff, A., Robalino, J. A., Sánchez-Azofeifa, G. A., Andam, K. and Ferraro, P. (2009) 'Park location affects forest protection: Land characteristics cause differences in park impacts across Costa Rica', *BE Journal of Economic Analysis and Policy*, vol 9, no 2, www.bepress.com/bejeap/vol9/iss2/art5

Robalino, J. A., Pfaff, A., Sánchez-Azofeifa, G. A., Alpízar, F., Leon, C. and Rodriguez, C. M. (2008) 'Deforestation impacts of environmental services payments: Costa Rica's PSA Program 2000–2005', Presentation at BioEcon Conference, 10 September, www.rff.org/RFF/Documents/EfD-DP-08-24.pdf

Rojas, M., Aylward, B. and International Institute for Environment and Development, Environmental Economics Programme (2003) *What Are We Learning from Experiences with Markets for Environmental Services in Costa Rica? A Review and Critique of the Literature*, International Institute for Environment and Development, London

Rosenbaum, P. R. and Rubin, D. B. (1983) 'The central role of the propensity score in observational studies for causal effects', *Biometrika*, vol 70, pp41–55

Sanchez, G., Pfaff, A., Robalino, J. and Boomhower, J. (2007) 'Costa Rica's payment for environmental services program: Intention, implementation and impact', *Conservation Biology*, vol 21, no 5, pp1165–1173

Sierra, R. and Russman, E. (2006) 'On the efficiency of the environmental service payments: A forest conservation assessment in the Osa Peninsula, Costa Rica', *Ecological Economics*, vol 59, pp131–141

Sims, K. R. E. (2010) 'Conservation and development: Evidence from Thai protected areas', *Journal of Environmental Economics and Management*, vol 60, no 2, September, pp94–114

15

Certification Process in the Coffee Value Chain

Achievements and Limits to Foster Provision of Environmental Services

Gabriela Soto and Jean-François Le Coq

Introduction

Various mechanisms have been promoted to foster the provision of ecosystem services. Product certification is one of the most promising and developed instruments to reward the socially and environmentally friendly practices of market producers.

This strategy started with organic production and Fair Trade, but in recent years has grown to a wide variety of certification labels, with the coffee sector experiencing the largest development. Before and during the coffee crisis of 2000 to 2003, different labels emerged, such as Smithsonian Bird Friendly coffee, Rainforest Alliance, Starbucks CAFE practices (Coffee and Farmer Equity practices), Utz Kapeh (now Utz Certified), the Common Code for Coffee Community (4C) and recently Nestlé's Nespresso AAA label. This trend is not unique to the coffee sector; similar certifications are being developed for sustainable cocoa, pineapple, cattle and palm oil. This strategy is growing and proving an important potential in changing how our food is produced.

With more years of implementation, the coffee sector offers a wide perspective to analyse the achievements and limits of this strategy in fostering environmental services. This chapter describes the development of the certified coffee market and the characteristics of the different certification strategies within Central America. We then review their achievements and limitations in promoting ecosystem services, particularly related to biodiversity conservation, and their reported socio-economic impacts. Finally, we propose areas of improvement to increase their potential as a tool to foster the provision of ecosystem services in the region.

Origin and Development of Sustainable Coffee Labels

There are three stages in the creation of sustainable coffee labels that explain the differences in objectives, methods and impacts. The first phase is linked to the global development of organic and Fair Trade (Fairtrade Labelling Organizations, or FLO, certified) production. This phase experienced exponential market growth during the 1980s, but did not enter into the Latin American coffee sector until the early 1990s, and it gained strength during the coffee crisis of 2000 to 2003 (Ponte, 2004). The second phase incorporates the development of labels with a biodiversity protection focus, during the late 1990s, such as the Smithsonian Institute Migratory Bird Centre's Bird Friendly certification and Rainforest Alliance's Sustainable Agriculture Network (SAN) (see Chapter 3 in this volume). The Rainforest Alliance label was first well known in the forestry and banana sectors, and later in the coffee sector. During and after the coffee crisis, the third group of sustainable coffee labels was developed, incorporating socially responsible coffee trading companies, such as Starbucks CAFE practices, Nespresso AAA, Utz Kapeh (now Utz Certified), promoted by Ahold (a collaboration between a supermarket in The Netherlands and coffee farmers in Guatemala), and 4C (a joint effort between coffee trading and producer organizations) (see Figure 15.1).

Organic production and Fair Trade initiated the awareness process of a consumer willing to fund the required changes at the farm level, in order to ensure a greater supply of ecosystem services and to improve producers' quality of life (Raynolds, 2002; Loureiroa and Lotade, 2005). Through these two labels the basis for third-party certification was also developed (Ponte, 2004).

Organic certification was launched in Europe and the US under the leadership of organic farmers and alternative consumers groups and associations, such as the Soil Association (England), Naturland and Bioland (Germany) and, in the US, the California Certified Organic Farmers (CCOF), the Organic Growers and Buyers Association (OGBA), the Organic Crop Improvement Association (OCIA), Oregon Tilth Certified Organic (OTCO) and Florida Organic Growers (FOG). In 1991, both Europe and the US enacted laws controlling the marketing of organic products (EU Regulation 2092/91 and US Organic Foods – Farm Bill Act 1990) because of the increase in public interest

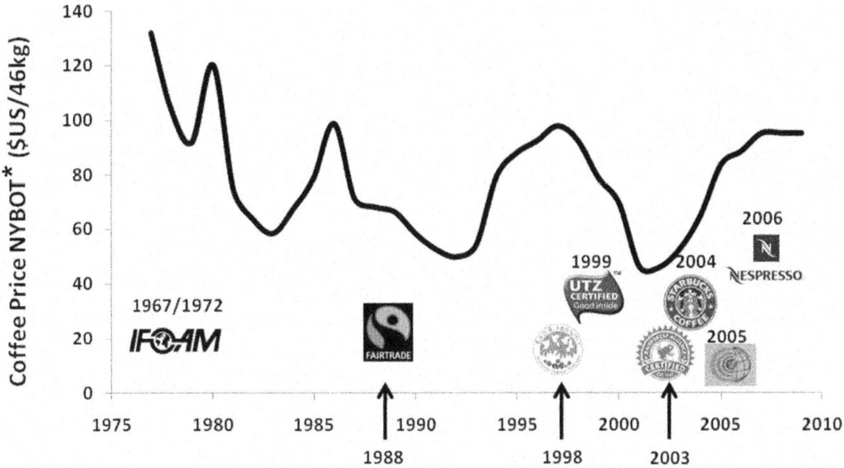

Figure 15.1 *World coffee prices and development of sustainable coffee labels over time*

Source: chapter authors, using information from ICO (2009)
* New York Board of Trade

in organic products. In 1999, the United Nations published the *Codex Alimentarius* for organic production. Among the sustainable coffee labels, only organic production has standards with legal status.

By 2000, the organic and Fair Trade coffee market was widely developed in large part due to the development of consumer consciousness and a legally established guarantee system. According to a *Coffee Sustainable Survey* of the US Coffee Specialty Industry, by the year 2001, 66 per cent of the coffee roasters sold at least one brand of organic or Fair Trade coffee and 77 per cent thought that the overprice of US$0.59 to $0.69 per pound of coffee was suitable (Giovannucci, 2001). But it was the coffee crisis that was the impetus of all coffee certifications (see Figure 15.2). The production of organic, Fair Trade, CAFE practices, Rainforest Alliance and Utz Certified coffee grew among farmers due to the better prices and the lower perceived risk (Ponte, 2004; Giovannucci and Potts, 2008). A survey conducted in 2010 showed that 16 per cent of all coffee entering the US market is certified, with an important primacy of Starbucks coffee, with around 2280,000 bags of 46kg in 2009. Rainforest Alliance and Nespresso AAA, although with a smaller volume, are showing the greatest average growth in the last four years (74 and 70 per cent, respectively) (Giovannucci, 2010).

Growth and development of coffee labels in Central America

In Central America, as in the rest of the world, organic and Fair Trade lead the way in coffee-sector certifications. Organic coffee was promoted by non-governmental organizations (NGOs) with a history of supporting agroecology

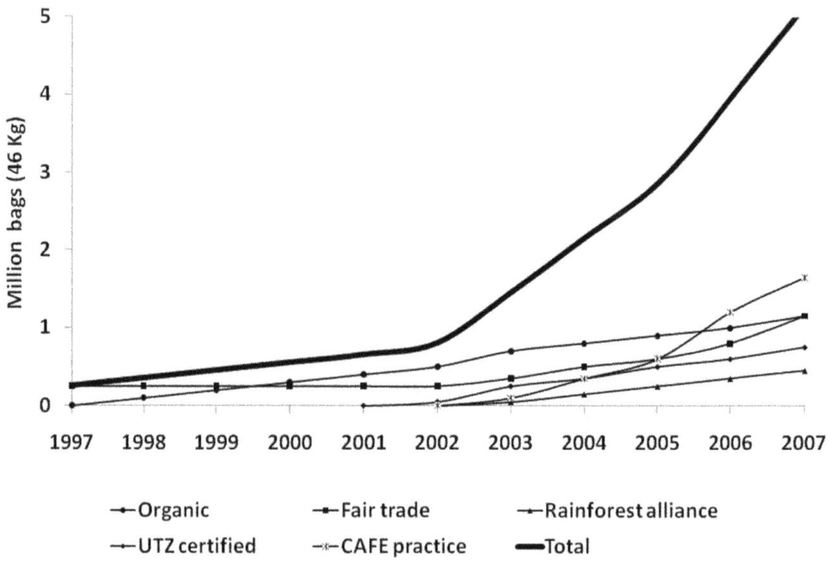

Figure 15.2 *Coffee certification growth in the world market, 1997–2007*

Source: chapter authors, using information from ICO (2009)

in co-operatives or small farmers' associations, as well as by foreign producers based in the region with strong ties to the US and European Union (EU) markets (Britt Coffee in Costa Rica in 1994, OCIA Chapter in Guatemala in 1996, etc.). Organic and Fair Trade production grew exponentially in the region during the coffee crisis and was an important strategy to support farmers in overcoming the coffee crisis (Lyngbaek et al, 2001; Ponte, 2004; Philpott et al, 2007; Cárdenas, 2008).

Fair Trade certification development was linked to the European market and funding agencies, such as Ebert Foundation and the Consortium of Cooperatives of Coffee Growers (COOCAFE) from Guanacaste and Montes de Oro, one of the pioneers who started marketing Fair Trade coffee in 1989 (Ronchi, 2002). The impact of Fair Trade certification in overcoming the coffee crisis was also crucial. Farmers with organic and Fair Trade labels were able to sell coffee at much better prices than conventional coffee (see Table 15.1).

Other labels, such as Rainforest Alliance, Starbucks CAFE practices and Utz Certified label, also grew rapidly during and post-crisis in Central America (see Table 15.1). Not all countries in Central America were able to differentiate between the amounts of exported speciality coffee as the Instituto Hondureño del Café (IHCAFE) could for Honduras.

Table 15.1 *Honduras speciality coffee exported (volumes and sale prices) from the 2005–2006 harvest*

Coffee seal	*Volume (bags of 46kg)*	*Average price (US$/bag)*
Organic	40,479	132.7
Utz Kapeh	17,578	105.8
Fair Trade/organic	10,395	138.2
Rainforest Alliance	9052	112.3
Fair Trade	8185	129.3
Organic/Fair Trade/Rainforest Alliance	2317	150.0

Source: IHCAFE (2006)

Strategies of Sustainable Coffee Certification

This section presents the characteristics of the various existing certification systems and analyses their limits and opportunities to foster ecosystem services (ES) provision.

The success of certification as a strategy to increase the supply of ecosystem services on farm depends on the different components of the certification structure:

- the objective and content of the standards that determine the level of intentionality towards ES provision;
- the certification structure which affects the liability of the requisite compliance control, its costs and the degree of access for farmers; and
- the market recognition that determines the economic incentive for the farmers' effort (investment) to comply with the normative and provide ES.

Certification must not only ensure service provision, but promote the profitability of the certified activity, such as coffee production. Thus, each component has the challenge of achieving a balance between these two main objectives: guarantee the ES provision for consumers' trust, and guarantee farmers' access and profitability. For example, the standards should be strict enough to guarantee the ES provision, but not so strict that farmers would not comply. Inspectors' farm visits should be sufficient to guarantee standards compliance, while maintaining affordability for the farmer (see Table 15.2).

The objectives and content of the standard

We analysed the standards in terms of objectives and contents, and discussed the implications in terms of potential effects on ES provision.

Objectives

It is important to understand that ecosystem services provision is not always the priority for all sustainable coffee standards (see Table 15.3). Specific labels

Table 15.2 *Challenges of the certification structure to guarantee the ES provision, farmer access and profitability*

Standard component	*Ecosystem services (ES) provision*	*Producer: Facilitate access and increase revenues*
Standards objectives and contents	Guarantee the provision of ES.	Enable cost-effective productivity. Promote a farmer strategy of continuous improvement.
Standards compliance control structure: accreditation body, certification body and inspectors	Guarantee equal compliance to the standards in all regions and among all farmers and farmers' organizations. Maintain a reliable guarantee system that is transparent for consumers, buyers and governments.	Adapt to local conditions. Be respectful of farmers' traditions and practices. Ensure that costs are accessible for farmers. Keep costs accessible to national and international agencies (these costs will eventually be transferred to the farmer).
Market recognition	Establish prices according to the ES provided. Give preference to products providing more ES.	Provide market prices (premiums) that compensate for the required investments and the decrease in productivity. Ensure stable prices, which will give confidence to the producer for long-term investments.

focus on social priorities rather than environmental, such as Fair Trade, while other certifications put more emphasis on ensuring that coffee quality meets their niche market requirements, such as CAFE practices and Nespresso AAA, who only certify coffee produced 800m above sea level. Seals are sometimes developed to promote the use of a baseline for sustainable coffee production (such as 4C).

Nevertheless, consumers do not perceive differences among labels, but maintain the perception that every sustainable coffee label guarantees environmental protection, appropriate social conditions and a fair price for farmers.

Table 15.3 *Main objectives of the different sustainable coffee labels*

Label	*Environment*	*Social*	*Cup quality*[1]
Organic	++++	+	–
Smithsonian Bird Friendly[2]	++++	+	–
Fair Trade[3]	+	++++	–
Rainforest Alliance	+++	+++	–
Utz Certified	++	++	–
Starbucks CAFE practices	+++	++	+++
4C	++	++	–
Nespresso AAA	++	++	+++

Notes: 1 Organoleptic characteristics. 2 Organic certification required. 3 In 2008, Fairtrade Labelling Organizations (FLO) added a detailed section on environmental standards.
Source: authors, based on interviews with auditors and certified co-operatives

Table 15.4 *Basic requirements of sustainable coffee certification standards and compliance control system*

Criteria/ requirements	*Organic*	*Smithsonian Bird Friendly*	*Fair Trade*	*Rainforest Alliance*	*Starbucks CAFE practices*	*Utz Certified*	*4C*
Criteria specific for coffee	No	Yes	No	Yes	Yes	No	Yes
Allow synthetic pesticides use	No	No	Yes	Yes	Yes	Yes	Yes
Transition period required before certification	3 years	Must be organic	No	No	No	No	No
Compliance assessment		Full compliance system			Scoring system		

Many labels have widened their scope of action to face this challenge, such as Fair Trade's environmental standards improvement or Rainforest Alliance's inclusion of climate change standards.

Content and design process of certification standards

Certification criteria vary among the different labels based on their objectives and the scope of standards. Some certifications are generic for all crops, while others are specific to coffee, allowing a greater degree of precision in aspects such as shade (see Table 15.5). Thus, organic, Fair Trade, Rainforest Alliance and Utz Certified are not coffee specific, whereas Smithsonian Bird Friendly, Starbucks CAFE practices and 4C are coffee specific (see Table 15.4).

The technical support behind each of the standards is also variable and depends on when the standards were developed and what methods were used to develop them. The first versions of the organic standards were written during the 1960s and 1970s by farmers and consumer associations in Europe or the US. These standards were later voted into the International Federation of Organic Agricultural Movements' (IFOAM's) General Assembly, with participants from around the world. While these methods were very democratic and participative, ecosystem services technical data was limited. In contrast to this strategy, the Smithsonian Bird Friendly seal developed its standards based on scientific data of the impact of coffee intensification (Rice, 1999) upon migratory bird behaviour in the Mesoamerican coffee landscape (Greenberg et al, 1997a, 1997b). As a result, these coffee-specific standards have a clear objective for a defined region. Bird Friendly seal research has since become the template for defining new criteria for other standards.

Variability of contents and possible practices regarding ES provision

Since shade structure and management are directly linked to biodiversity within the coffee system (Perfecto et al, 1997; Moguel and Toledo, 1999; Mas

Table 15.5 *Coffee shade requirements in sustainable coffee labels*

Requirement	*Organic*	*Smithsonian Bird Friendly*	*Rainforest Alliance*	*Utz Certified*	*Starbucks CAFE practices*	*4C*
Regulation version	NOP-USDA, 834/2007 889/2008	April 2002	February 2009	Version 1.1 January 2010	November 2009	May 2009; generic indicators February 2010
Must have shade in the coffee plantation	No mention of shade	Yes	Only for crops usually managed in agroforestry systems or in a natural forest region	If compatible with local production practices and considering productivity	Shade required where the natural vegetation was forest	No mention of shade
Diversity (number of species ha^{-1})		10	12	–	'Several species'	–
Minimum height of main species (m)		12	–	–	–	–
Strata		3	2	–	–	–
Percentage minimum shade year round		40	40	–	Additional points for 10%, 40% or 75% shade	
Native species		'Top strata'	'Preferable'	–	Additional points if only native species are used	

and Dietsch, 2004), they provide a good example of understanding standards' variability regarding impacts upon ES provision. Standards that are not coffee specific, such as organic, do not mention shade structure in their requirements; but shade must be implemented to control weeds, promote biodiversity and manage coffee nutrition within the farm system. On the other hand, standards such as Smithsonian Bird Friendly clearly define the number of trees per hectare, the height of the trees and a minimum shade percentage (see Table 15.5).

The lack of shade criteria or the fact that shade is optional within the scoring system has made it possible to have Utz Certified, 4C, Starbucks CAFE practices, Fair Trade, organic and Rainforest Alliance certified farms with few or no shade trees. The implications of this for ecosystem services provision will be discussed further in this chapter (see also Chapter 3 in this volume).

The scoring system used in Rainforest Alliance, Starbucks CAFE practices or Utz Certified offers the farmer the possibility of being certified and receiving consumer recognition (potential premium) at the initial stages of implementing criteria while improving farm management. Full compliance with the standards is a requirement to be a certified farm for organic and Smithsonian Bird Friendly programmes. The scoring system strategy risk comes from the consumer's perception of certified farms. Most consumers are not aware of the different standards, which potentially could result in lost confidence when they see a full sun farm certified as a sustainable farm.

Adaptations of content and variability of certification application and practices

Some labels have made an effort in adapting standards to regional conditions. In 2009, Rainforest Alliance hosted workshops to discuss the coffee standard of each Mesoamerican country, with participation from farmers, co-op technicians, government extension agents and the academic sector, to guarantee that the standards are adapted to local conditions and to define training issues for local inspectors.

As a result of this consulting process, some standards were adapted to local conditions. For example, Rainforest Alliance and Starbucks CAFE practices define the shade requirement based on the natural growth of the area before agriculture. Therefore, if the natural growth in an area was forest, shade is required, but if it was prairie, shade is not necessary. One of the concerns with this 'optional and gradient' standard system is that there is more room for interpretation by inspectors. Adequate training for inspectors or auditors is fundamental for the success of the programme. Farmers often complain about the interpretation variability of the standards pending the inspector's visit each year.

The regional standards adaptation process has been analysed by developing standards committees worldwide for many years. On the one hand, adaptation has the advantage of considering different biophysical and socio-

Table 15.6 *Components and characteristics of the standards compliance control structure of sustainable coffee certification*

Standards compliance control structure	*Functions*
Accreditation body	Controls the operation of the certification agencies based on ISO 65 and ISO 19011 requirements, as well as each specific standard. The accreditation body can be a private company or a governmental institution such as the National Organic Programme of the US Department of Agriculture (USDA).
Certification agency	Certifies coffee production and processing based on: • the farm management plan (FMP) provided by the producer; • the inspection report, which establishes potential non-compliances.
Inspector verifier	The inspector receives a copy of the FMP from the certification agency. The inspector then visits the farm and/or the processing plant. A detailed report is sent to the certification agency with the potential non-compliances observed in the field.
Farmer or farmers' organization	• Develop a FMP and send it to the certification agency to apply for certification. • The inspector visits the farm or processing plant. • A corrective plan of action is developed to comply with the non-compliances found on the farm.

Table 15.7 *Application structures for different labels in coffee certification in Costa Rica*

Who?	*Organic*	*Smithsonian Bird Friendly*	*Fair Trade*	*Rainforest Alliance*	*Utz Certified*	*4C*	*Starbucks CAFE practices*	*Nespresso AAA*
Defines the standards	National and international legislation	Smithsonian Institute of America	International advisory committees coordinated by each of the headquarters of these labels				Starbucks in collaboration with Conservation International	Designed by Rainforest Alliance*
Controls private agencies' activities (accreditation)	Governments: NOP-USDA** in the US; Plant Health and Quarantine Service in Central American countries	Must be ISO 65 accredited	They certify themselves or work in collaboration with other certification agencies	Most of the certification is done by themselves or in strategic alliances with NGOs	ISO 65 accredited agencies; they also monitor the certification agencies	Certify themselves	ISO 65 accredited agencies and Scientific Certification Systems†	Rainforest Alliance
Controls standards compliance (certification function)	Private agencies or government offices	Independent private agencies	Fairtrade Labelling Organizations (FLO) certified	Control division of Rainforest Alliance	Independent private agencies	Internal personnel	Independent private agencies	Rainforest Alliance auditors
Rule of compliance assessment	Full compliance system			Scoring system				

Notes: †Private company contracted by Starbucks to develop its certification system.
* The Nespresso AAA are not public standards.
** National Organic Program of the United States Department of Agriculture.

economic conditions of each region; on the other, it could increase the variability of implementation and, by doing so, risk losing consumers' credibility. For example, a coffee farmer in Central America who invests in shade management and accepts a decrease in productivity may feel it is unfair to have the same label as a full sun coffee from the Cerrado in Brazil.

Structures of the Control System

The compliance control system structure of sustainable coffee certification is critical to the certification strategy because it is the mechanism that gives credibility to the eco-label strategy. Thus, it should be transparent, fair and strict enough to be trustworthy. Nevertheless, it should be cost effective and adaptable in order to ensure farmer adoption. In this section, the different control structures and their implications in terms of system liability and efficiency are presented.

The basic structures of the certification control system

Certification's compliance control system is essentially composed of three main actors who assume three specific functions (see Table 15.6).

Each label has developed different certification structures from the field visits selection criteria (see Table 15.7). The fact that organic standards are enforced by public regulation makes the standards definition a very structured, open process, relying on government official implementation. However, in most cases, modifications are difficult and time consuming for all stakeholders, while private standards are easier to modify.

Differences are also observed in the accreditation system. While organic certification is mainly government controlled (private accreditations are also available, such as the Organic Accreditation System (OAS) from IFOAM), there are standard-setting bodies that conduct the certification themselves, where accreditation by a third party is not required (e.g. Rainforest Alliance certification). The advantage of having an accreditation system is that the division of roles between the standard-setting and standard-controlling bodies increases the transparency of the process. However, it also increases the certification cost. Nonetheless, Utz Certified and Starbucks CAFE practices have implemented accreditation systems through their regional offices with no additional cost for the certification agency or the farmers.

Cost control, monitoring and farmers' accessibility

One of the highest costs in the promotion of ecosystem services through certification is compliance inspection and monitoring. The certification process moves about US$200 million worldwide, from field inspectors and agency coordination and certification decisions, to agencies in accreditation processes with governments as well as private agencies (e.g. ISO 65). Sooner or later these costs are transferred to the producer or the consumer, and have become a

Table 15.8 *Organic, Fair Trade and Rainforest Alliance certification costs of eight case studies in coffee production in Costa Rica, 2007*

Label	*Zone*	*Form of certification*	*Coffee area (ha)*	*Certification cost (US\$ ha^{-1})*
Organic	Los Santos (Tarrazú)	Individual	3.5	43.8
	Central Valley	Individual	56.4	33.4
	Pérez Zeledón	Association	6.6	12.6
Fair Trade	Los Santos	Co-operative	4.2	0.7
	Pérez Zeledón	Co-operative	2.1	0.5
Rainforest Alliance	Los Santos (Tarrazú)	Co-operative	2.8	11.8
	Central Valley	Individual	49.7	40.1
	Turrialba	Individual	675	9.6
Conventional	Los Santos	Individual	7.1	0
	Pérez Zeledón	Individual	8.5	0

Source: adapted from Moreno et al (2009)

growth constraint. Case studies in Costa Rica show high variability in certification costs (Moreno et al, 2009) (see Table 15.8), depending on certification type (individual or group), agency, size of and access to farms, number of inspection days, etc. The size of the organization (number of certified producers) is a factor that strongly affects the certification cost per hectare (fixed costs versus variable costs). In the study sample, we worked with organizations of different sizes (Coopetarrazú with 2600 producers; CoopeAgri – Fair Trade certified with 16,000 producers; and associations of organic producers with 15 to 20 producers), which justified the variation in costs per hectare. Surveys with cocoa producer organizations in Central America reported similar certification costs for organic and Fair Trade certification (PCC, 2010).

Currently, many efforts have been made to reduce these costs at the producer level, including collective certification of small producer groups; funding agencies supporting small producers to cover certification costs; training of local inspectors instead of working with international inspectors; local certification agencies; reducing the frequency of visits; etc. The reduction of these costs should be a constant quest of the certification programmes themselves. For example, the Smithsonian Bird Friendly label achieved cost reduction through its union with organic certification, which enables it to reduce the number of inspection visits to one for both certifications, to reduce the frequency of audits from one per year to one every three years, and to avoid accreditation cost for agencies. Other labels such as Starbucks CAFE practices and Rainforest Alliance have developed similar efforts. Farmers' organizations that adopt multi-certification and a common system of internal control systems are able to reduce the costs of investment in training, record-keeping, etc. at the farm level.

Multiplicity of standards and access to farmers

Another difficulty that farmers encounter in the organic sector is to have different standards for different markets. The current structure of the organic market states that no matter where the products are produced in the world, they must be produced following the standards of the regions where the products are sold. Producers selling to the US market must meet the NOP standards (Part 205 of Title 7 of the Code of Federal Regulations USDA), while producers exporting to Europe must comply with European Union Regulations 834/2007 and 889/2008, and exports to Switzerland or Japan must comply with Biosuisse or Japanese Agricultural Standard (JAS) regulations, respectively. As a result, producers from exporting countries who want to maintain their access to diversified markets have more constraints than producers from developed countries, such as Europe and the US, in selling to their local markets.

This additive effect of requirements has made farmers feel that the regulations are constantly changing and becoming stricter with time. This was mentioned as one of the reasons why farmers are stepping out of organic coffee in Central America (Haggar and Soto, 2010).

Variability of rules of compliance and access to farmers

Another aspect relevant to farmers' access is the method of evaluation of compliance. Different labels are currently using two systems: the full criteria compliance method used by organic, Smithsonian Bird Friendly and Fair Trade, and the scoring system. In the full compliance system, the producer must ensure that all requisites are certified. In the scoring scheme, most of the requisites have a score and a minimum score certifies a producer (e.g. 60 per cent of the total score for CAFE practices). However, a balance between the three main topics (environmental, social and transparency factors) is required. Farmers, for example, cannot have high scores in social issues that will balance low performances in environmental issues. In addition, there are compulsory 'critical criteria'. Among these criteria are aspects such as minimum salaries, child labour, anti-discrimination, etc. This scoring assessment system is used by Rainforest Alliance, Starbucks CAFE practices, Utz Certified and 4C.

There are two consequences regarding the differences in the assessment system. First, full compliance systems are stricter than scoring systems and tend to be more clearly understood by consumers; but the scoring system offers a more inclusive pattern from the producers' point of view. It enables facilitation of a continuous improvement process within the framework of the certification, whereas in a full compliance system, farmers have to develop the compliance by themselves. Second, the existence of various assessment systems tends to complicate the comparison between certifications regarding their impact upon provision of ecosystem services; certified farmers with a scoring system may have a variety of farmers' practices.

Incentives for Certification and Farmers' Interests

Producers adopt certification for a variety of reasons: an interest in protecting the environment and family health, access to niche markets, better prices or price stability, or (as a certified producer said in Nicaragua) 'I would do anything not to go through the same anxiety that I suffered during the coffee crisis.' But the reason why they decided to become certified does not matter; they will not remain unless they are recognized for their efforts. We describe here the modalities of economic rewards to producers and analyse their results and limitations.

Characteristics of economic incentives

Economic rewards to compensate the certified producers' efforts take various forms according to various labels. The common perception is that certification will lead to a premium over the conventional market price. Yet, the reality is more complex.

The only certification which establishes a premium as part of the standard is Fair Trade, which explicitly regulates commodity prices, and obligates traders to pay a minimum price as well as a premium for development of US$5 to $10 kg^{-1} of coffee. No other certification has control over the certified coffee price. However, some of the labels developed by coffee businesses have established a reward system. For example, Starbucks gives a one-time premium to CAFE practices producers who make improvements during the initial years. Producers of Utz Certified coffee or Nespresso AAA are offered a fixed

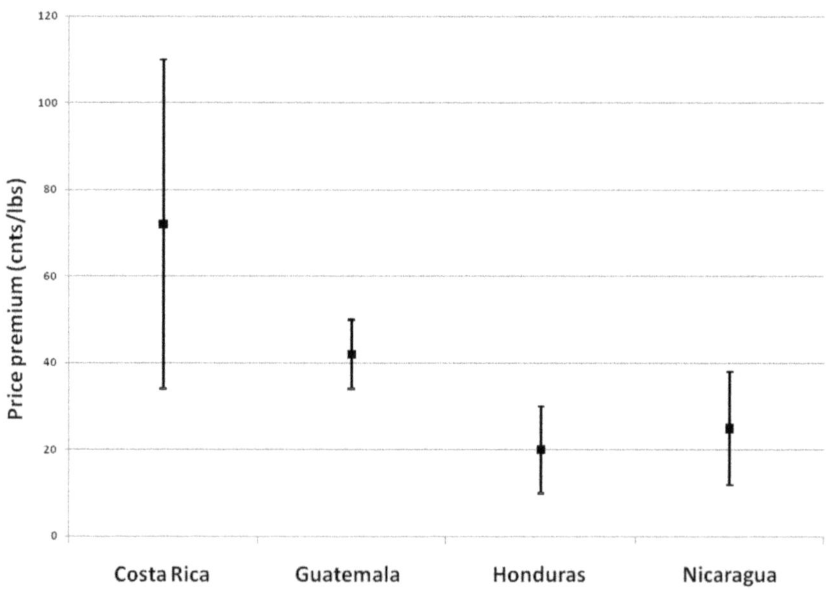

Figure 15.3 *Price premium for organic certified coffee in Latin America harvest 2002/2003*

Source: adapted from CIMS (2004)

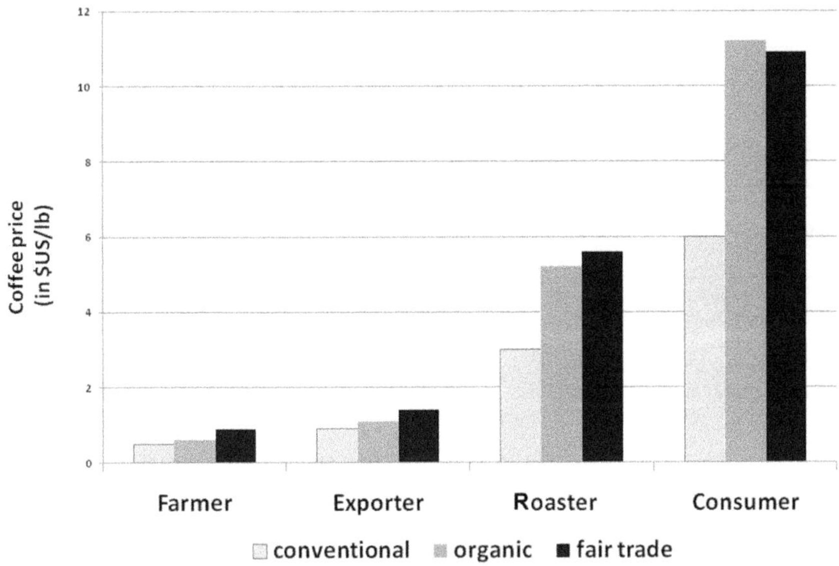

Figure 15.4 *Coffee price distribution in the coffee value chain*

Source: CIMS (2004) and ICO (2004)

premium (from US$2 per 46kg for the former, to US$5 per 46kg for the latter, according to producers interviewed by the authors in Costa Rica in 2007). For organic and Rainforest Alliance, the premium is an element of price negotiation between producers and traders and, thus, depends on market rules. Whereas Rainforest Alliance association is constantly and actively promoting its label with traders and roasters, there is no specific promotion for organic products. Organic certification agencies are specifically forbidden to do so by ISO 65 accreditation requirements. Nevertheless, since it is the better-known label in the market, the premium for organic is usually the largest one (see Table 15.1 and Figure 15.5). Other labels, such as 4C, do not promote a premium system. Therefore, the premium linked to certification, with the exception of organic, is generally very limited in comparison to conventional price. According to our estimation, in 2007 in Costa Rica, the average premium level, except for organic, represented between 1.5 and 7.5 per cent of the conventional price.

A second characteristic of the economic reward is that there is no guaranteed reward level. The reward level is variable and depends on offer and market demand in this market segment and on the price level in the conventional market. For example, the premium for organic/Fair Trade production was around US$70 to $100 per 46kg compared to conventional production during the coffee crisis during early 2000; however, it was only US$5 to $10 per 46 kg in 2009 during a high price conjuncture in international markets (Haggar and Soto, 2010).

Table 15.9 *Evolution of the number of organic producers in Central America*

	Number of organizations participating in workshops	Number of organic producers in these organizations* 2004–2005	2009	Percentage change in the total number of organic producers Between 2004–2005 and 2009
Guatemala	5	1277	738	–42%
Nicaragua	7	2718	2485	–8%
Costa Rica	7	897**	388**	–57%

Notes: * Information based on workshops held in 2010 in Nicaragua, Costa Rica and Guatemala. Data provided by producers participating in the workshops. These are not country averages but averages of the organizations that they represent.
** Data collected by students from CATIE (Quispe, 2007; Ramirez, 2010). This is a country average.
Source: adapted from Haggar and Soto (2010)

As the economic reward is linked to the coffee commodity market, the price of coffee depends not only on certification, but also on other factors such as quality, technology and organoleptic characteristics, as well as marketing of the product or how well the region is known (Tarrazú in Costa Rica; Antigua or Huehuetenango in Guatemala). Thus, the premium reflects not only the ES provision, but also commercial attributes. Organic producers in countries such as Costa Rica and Guatemala benefited from better organic production premiums than other countries in the region for their quality and origin of fame. For example, during the 2002 to 2003 harvest, they received an average premium of US$70 per 46kg of green coffee, while countries such as Nicaragua and Honduras received a premium of US$20 to $30 (see Figure 15.3) (Kilian et al, 2004).

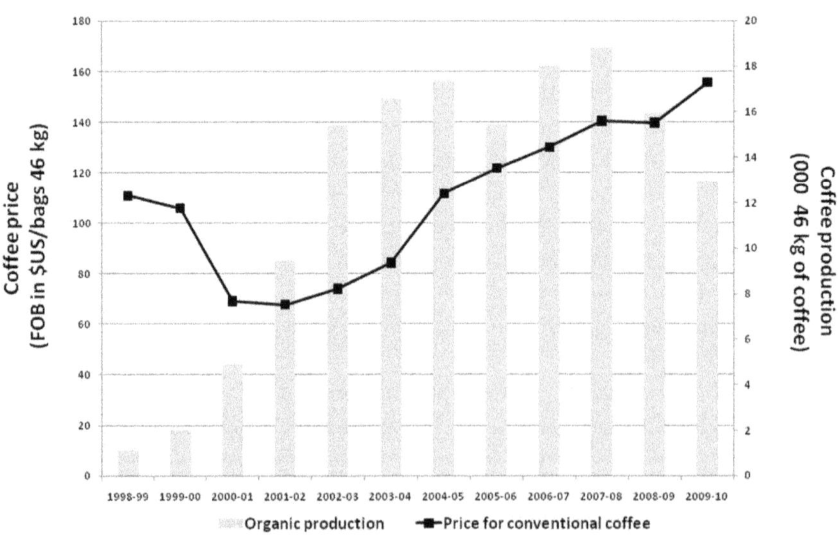

Figure 15.5 *Organic coffee production (46kg bags) in Costa Rica from 1989 to 2009–2010 harvest*

Source: authors based on ICAFE data from 2010

Another characteristic of economic rewards is no premium guarantee since certification does not give a guarantee to effectively sell the coffee as certified coffee. For most eco-labels, it is common that producers have to sell a part or sometimes all of their certified production as conventional production because they cannot find a trader interested in the product. This has been particularly the case for Fair Trade, Utz Certified and Rainforest Alliance during the last few years since certified production exceeded demand. Indeed the volume effectively sold as Fair Trade, Utz Certified and Rainforest Alliance at the Central American level was only 14, 32 and 32 per cent of the certified production, respectively (authors, based on Kilian and Pratt, 2009).

Finally, the rewards distribution along the commodity chain is not regulated by the certification standard. Thus, the producers have no guarantee of receiving the entire premium that the consumers paid for the product. Indeed, the price premium paid by the consumer is distributed among all of the actors of the commodity chains. In many cases, the additional price paid by the consumer is higher than the additional price received by the producers (see Figure 15.4) (CIMS, 2004).

Economic rewards and benefits for producers: The importance of productivity

In spite of economic rewards, the number of organic coffee producers in Central America has suffered a decrease during the last few years (see Table 15.9). Moreover, according to personal communications with leaders of co-operatives and producers in the region, the producers' interest in other coffee labels is also declining. The common reason to explain this tendency is that economic rewards do not cover the producers' efforts to comply with certification standards. The benefits for producers are a critical factor in the sustainability of the certification strategy.

Organic certification shows that because recognition for the provision of ES is paid by quintal of coffee, what is important is not just the premium received per quintal, but also the number of quintals sold. During periods when the difference between organic/Fair Trade production compared to

Table 15.10 *Comparison of productivity in organic and conventional farms in Central America, 2009*

	*Average productivity in organic farms (quintals ha^{-1})**	*Average productivity in conventional farms (quintals ha^{-1})**	*Reduction of productivity between organic and conventional (%)*
Guatemala	10	13	23
Nicaragua	10	14	29
Costa Rica	12	25	52

Notes: * Data provided by producers participating in the organic coffee crisis analysis workshops. There are no country averages, but averages from the regions that they represent.

1 quintal = 46kg of green coffee.

Source: adapted from Haggar and Soto (2010)

conventional production was important (e.g. US$70 to $100 during the coffee crisis years), the producer felt rewarded and the amount of organic coffee production increased (see Figure 15.5); but during periods of high conventional coffee prices, the organic or Fair Trade premiums do not pay for the differences in productivity (as in 2009, where the differential was US$5 to $10). This reduction of productivity results from the density and management of shade trees, and the limited use of organic fertilizers.

The differential between organic and conventional productivity is not equal in all Central American countries (see Table 15.10); countries with higher productivity in conventional coffee, such as Tarrazú in Costa Rica or Huehuetenango in Guatemala, are regions where most organic farmers have already converted to conventional or some other sustainable certification. Moreover, in areas with recognized quality coffee and high coffee price, such as Tarrazú, there is no interest in organic production since organic coffee premiums do not compete with the premium obtained for quality.

For other labels, the situation seems less stringent. Although the economic rewards are lower than for organic coffee, the reduction of productivity when complying with requisites seems to be less than for organic coffee. For example, CIMS (2006) shows that in the region, Rainforest Alliance and Utz Certified coffee productivities ranged from 38 quintals to 40 quintals ha^{-1}, which was comparable to conventional production levels.

Finally, producers' perceptions are also an important element for the sustainability of the eco-label mechanism. A survey showed that while producers' satisfaction was initially high, it has decreased over time because the economic reward is less than originally offered (Giovannucci and Potts, 2008).

Impact of Sustainable Coffee Certification upon the Provision of Ecosystem Services and Farmers' Welfare

Provision of ecosystem services

The provision of ecosystem services from agroforestry systems has been widely documented (Schroth et al, 2004; Montagnini, 2006; Jose, 2009), and there is clear evidence that the two main factors that will increase biodiversity and the provision of ES are shade tree diversity and distance to forest patches (including impact of riparian forest, live fences, etc.) (see also Chapter 3 in this volume). However, farmers have little control over the organization of the landscape outside of their farm, so shade management is the area where certification could have an impact in improving the provision of ES. But after reviewing the variability that exists in the shade criteria and in the implementation of these criteria in the field (see Table 15.5), one may wonder about the real impact of certification upon the provision of ES. Unfortunately, there is little scientific evidence which compares the impact of the different seals upon the provision of ES, with the exception of organic and Smithsonian Bird Friendly (see Table 15.11).

Table 15.11 *Studies comparing the provision of ecosystem services in certified coffee farms in Mesoamerica and Brazil*

Ecosystem service	*Indicator*	*Region*	*Certification*	*Main results*	*Source*
Biodiversity and pests and disease	Fruit-consuming butterflies and forest birds	Chiapas, Mexico	Organic, Rainforest Alliance, Bird Friendly and conventional	More variation with shade structure than with certification label. Farms with more complex shade structures were more diverse.	Mas and Dietsch (2004)
	Trees, epiphytes, birds, ants and yield	Chiapas, Mexico	Organic, Organic + Fair Trade, conventional and 'similar to' Rainforest Alliance and Bird Friendly (no certified farms in the region)	The ideal for biodiversity is standards more specific to tree requirements.	Philpott et al (2007)
	Tree diversity	Costa Rica	Organic, conventional, Fair Trade, Rainforest Alliance, CAFE practices and Utz Certified	CAFE practices, Rainforest Alliance and Utz Certified no difference in percentage shade from conventional. Rainforest Alliance, organic and Fair Trade were different from conventional in the biodiversity of trees.	Quispe (2007)
	Ants	Turrialba, Costa Rica	Organic and conventional (including a diversity gradient of high diverse to low diverse)	Organic farms had higher species richness.	Barbera et al (2004)
	Cicadellidae			Higher diversity of Cicadellidae on organic shaded systems.	Ramos (2008)
				Less nests in organic shaded systems.	Varon et al (2007)
Hydrological services	Native species number and conservation of water resources	Minas Gerais, Brazil	Rainforest Alliance and conventional	Rainforest Alliance had higher numbers of native species and better water conservation.	Palmieri (2008)
Soil quality	Soil carbon	Cartago, Costa Rica	Organic and conventional	More homogeneous distribution of soil carbon in organic farms.	Payan et al (2009)

Table 15.11 *continued*

Ecosystem service	*Indicator*	*Region*	*Certification*	*Main results*	*Source*
	Organic matter, mycorrhizae and nematodes	Guatemala and Brazil	Organic and conventional	Higher contents of organic matter, mycorrhizae, bacteriophages and nematodes in organic production.	Alfaro (2004)
	30 indicators of soil quality	Turrialba, Costa Rica	Organic and conventional with and without timber trees and bananas	Higher additive Index of Soil Quality (ICSA) (combination of different variables of soil) in organic farms.	Porras (2006); George (2006)
	Ground cover	Costa Rica	Organic, conventional, Fair Trade, Rainforest Alliance, CAFE practices and Utz Certified	Significant differences in the ground cover of organic farms.	Quispe (2007)
	Earthworms and microbial biomass and yield	Turrialba, Costa Rica	Organic and conventional with different shade trees	Similar yields and microbial biomass between organic and conventional farms. More earthworms in organic.	Sanchez de León et al (2006)
	Soil fertility	Turrialba, Costa Rica	Organic and conventional with different shade trees	Higher P, Ca and K, and lower acidity after four years of organic management.	Soto et al (2007)
Provision	Other products extracted from the organic farm	Turrialba Costa Rica	Organic and conventional	Organic farms were more profitable when considering the other food products produced on-farm; but coffee productivity was lower on organic farms.	Cárdenas (2008)
	Profit and coffee yield	Costa Rica	Organic and conventional	22% less yield in organic, 5% more profitable.	Lyngbaek et al (2001)
Carbon sequestration	Carbon footprint	Turrialba, Costa Rica, and Masatepe, Nicaragua	Organic and conventional with different shade trees	Higher CO_2 kg^{-1} ha^{-1} on conventional farms; higher CO_2 kg^{-1} of coffee on organic farms due to lower yields.	Noponen et al (2010)

Comparative studies of the impact of certification upon ES provision have several constraints. The first and most common is the definition of the categories of management systems to be compared (e.g. organic versus conventional), when management practices within each of these categories can be variable (trees or no trees, pruned trees or free growth, etc.). Other studies compare before and after certification; but most farmers did not keep records before certification, so the study must rely on the farmers' memories, which are subject to error. Another strategy is to compare certified and uncertified farms in a specific time, matching and comparing socio-economic and biophysical characteristics. The drawback of this methodology is limited access to databases to identify certified and uncertified farms with the same characteristics, or, alternatively, the high cost of a sampling effort. Facing these difficulties, the ISEAL alliance developed a code of good practices to conduct impact studies for this type of analysis (see http://community.isealalliance.org/content/ Impacts-code).

Other aspects to be considered when interpreting the results are regional differences, such as the existence of strong environmental national regulations, which could alter the results (Alonso and Jiménez, 2009). Preliminary data on the impact of organic, Fair Trade, Rainforest Alliance, Utz Certified and Starbucks CAFE practices in Honduras, Nicaragua, Costa Rica, Peru and Kenya collected within the COSA project (a multi-criteria cost–benefit analysis of sustainable practices in coffee) show a wide range of economic and biodiversity impacts of the same labels in different countries (Giovannucci and Potts, 2008).

A review of ecosystem services provision in certified coffee farms in the Mesoamerican region and Brazil (see Table 15.11) shows a trend towards positive impacts of organic, Smithsonian Bird Friendly and Rainforest Alliance. More research is required to determine the impact of the most recent labels, such as Utz Certified, CAFE practices, Nespresso AAA or 4C.

The challenge that the standard-setting bodies are facing is how to develop standards to improve the provision of ES, but to be understood and implemented by farmers. Should the standards refer to the provision of ES (e.g. to avoid erosion), or should the practice to avoid erosion be requested (e.g. construct terraces in the field). It is clear that for the inspector visiting the farm once a year for two or three hours (depending on the size of the farm and access to all fields), it would be easier, for example, to verify the presence or absence of the terraces than to measure laminar erosion.

The other question that remains is: are the standards strong enough to make the necessary changes in farm practices to improve provision of ES? Quispe (2007) compared changes in farming practices before and after certification in Costa Rica, observing limited changes on the Utz Certified, Rainforest Alliance and Starbucks CAFE practices certified farms (reduction of one herbicide application, no changes in fungicides or fertilizers used), and observing no change in the percentage of shade, even though Rainforest Alliance producers increased tree plantings in the plantation (too small at the time of the study to see their impacts reflected in shade percentage). The only producers with radical changes in management practices were organic producers who modified most of their practices (e.g. removing herbicides, fungicides and synthetic fertilizers).

On the other hand, auditors and verifiers monitor regulation compliance for changes in plantation management practices. But they should also use indicators to quantify the provision of ecosystem services without increasing the costs of certification (two- to three-hour visits per farm depending on farm size) (see Chapter 3 in this volume).

Social impacts of certification

Some coffee certifications have important social implications (De Lima et al, 2008; Rivera, 2008), especially Fair Trade certification (Ronchi, 2002; Bacon, 2005). Although social impacts are not covered in this chapter, they should not be ignored as a fundamental component of the strategy's success: they are an important part of consumers' preference criteria.

In Central American regions where education or health access was limited, the impact of CAFE practices and Rainforest Alliance certification on large farmers has made an important difference in farmers' communication. In Costa Rica, where social security and access to education is available in most coffee areas, the major impact has been in workers' housing, especially harvesters, who often come from neighbouring countries. The strongest economic impact of implementing these standards has been felt on medium-sized farms (5ha) (Moreno, 2008).

Conclusions

The growth in recent years of the green label sector is a promising strategy to promote the required changes to foster ecosystem services provision through market mechanisms. One of the main achievements of sustainable certification processes has been to improve the link between the producer (family and production system) and the consumer. Consumer preference in the market is a tool to obtain changes at the farm level. If this link is valuable and powerful, it is also extremely fragile and subject to market rules. Certification development experience in Central America highlights some limitations and lesson learned:

- As a result of consumers' concern about reduced knowledge on certification issues, the different certifications tended to converge and make more room for environmental concerns.
- Continuous evolution of the standards has tended to fine-tune the criteria of the norm. Nevertheless, there is still room for improvement in the mobilization of technical knowledge to improve ES provision guarantees in the criteria of the norm.
- The compliance standard control structure of existing certifications offers a good level of guarantee to consumers. However, this control system is

Table 15.12 *Areas of improvement of certification to promote the provision of ES*

Areas of improvement	*Possible actions*
Improve the guarantee for provision of environmental services (ES)	Develop inspection methodologies that allow the use of more indicators to quantify the provision of ecosystem services. Evaluate changes to regulations that encourage greater provision of ES. Adapt the rules to local biophysical and socio-economic conditions. Harmonize criteria for interpretation of standards by auditors and inspectors in the field.
Cost reduction	Various actions are possible to reduce certification costs, such as more government involvement in the certification process (a test is being carried out in Costa Rica with a mixed private–state certification; inspections will be done where government extension agents working in different regions will conduct the inspection and send the report to the private certification agencies; inspection costs are covered by the government). Participatory certification for local (and international) markets, as well as alliances between certification programmes to reduce inspection costs, should also be encouraged. Sell certified coffee in local markets.
Improve the recognition of producer investment in sustainable production	In the case of organic production, harmonize regulations in the various export markets. Establish better distribution of the 'premium' among the commodity chain. Various options at the institutional or standard level can be used, including standard regulations (such as Fair Trade) and state regulations. Modify the balance of power between producers and their organizations and other actors of the commodity chain. Develop information and promotion campaigns on certification to ensure consumers' preferences.
Improve productivity (especially for organic farms)	Promote intensification of production under certification commitment to maintain or upgrade profitability of certified production. Identify risk management practices under certification commitment to secure the incomes of producers. Promote technical assistance support.

costly for individual smallholder farmers; as a result, several efforts have been made to reduce these costs.

- Certification has led to various forms of remuneration to compensate farmers' efforts. Nevertheless, the balance between remuneration and effort is not sufficient to develop sustainable economic interests for producers, especially in organic farming.

This strategy to motivate changes on farm through market incentives has great potential but also great challenges ahead. Some improvements are necessary to promote ES provision through the certification strategy (see Table 15.12). Improvements may be difficult since the certification strategies have inherent tensions and trade-offs, such as between the complexity of requisites and the capacity to evaluate; between flexibility of requisites (adaptation) and consumers' credibility perception; between the accuracy of control and its costs; and between levels of effort asked of farmers and the compensation provided by the market. In order to develop this strategy, further support from governments is required, as well as further identification of robust and easily tested criteria for standard requisites. Alternative propositions are therefore needed for better market recognition and to improve farmers' productivity under different certification schemes.

Acknowledgements

We are grateful for the financial support provided by the European Union via CAFNET (Connecting, Enhancing and Sustaining Environmental Services and Market Values of Coffee Agroforestry in Central America, East Africa and India) collaborative project and by the French Agence Nationale de la Recherche (ANR) through the SERENA (Environmental Services and Rural Land Uses) project. Support was also provided by Pôle de Compétences en Partenariat (PCP): Agroforestry Systems with Perennial Crops, CIRAD–CATIE–INCAE–Bioversity–CABI–Promecafé.

References

Alfaro, V. T. M. (2004) *Matéria Orgânica e indicadores biológicos das qualidades do solo na cultura do café sob manejo agroforestal e orgânico*, PhD thesis, Universidade Federal Rural do Rio de Janeiro, Brazil

Alonso, S. and Jiménez, G. (2009) 'Impacto de las regulaciones ambientales en las estrategias de comercialización de café costarricense', *Revista Iberoamericana de Economía Ecológica*, vol 10, pp29–43

Bacon, C. (2005) 'Confronting the coffee crisis: Can fair trade, organic, and specialty coffees reduce small-scale farmer vulnerability in northern Nicaragua?', *World Development*, vol 33, no 3, pp497–511

Barbera, N., Hilje, L., Hanson, P., Longino, J., Carballo, M. and de Melo, E. (2004) 'Diversidad de especies de hormigas en un gradiente de cafetales orgánicos y convencionales', *Manejo Integrado de Plagas y Agroecología*, vol 72, pp60–71

Cárdenas, A. (2008) *Incentivos económicos para la producción ecoamigable en fincas cafetaleras en el Corredor Biológico Volcánica Central – Talamanca, Costa Rica*, MSc thesis, CATIE, Costa Rica

CIMS (2004) *Prices of Sustainable Coffee from Latin America*, CIMS, Costa Rica

CIMS (2006) *Opciones de Mercado para el café sostenible: Resultados consultoría realizada por CIMS para COOCAFE, Costa Rica*, CIMS, Costa Rica

De Lima, A. C., Novaes Keppe, A. L., Palmieri, R., Correa Alvarez, M., Maule, R. F. and Sparovek, G. (2008) *Impact of Sustainable Network (SAN) Certification on Coffee Farms: Case Study in the Southern Region and Cerrado Areas of the State of Minas Gerais, Brazil*, Instituto de Manejo e Certificacao Florestal e Agrícola (IMAFLORA), Brazil

George, A. (2006) 'Estudio comparativo de indicadores de calidad de suelos en fincas de café orgánico y convencional en Turrialba, Costa Rica', masters thesis, CATIE, Costa Rica

Giovannucci, D. (2001) *Sustainable Coffee Survey of the North American Specialty Coffee Industry*, Report for the Word Bank, Washington, DC

Giovannucci, D. (2010) 'Take this personally', Presentation, Speciality Coffee Association, Anaheim, US

Giovannucci, D. and Potts, J. (2008) *Seeking Sustainability: COSA Preliminary Analysis of Sustainability Initiative in the Coffee Sector*, IISD, http://papers.ssrn.com/sol3/papers.cfm?abstract_id=1338582

Greenberg, R., Bichier, P., Cruz, A. and Reitsma, R. (1997a) 'Bird populations in shade and sun coffee plantations in Central Guatemala', *Conservation Biology*, vol 11, no 2, pp448–459

Greenberg, R., Bichier, P. and Sterling, J. (1997b) 'Bird populations in rustic and planted shade coffee plantations of Eastern Chiapas, Mexico', *Biotropica*, vol 29, no 4, pp501–514

Haggar, J. and Soto, G. (2010) *Análisis del Estado de la Caficultura Orgánica*, Consultoría para la Coordinadora de Comercio Justo en América Latina, Turrialba, Costa Rica

ICO (International Coffee Organization) (2009) 'Coffee prices', January, International Coffee Organization, London, www.ico.org/frameset/priset.htm

IHCAFE (Instituto Hondureño del Café) (2006) *Informe de actividades cosecha 2005–2006*, Instituto Hondureño del Café, Honduras

Jose, S. (2009) 'Agroforestry for ecosystem services and environmental benefits: An overview', *Agroforestry Systems*, vol 76, pp1–10

Kilian, B. and Pratt, L. (2009) 'Challenges and perspectives of the Central American coffee sector', Presentation at SCAA, Atlanta, GA, 15–16 April

Kilian, B., Pratt, L., Jones, C. and Villalobos, A. (2004) 'Can the private sector be competitive and contribute to development through sustainable agricultural business? A case study of coffee in Latin America', *International Food and Agribusiness Management Review*, vol 7, no 3, pp1–25

Loureiroa, M. L. and Lotade, J. (2005) 'Do fair trade and eco-labels in coffee wake up the consumer conscience?', *Ecological Economics*, vol 53, no 1, pp129–138

Lyngbaek, A. E., Muschler, R. G. and Sinclair, F. L. (2001) 'Productivity and profitability of multistrata organic versus conventional coffee farms in Costa Rica', *Agroforestry Systems*, vol 53, pp205–213

Mas, A. H. and Dietsch, T. V. (2004) 'Linking shade coffee certification to biodiversity conservation: Butterflies and birds in Chiapas, Mexico', *Journal of Applied Ecology*, vol 14, pp642–654

Moguel, P. and Toledo, V. M. (1999) 'Biodiversity conservation in traditional coffee systems of Mexico', *Conservation Biology*, vol 13, pp11–21

Montagnini, F. (2006) *Environmental Services of Agroforestry Systems*, Food Products Press, US

Moreno, C. (2008) *Aplicabilidad de la legislación y las normas de certificación en sistemas agroforestales de café (SAFC) en Costa Rica y sus efectos en la rentabilidad del productor*, MSc thesis, CATIE, Costa Rica

Moreno, C., Navarro, G., Le Coq, J. F. and Soto, G. (2009) 'Farmers' perception and economic constrains in the implementation of the legal framework and voluntary certification systems influencing coffee AFS in Costa Rica', Paper presented to the 2nd World Congress on Agroforestry, Nairobi, Kenya

Noponen, M., Healey, J., Edwards-Jones, G., Haggar, J. and Soto. G. (2010) 'Coffee agroforestry systems in Costa Rica: Carbon emissions vs. sequestration', Poster

presentation, SENRGY, Bangor University, Wales, and CATIE, Costa Rica
Palmieri, R. (2008) *Impactos socioambientais da certificação Rainforest Alliance em fazendas produtoras de café no Brasil*, MSc thesis, Universidade de São Paulo, Piracicaba, Brazil
Payan, F., Jones, D. L., Beer, J. and Harmand, J. M. (2009) 'Soil characteristics below *Erythrina poeppigiana* in organic and conventional Costa Rican coffee plantations', *Agroforestry*, vol 76, no 1, pp81–93
PCC (Proyecto Cacao Centroamérica) (2010) *Informe de taller no* 2, Enero 2010, CATIE, Costa Rica
Perfecto, I., Vandermeer, J., Hanson, P. and Cartín, V. (1997) 'Arthropod biodiversity loss and the transformation of a tropical agroecosystem', *Biodiversity Conservation*, vol 6, pp935–945
Philpott, S., Bichier, P., Rice, R. and Greenberg, R. (2007) 'Field testing ecological and economical benefits of coffee certification programmes', *Conservation Biology*, vol 21, no 4, pp975–985
Ponte, S. (2004) 'Standards and sustainability in the coffee sector: A global value chain approach', United Nations Conference on Trade and Development, International Institute for Sustainable Development, Winnipeg, Canada, p49
Porras, C. M. (2006) 'Efecto de los sistemas agroforestales de café orgánico y convencional sobre las características de suelos en el Corredor Biológico Turrialba – Jiménez, Costa Rica', masters thesis, CATIE, Costa Rica
Quispe, J. (2007) *Caracterización del impacto ambiental y productivo de las diferentes normas de certificación de café en Costa Rica*, MSc thesis, CATIE, Costa Rica
Ramirez, A. (2010) 'Impacto económico de las diferentes certificaciones de café sostenible en Costa Rica', masters thesis, CATIE, Costa Rica, in preparation
Ramos, M. (2008) 'The effects of local and landscape context on leafhopper (Hemiptera: Cicadellinae) communities in coffee agroforestry systems of Costa Rica', PhD dissertation, University of Idaho-CATIE, Turrialba, Costa Rica
Raynolds, L. (2002) 'Consumer/producer links in fair trade coffee networks', *Sociologia Ruralis*, vol 42, no 4, pp404–424
Rice, R. (1999) 'A place unbecoming: The coffee farm of Northern Latin America', *Geographical Review*, vol 89, no 4, pp554–579
Rivera, L. (2008) *Una aproximación al análisis de provisión de capitales como determinante de la adopción de sistemas agroforestales de café certificado en Costa Rica*, MSc thesis, CATIE, Costa Rica
Ronchi, L. (2002) *The Impact of Fair Trade on Producers and Their Organisations: A Case Study with COOCAFE in Costa Rica*, Poverty Research Unit, Sussex, UK
Sanchez De León, Y., De Melo, E., Soto, G., Johnson-Maynard, J. and Lugo-Pérez, J. (2006) 'Earthworm population, microbial biomass and coffee production in different experimental agroforestry managed systems in Costa Rica', *Caribbean Journal of Science*, vol 42, no 3, pp397–409
Schroth, G., da Fonseca, G. A. B, Harvey, C. A., Gascon, C., Vasconcelos, H. and Izac, A. N. (2004) *Agroforestry and Biodiversity Conservation in Tropical Landscapes*, Island Press, Washington, DC
Soto, G., García, L., Haggar, J., de Melo, E., Munguía, R. and Staver, C. (2007) *Efecto del sistema de manejo del café (*Coffea arabica*) orgánico y convencional, con diferentes árboles de sombra sobre las características de suelo en un andisol en Nicaragua y un ultisol en Costa Rica*, Congresos Nacional de Suelos, Costa Rica

Varon, E., Eigenbrode, S. D., Bosque-Perez, N. and Hilje, L. (2007) 'Effect of farm diversity on harvesting of coffee leaves by the leaf curring ant *Atta cephalotes*', *Agricultural and Forest Entomology*, vol 9, pp47–55

Zhang, W., Ricketts, T. H., Kremen, C., Carney, K. and Swinton, S. M. (2007) 'Ecosystem services and dis-services to agriculture', *Ecological Economics*, vol 64, no 2, pp253–260

16

Securing the Continuous Supply of Drinking Water in a Territory Requires Concerted Actions and Integrating Intervention Strategies

A Case Study in Copán Ruinas, Honduras

Cornelis Prins and Josué León

Introduction

Too often, conflicts of interest between service providers and their users (organizations) are not resolved and stand in the way of proper protection and wise use of natural resources. On the other hand, farmers' efforts to conserve water replenishment zones are often not properly encouraged and rewarded.

Conflict management is inherent to natural resource management. If resources are scarce and used for different purposes, differences of interest between stakeholders must be channelled in a constructive way and by different means, such as negotiation, compensation, incentives and sanctions, rules of the game (national law, municipal ordainment, direct agreements), and technological devices (green and profitable ways of land use). In this way, conflicts of interest among different resources can be transformed into platforms of joint massive action for effective water protection (Prins, 2008; Prins and Kammerbauer, 2009).

Generally, agricultural space is divided into numerous farms, some of them managing small or very small areas. In order to produce ecosystem services, cooperation platforms among farmers need to be created at a sufficient intervention scale (Pretty, 2003; Goldman et al, 2007). Common pool resources can be conserved if collective action and rules of the game among the resource users are properly designed and implemented (Ostrom, 1990; Meinzen-Dick and Di Gregorio, 2004).

Water is a scarce and finite resource, an economic good, and an essential dimension of public health and social welfare. Its value as a common resource is recognized, but not properly accounted for in economic terms. A culture and habit still prevail of paying little for water, especially if it is perceived as abundant. Water tariff incomes, in most cases, are just enough to pay for the costs of systems operation and physical infrastructure maintenance. In most rural Latin American towns, the lack of water meters does not allow for real water use measurements, which, in turn, creates a major obstacle to sound water management practices and results in a lack of incentives to economize water use.

Finally, there is a gap between formal and real world protection of natural resources. Zones of formal legally protected forests and water resources are often not protected due to a lack of social backing and organized local vigilance. Therefore, the management of natural resources through law alone may not be sufficient.

In this chapter, we argue that payment for hydrological ecosystem services (PHES) can promote good water management practices; but to be effective it must be embedded in a broad global framework that includes other actions and regulations which go hand in hand and at a particular pace. When the objective is water conservation, PHES can be a means of reaching an objective; but if used alone, it may be inefficient.

The argument is presented using the example of Copán Ruinas, Honduras, at the border of Guatemala. This centre of the Mayan culture receives hundreds of thousands of tourists every year. The increasing demand on water resources by the local population and tourists is currently poorly managed and capacity may be threatened in the near future. Many of the Copán Ruinas water reservoirs and replenishment areas are situated within the Quebrada de Marroquín, which is part of the Carrizalón Mountain region (see Figures 16.1 to 16.3). A joint venture between local government and the Tropical Agricultural Research and Higher Education Center (CATIE)–FOCUENCAS (Innovation, Learning and Communication for the Adaptive Co-Management of Watersheds) programme is under way to ensure continuous water flow in this area through a broad variety of measures, such as precise delimitation of water replenishment areas; a PHES pilot scheme and promotion of friendly technologies by producers settled in these areas; negotiations and agreements between producers and water users; backing by local authorities and the legal framework; and adjustment of water tariffs. Through this effort, different actors interact concertedly and a critical mass is forged in order to create a solid social and institutional base for actual and future water protection.

The Context of the Experience

Geographical location

The Copán Ruinas municipality is located in the Copán River watershed, which covers four municipalities (Copán Ruinas, Santa Rita, Cabañas and San Geronimo), associated with the Association of Municipalities of the Maya Route of Copán Ruinas, Santa Rita, Cabañas and San Jeronimo (MANCORSARIC). The 1988 Hurricane Mitch disaster, which caused enormous havoc in Honduras and showed the vulnerability of natural resources, was the impetus for the formation of MANCORSARIC as an organizational basis for joint municipal action in natural resource management in the Copán River watershed.

This natural event was also the basis of the CATIE–FOCUENCAS (first phase) programme supported by the Swedish International Development Cooperation Agency (Sida), whose intent was to respond to the effects of the Hurricane Mitch disaster, which is reflected in the programme's name: Strengthening the Local Capacity for Watershed Management and Prevention of Natural Disasters. The programme gave substantial support to the development of MANCORSARIC.

MANCORSARIC covers 871 square kilometres with a population of 70,000 inhabitants. Temperatures range between 18°C and 24°C, and annual

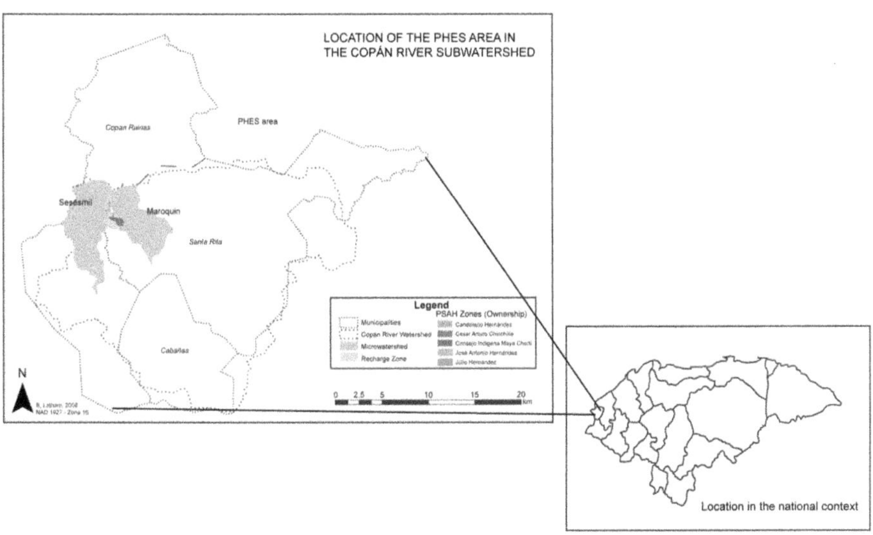

Figure 16.1 *The municipality of Copán Ruinas, Honduras: Case study location*

Source: MANCORSARIC

average rainfall ranges between 1200mm and 1800mm. Predominant economic activities include livestock, coffee, grains (maize and beans) and tourism (for the Copán Ruinas and MANCORSARIC case study location, see Figure 16.1).

Legal framework

Three recent laws are most relevant to this case: the 2003 Territorial Ordainment Law, the 2004 Drinking Water and Sanitation Law and the 2007 Forest and Protected Area Law. The Territorial Ordainment Law guides spatial policy and planning processes utilizing environmental, economic and welfare dimensions, and provides for civil participation in the elaboration of plans. In 2008, the Ministry of Governance and Justice (responsible for implementation of the Territorial Ordainment Law), in collaboration with the Copán Ruinas local government, carried out a pilot territorial ordainment project which built on the experience and results achieved in this action research case.

The Drinking Water and Sanitation Law orders the gradual transference of the drinking water administration from the National Autonomous Service of Aqueducts and Sewers (SANAA) to the municipality, which, in turn, may delegate part of this responsibility and operation to local drinking water organizations. Copán Ruinas is currently in the transference process. The process is significant because it will readjust the water tariff and integrate water consumption, production and conservation.

The recent Forest and Protected Area Law is of utmost relevance because it ordains a variety of measures and procedures to protect forests in water replenishment areas. Many of these actions are already carried out in the project and process areas of this case study. The Forestry Law mentions the necessity of hydrological ecosystem payment schemes; however, it does not indicate how to finance and organize them. The newly created Institute for Forest Conservation (IFC) implements the law and oversees the creation of protected areas. IFC replaces the Honduran Forest Development Corporation (COHDEFOR), which we will discuss later in the chapter.

The local government plays a central role in implementing these laws, which involves municipal capacity-building. FOCUENCAS I and II have played an integral role in municipal capacity-building.

The FOCUENCAS II (second phase) programme

The case study must be understood within the context of the FOCUENCAS II programme. Even more than during the first phase, the emphasis is on trying new approaches, methods and tools in order to produce inputs for scaling out and up. This philosophy and aim is expressed in the programme's official name: Innovation, Learning and Communication for the Adaptive Co-Management of Watersheds. The programme operates four watershed laboratories, two in Nicaragua and two in Honduras (one in the Copán River watershed), and combines action, investigation, learning and communication, resulting in an action research approach.

The FOCUENCAS programme supports the creation of platforms for concerted action and decision-making through watershed management boards; capacity-building; arrangements between various stakeholders and resource users in the watershed and its smaller territories; protection of water sources and replenishment areas; and stimulation of environmentally friendly production technologies and cost compensation.

In the Copán River watershed, MANCORSARIC is the FOCUENCAS counterpart and the Board of Environment and Production (MESAP) its working partner. In MESAP, a great variety of actors (local and national government, civil societies and international agencies) join efforts to share responsibilities, capacities and resources, with some promising results, although initial transaction costs are high.

Due to the scaling out and up objective of the programme, the central programme activity consists of systematization and documentation of the lessons learned in the laboratory watersheds. This process enables the establishment of learning alliances with similar programmes to raise political decision-making leverage and to enhance the chance of effective implementation of national policies and legal frameworks, such as the Honduran Drinking Water and Forestry Laws.

Antecedents of Water Protection in the Carrizalón Mountain

During the 1990s, seven semi-urban communities of Copán Ruinas joined hands and created a two-tier organization in a huge effort to conduct water from a distance of more than 30km to their communities and houses (see Figure 16.2). It is a lively tale of *endogenous water organization and forest protection* around water sources. It took ten years for these communities to see their dreams realized and efforts rewarded, not just in terms of water works and access to good-quality drinking water, but also in organizational strengthening, internal regulation, conflict management and experiential learning. Their vision widened during the process to include protection of their water sources to ensure a water supply for their growing neighbourhood populations. Therefore, the Central Water Board organized water reservoir maintenance and protection of the Quebrada Marroquín vegetation against fire and illegal logging. Because of its history and acquired knowledge, the Central Water Board of the seven communities has become a proactive actor in the concerted action and protection of the Carrizalón Mountain .

On the contrary, the 1988 Carrizalón Mountain *formal forest protection declaration* by COHDEFOR is an instructive example of no protection. The declared area encompasses 5921ha of mostly forest vegetation (although partly degraded and converted into agricultural and animal husbandry land), located within the municipalities of Copán and Santa Rita (see Figure 16.3). The primary reason for the declaration was protection of the water supply. Nevertheless, this aim was not well thought out because the zone consists of a

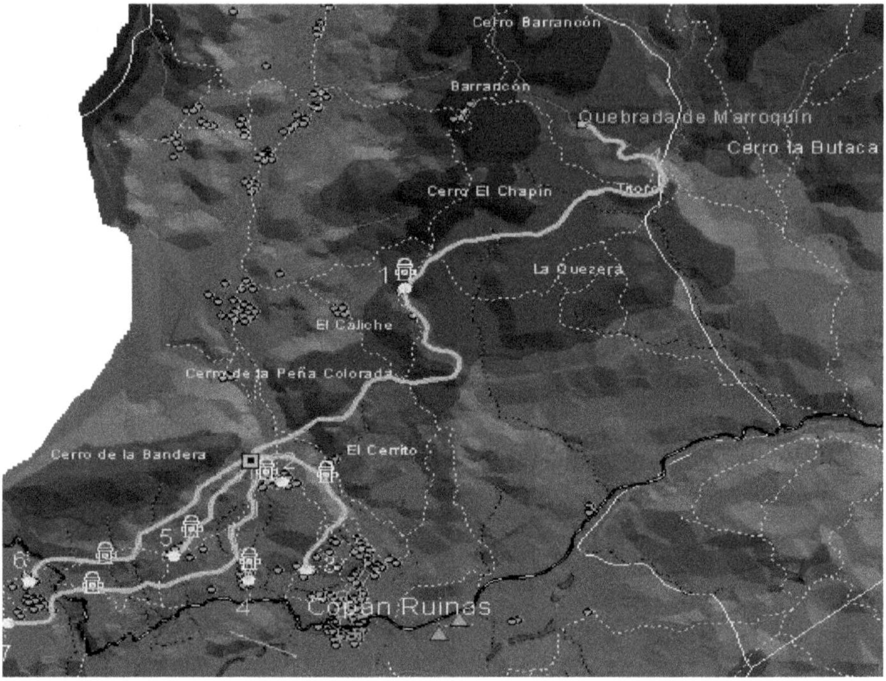

Figure 16.2 *Location of the water project of the seven communities of Copán Ruinas*

Source: Noel Chica

wide variety of micro-watersheds and deep ravines, some more or less critical for water production and protection.

On the other hand, the Honduran Forest Development Corporation (COHDEFOR) declaration remained the letter of the law because of a lack of openness and backing by the local authorities and civil society. In fact, parts of the forest were gradually converted into land for animal husbandry and agriculture, and many municipal lands became private property through land titling by the National Agrarian Institute (INA). Hence, reality was not in line with the COHDEFOR ordainment, and forest and water protection was not achieved. Still, development is not a linear process. Political and social conjuncture differs from that during the 1980s and widens the scope for joint action. Currently, there is a heightened environmental consciousness among the population (caused by, among other things, the Hurricane Mitch disaster), increased involvement by local government because of MANCORSARIC and the FOCUENCAS programme, additional backing and interest by civil societies and development agencies on how to advance natural resources protection, and prevention and mitigation of natural disasters. FOCUENCAS II's action research approach helped to design a road map to address and solve the problem (see Figure 16.4 for a timeline of the main historical events and processes).

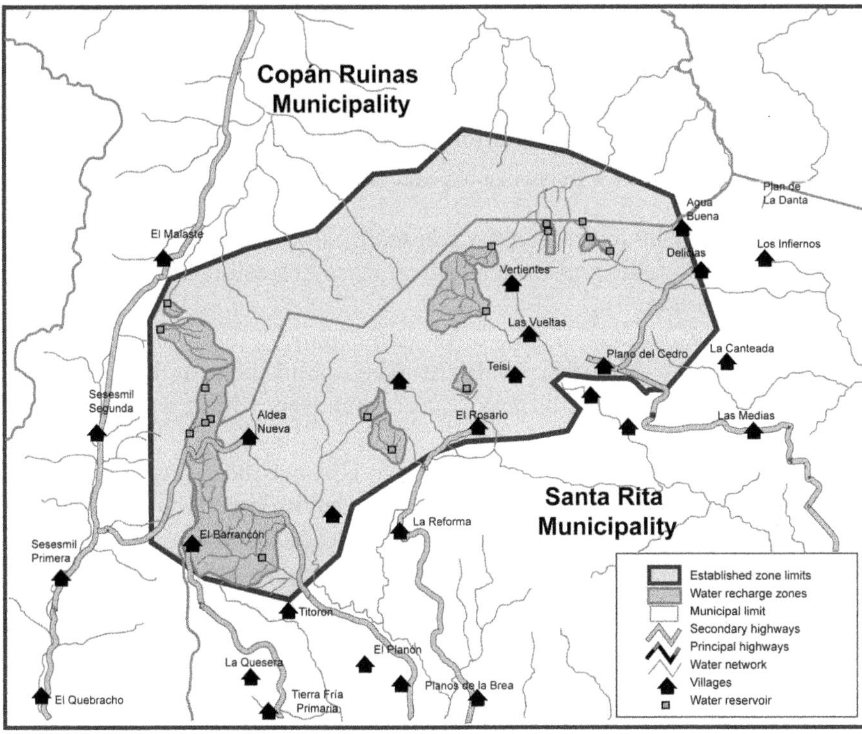

Figure 16.3 *Map of the Carrizalón area*

Source: FOCUENCAS

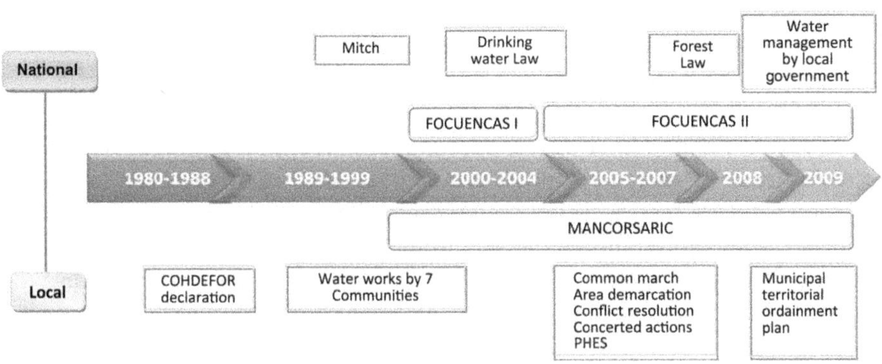

Figure 16.4 *Timeline of main events and processes*

Source: Cornelis Prins

The Intervention Approach

In order to tackle the water problem in its multiple dimensions, a set of integrated working hypotheses was formulated and tested. According to these assumptions, the following conditions must be met in order to begin solving the water problem:

- Link a variety of actors; build a common vision that considers particular interests and views, and form a critical mass to feasibly attain the desired water protection.
- Arrange and agree upon rules of the game in order to protect natural and water resources effectively, while creatively channelling conflicts of interest and perspectives of the different actors and users within a circumscribed territory.
- Integrate a variety of ways and means to advance the protection of water resources and continuous hydrological ecoservices production (in quality, quantity and opportunity), such as precise delimitation of water replenishment areas; communication and education; conflict management, negotiation and agreement with producers; PHES and other forms of compensation; and environmentally friendly land use in the precise areas of water replenishment.
- Build awareness and gradual change of consumer habits; put water meters into place; increase water tariffs to pay for the costs of water production, among other measures.
- Secure the backing of local authorities and the national legal framework; create a solid social and institutional basis for the effective and sustainable protection of water resources.
- Apply a cautious, incremental and learning approach, as well as intervention adaption, according to accumulated experience and knowledge.

In line with the action research approach, different strategies to secure the conservation of water resources were integrated and examined. Working hypotheses were tested through experimental action and joint reflection and learning, and the action strategy was adjusted to meet the desired action goal (Greenwood and Levin, 1998). This approach was a key factor because the programme operated in uncharted territory where the road towards its vision was 'undiscovered', and adjustments in accordance with the experience and learning were necessary.

Implementation Process

Gaining momentum

In March 2005, MESAP, COHDEFOR and FOCUENCAS, in participation with Copán and Santa Rita water organizations, the Marroquín and Sesesmile municipal environmental organizations (CAM), and the functionaries of the

COHDEFOR and FOCUENCAS municipal environmental offices organized a visit to observe and discuss the degree of vulnerability of forest and water resources in the Carrizalón zone. The attention focused on the higher regions of the Marroquín and Sesesmile micro-watersheds, where both watersheds converge and where most of the Copán Ruinas and Santa Rita water sources are located. An open-air meeting was held to read and discuss thc 1988 COHDEFOR declaration and to compare its content with the observations and results in place. It was obvious that most of the measures had remained on paper and that the situation on the ground had worsened. There was a shared feeling that a group effort was needed to reverse the process. The assembly decided to start demarcating the zone.

Rethinking the plan

In a groundbreaking and eye-opening workshop later that year, the original idea and plan was adjusted. A participatory inventory of the producers within the boundaries of the area to be demarcated informed the assembly that there were numerous producers within the area and nearby water reservoirs, many of whom had proprietary title. The assembly further grouped the producers as follows: hardliners (red), producers with an interest in environmental matters and open to negotiation (green), and producers in between (orange).

It was concluded that physical demarcation was not the proper way to start the process because it would generate unnecessary paralysing conflicts with the producers and could even be counterproductive. Producers, instead, should be educated on the purpose and benefit of conservation. Hence, demarcation should not be the starting point, but a result of an education and negotiation process with producers. Action should be initiated in the most vulnerable water reservoirs and replenishment areas, and where producers were most approachable and amenable to change.

First steps

A first step was a precise delimitation of the most critical water replenishment areas. The area declared by COHDEFOR consists of a variety of micro-watersheds and deep ravines; however, the most important water replenishment areas were not properly identified. The scope of demarcation and protection was, for the time being, too extensive.

Hence, within the 5921ha of the Carrizalón Mountain, 468ha were prioritized for delimitation, demarcation and protection through an integral approach and joint action. The main criteria for selecting this smaller area (Quebrada de Marroquín) (see Figures 16.3 and 16.5) and initiating protective actions were the quantity of water reservoirs and the occurrence (or risks) of degrading practices in their replenishment areas.

The area was delimitated by means of georeferencing, cartography, identification of various water points and estimates of replenishment area limits. Within the selected area, tenancy, soil use and economic activities were identi-

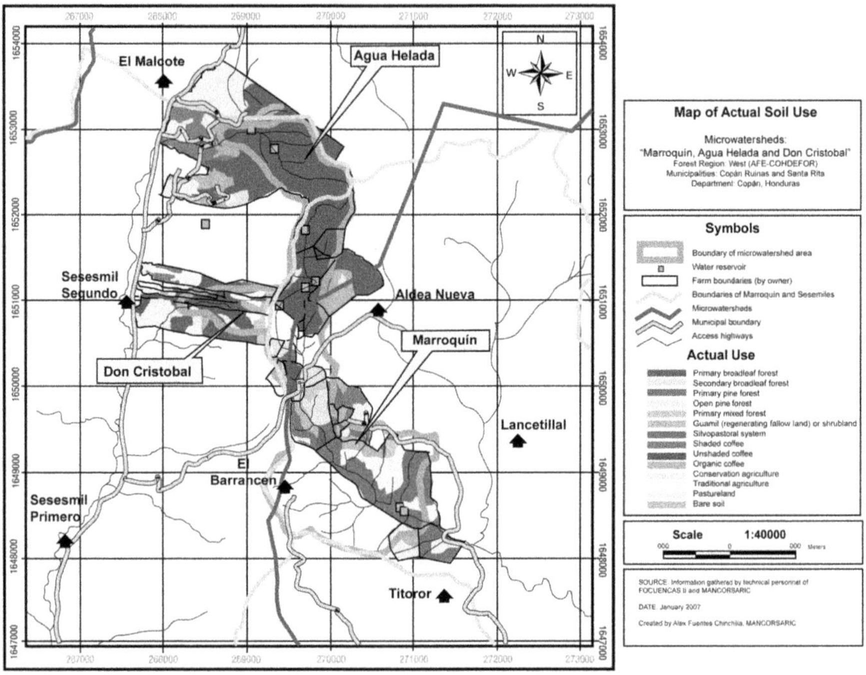

Figure 16.5 *Land-use map of Quebrada de Marroquín*

Source: FOCUENCAS

fied through direct observation and comparison with real property tax list data (*catastro*).

Producers in the area specialize in coffee, grains and cattle. Most producers have forest patches within their property and are smallholders, although there are also some medium-sized producers. One example of the extremes between producers is the lord mayor of Copán Ruins who owns 100ha of coffee plantations and forests, while a group of 25 Indian Chorti families share 130ha, primarily producing grains.

The demarcation of the prioritized la Quebrada de Marroquín area began slowly while approaching the producers. With the help of the producers (see Figure 16.6), the area is now completely demarcated and serves as a sign of agreement. The fact-finding, observation and demarcation activity stimulated learning and organization, and resulted in a growing integration of water users and producers.

A pilot PHES project also played an important role in stimulating concerted action and a common vision between these central actors.

Implementation of PHES

An exploratory study was carried out to determine the willingness and capability of the producers in the prioritized protection area to adapt their land uses to

Figure 16.6 *Joint and agreed-upon demarcation of the la Quebrada de Marroquín area*

Source: Josue Leon

water protection (Retamal et al, 2008). The study laid the basis for the design of the pilot PHES. In 2006, a municipal ordainment gave the green light to initiate the pilot project, and FOCUENCAS and another donor provided the seed money.

Costs of land-use change were estimated and PHES contracts designed with the following conditions. The amount paid to a producer corresponds to the contribution of the land-use change (or, in the case of forests, no change) to water protection (quality and quantity) and the costs of its implementation and maintenance. Therefore, natural primary and secondary forests are worth more than coffee. Organic coffee (or shade coffee with ground cover and free from pesticides) receives higher marks than conventionally grown coffee. The amount received also corresponds to the cost of the agreed-upon changes. Nevertheless, because of the limited funds available, an upper limit of US$1000 per contract was established (for more detailed information on the design of the PHES scheme and the calculation of cost and benefits, see Chapter 7 in this volume).

The results are still limited in quantitative terms, with some 175ha and six contracts (of which one is a group contract with the Chorti Indian community). Some other producers voluntarily apply conservation practices, but do not want to bind themselves to a formal contract.

The experience of the Chorti Indian community is enlightening. In October 2003 they started to build houses on the property of an idle hacienda above the water reservoir of a nearby hamlet. Mediation between the municipal authorities, COHDEFOR and the Chorti Maya Board was needed to relocate the Chorti Indian community to a more appropriate site. Their land covers 130ha and lies within the prioritized area. The group obtained a contract within the PHES scheme and has become a reliable ally for joint conservation activities, such as controlling forest fires and illegal logging. The community protects the water reservoir that it threatened to contaminate five years earlier. A fundamental aspect of this change in behaviour has been the trade-off between conservation and improvement of their livelihoods.

Several bigger producers considered the offered sum insufficient to compensate for the opportunity costs. For the lord mayor of Copán Ruinas, the economic incentive was insufficient to compensate for his 100ha of coffee and forests. However, it was a worthwhile political asset to be considered a responsible producer and a progressive lord mayor.

On the whole, it appears that the importance of PHES lies more in its example than in its scope of application. It must be emphasized, however, that PHES is not the only incentive used in the FOCUENCAS programme.

Other incentives

The Barrancón community, living on the border of the Quebrada de Marroquín area, did not participate in the PHES scheme but received support from other funds with similar conservation effects. The community is organized in a rural saving fund association (*Caja Rural*) and receives credit and technical assistance from a non-governmental organization (NGO). This enabled the members to introduce (based on grain production) more profitable crops such as pineapple and sugar cane, while applying agroforestry technology. The community also received modest support from the FOCUENCAS Environmental Fund to reward their participation in joint protection activities with neighbouring groups, such as fire control of the surrounding woods.

Good technical assistance may be an effective means of changing the production mode and behaviour of small and medium-sized producers. This was demonstrated in a green animal husbandry project. The project consisted of better pasture management promotion and diversifying forage production through the farmer's field schools approach, which showed the feasibility of combining protection of natural vegetation around water sources and intensifying forage production.

For water users, the main driving force was to ensure ongoing water flow, both in terms of quantity and quality, and to protect the natural vegetation around their water sources. In 2008, the water board of the seven communities obtained 100ha within the selected 468ha area, which gave them direct control over the area around their water reservoirs.

Tangible and intangible results

Tangible actions and results of the intervention approach went hand in hand with less visible results, such as joint fact-finding and reflection; learning and adjusting; negotiation; compensation; agreements on new rules of the game; concerted action by different actors; and a combination of grassroots action and respect for the legal framework. This 'software' has been a fundamental asset in the implementation process.

Challenges and Opportunities Ahead

The process is headed in the right direction and the results so far signal that it is on the right track.

Scaling out and up

The intended scaling out and up must go on. The results of the clearance and demarcation of the Quebrada de Marroquín stimulated other communities and water organizations within the Carrizalón Mountain area to initiate similar processes (see dark areas on Figure 16.3). Within the La Fortuna mountain area in the neighbouring municipality of Cabañas, another water protection process was started. In this area, the water sources of lower-lying communities are situated within an animal husbandry farm of 500ha. The owner was willing to protect the natural vegetation around these sources through improving and intensifying forage production. This was an interesting example of a trade-off between profitability and conserving an ecosystem service.

In order for interventions to be effective, they must take place on different scales and be complementary. A farm plan is a mini ordainment of the farm in space and time by the farming household. Spatial arrangements in farms and minor territories should be incorporated in a broader territorial plan, and the experience acquired from minor plans facilitates the implementation of more detailed plans. Therefore, as a follow-up of the Quebrada de Marroquín experience, all water replenishment areas in Copán Ruinas have been mapped and incorporated within a pilot municipal plan.

On the other hand, the new Forest and Protected Area Law offers an excellent opportunity to influence national policy-making. Good practice helps to create conditions conducive to implementing the law, as well as effective forest and water protection. In spite of good intentions for implementation, many laws and regulations have remained idle because the conditions on the ground were not put into place.

Pending questions and actions

There are still some questions pending and further actions to be undertaken. On the *supply side* of the equation, there still remains the need for more intensive application of environmentally friendly technologies, such as for coffee, animal husbandry, grains and fruits. Facilitation of feasible green technology and economic incentives must go together. Producers need time to adapt to the new requirements and opportunities. It is preferable to have a mix of incentives and forms of compensation, as shown earlier.

The supply-side approach of hydrological ecosystem services production will not be effective in the end if the *demand side* is not properly taken care of. This would be like walking on one foot. An effective service demand requires focused attention and a clear strategy since it is the weak link in the whole system

How do you continue paying producers for water resources protection in such a way that it is a real incentive for land-use change and environmentally friendly technology application, while reaching enough producers to form a critical mass? Currently, external donors replenish the PHES scheme fund; however, these fund sources are finite. As a result, financial compensation could be locally mobilized, if the correct strategy is followed, by the presence of the many hotels and restaurants in the area.

The transfer of the water administration from SANAA to the municipal government in 2010 is a golden opportunity to put into place structural and lasting improvements to the water system such as the gradual instalment of meters, payment and progressive tariffs according to the quantity of water used, and capacity of payment. Efficiency and equity must be the guiding principles.

There are promising signs that this is not wishful thinking, but a window of opportunity in the near future. The local authorities of Copán Ruinas, for example, have become focused on the drinking water issue and are convinced of the importance of proper water protection devices and policies. There is also a growing consciousness among local hotel owners that water is a fundamental element of service to the thousands of clients who visit Copán each year; therefore, they cannot go on paying so little for water. It would be an ecological and an economic disaster if, some day, there was not enough high-quality water to prepare meals and provide proper sanitary conditions to tourists.

In 2005, a study was carried out to enquire into the willingness of the Copán Ruinas population to pay an additional sum on their water bill for water protection purposes. On average, people were prepared to pay nearly US$1 extra for conservation purposes. Nevertheless, the leading slogan should become 'slowly but surely': water is a commodity, like electricity, and therefore payment should compensate for the costs of its production, protection and conservation, whether in accordance with the water quantity used or service received, and take into account considerations of equity.

In 2008, water tariffs were raised by 50 per cent to cover operation costs. During the same year, a *Cabildo Abierto* (a meeting of local authorities and local people) was held in which it was agreed, in principle, to begin the gradual installation of water meters and to apply different tariffs according to the quantity of water consumed. There was also consensus in this meeting that the increase in tariffs must go hand in hand with the improvement of water quality and service supply.

Key Lessons Learned

The results validate the approach applied, and the lessons learned transcend the case study and shed light on the road towards effective protection and proper use of water resources:

- Determining the willingness to provide ecosystem services is as important as determining the willingness to pay for it. Therefore, it is necessary to work on both the supply and demand side of the water conservation problem to really start resolving it.
- Conflicts of interest and vision between agricultural producers and water users in these crucial water conservation areas must be channelled in a creative and constructive way to obtain social cooperation. Compensation for environmentally friendly land use must be seen as a means of conflict management and large-scale cooperation for the purpose of water conservation.
- The effect of PHES on water protection must be analysed in relation to other types of incentives and the overall intervention strategy. PHES can promote good water management practices; but it must be embedded in a broad global framework that includes other actions and regulation. The objective is water conservation and not PHES.
- One recipe does not work for all dishes. PHES is important; but it is not the only form of compensation. Other complementary incentives are needed according to varying circumstances and types of producers.
- For analytical and practical purposes, we must take into account that different types of producers have different profiles and logics. For poor producers, a relatively modest amount of support is a sufficient incentive to accept the desired and required ecological conduct if the compensation is accompanied by assistance to improve their production and welfare. For larger producers, the offered sum may not be enough to compensate for the opportunity costs involved. For the lord mayor of Copán Ruinas, the economic incentive was insufficient to compensate for his 100ha of coffee and forests. However, it was a worthwhile political asset to be considered a responsible producer and a progressive lord mayor.
- Both poor and rich producers must be involved in conservation schemes. Poor producers are often obliged, due to their poverty, to utilize degrading practices (many small degrading practices add up to a big impact). The impacts of rich producers are often quite high due to the scope of the environmental externalities of degrading farming or animal husbandry on

medium or big properties. Hence, both types of producers must be included in conservation objectives through principled, although differentiated, and pragmatic approaches.

- An alternative set of rules, beyond those pertaining to PHES, are necessary in case the providers and consumers of water services belong to the same group. This occurs when a water user organization owns the land around its water sources and manages the entire system, including protection and maintenance of the vegetation around the water reservoirs; conduction and repartition of the water stream; agreement on and collection of water tariffs; organization of labour to maintain the system; and agreement on the internal rules of the game.
- A clear analysis is necessary to assess the particular situations of the different actors in order to carry out a differentiated protection strategy and to build the critical mass needed for an effective water protection policy within a well-defined territory and with a clear understanding of the ecological, economic and social relations and processes.
- Respect for the legal framework and carrying out innovative good practice must go hand in hand. As shown in the COHDEFOR declaration, a formal protection decree remains a 'paper tiger' if it is not rooted in civil society. On the other hand, local action that is not backed by national or local regulations will also be limited in its effects.
- Regular reflection helps to readjust and fine-tune actions to be taken and methods to be applied in order to be more effective in advancing towards a shared vision. The FOCUENCAS philosophy refers to this as the wheel of learning, part and parcel of adaptive management and action research. It is a cautious and effective way of planning and acting when entering into uncharted territory and having to find your own way.

These are not the Ten Commandments – just a few proven and useful guidelines to orient actions in similar, although always particular, circumstances. A successful replication is never a photocopy!

References

Goldman, R., Thompson, B. and Daily, G. (2007) 'Institutional incentives for managing the landscape: Inducing cooperation for the production of ecosystem services', *Ecological Economics*, vol 64, pp333–343

Greenwood, D. and Levin, M. (1998) 'Introduction to action research', in *Social Research for Social Change*, Sage Publications, London

Meinzen-Dick, R. and Di Gregorio, M. (eds) (2004) *Collective Action and Property Rights for Sustainable Development*, IFPRI 2020 Vision Focus Briefs, Washington, DC

Ostrom, E. (1990) *Governing the Commons: The Evolution of Institutions for Collective Action*, Cambridge University Press, New York, NY

Pretty, J. (2003) 'Social capital and the collective management of resources', *Science*, vol 302

Prins, C. (2008) 'Abordaje de conflictos para la efectiva protección de los recursos hídricos. Camino hacia la gobernabilidad', Foro internacional de gestión de cuencas hidrográficas, CATIE, Costa Rica

Prins, C. and Kammerbauer, H. (2009) *Análisis y abordaje de conflictos en cogestión de cuenca y recursos hídricos*, Serie técnica, CATIE, Costa Rica

Retamal, R., Madrigal, R., Alpízar, F. and Jimenez, F. (2008) *Metodología para valorar la oferta de servicios ecosistémicos asociados al agua de consumo humano*, Copán Ruinas, Honduras

17

Payment for Environmental Services

Perfecting an Imperfect Market by Building up Environmental Solutions

Carlos Manuel Rodríguez

Introduction

Never before in the history of our countries have we had as much knowledge about our natural environment, as many institutions for sustainability and research, as many environmental standards, organizations, professionals and efforts to protect the environment, as now. Paradoxically, we have never had as many serious environmental problems as at present. Ironically, as worldwide governments accept and preach the discourse of sustainability, few concrete and sustainable developments reflect the necessary changes needed to approach the problem of the destruction of our natural environment. Normally, environmental challenges are simplified and focused around climate change; however, recent scientific information suggests that the problems are significantly more complex and are a serious threat to environmental sustainability. Loss of biodiversity, contamination by persistent organic substances, soil degradation and desertification have exceeded tolerable limits and their restoration or care requires knowledge and resources that may not exist or be available. However, positive examples of effective benefits of systematic and coordinated global actions towards a solution, such as in the depletion of the ozone layer, could benefit other approaches towards environmental stability.

Without a doubt, the current state of our environment is the direct result of our development model, with consumption and production patterns of particular concern. These patterns respond exclusively to the political goals of economic growth, where sustainability remains high on the value scale, but is not reflected in public policy. At most, sustainability goals trigger actions that do not exceed those of lower-value scales and therefore do not require structural reforms. Our current consumption and production patterns focus on a fossil fuel economy and on the manufacture of disposable non-returnable goods. On top of this, the current global population of 7 billion people is likely to expand to 9 billion within the next 50 years, with the vast majority of the population living a modern Western lifestyle. During the next few decades, if countries such as China and India continue in their efforts towards fast unsustainable economic growth, there will not be enough oil, aluminium, wood, paper, iron and water to meet the resource demands of a consumer society that is not efficient in the production and recycling of goods.

Since global policy goals are mainly oriented towards economic growth, government institutions act accordingly by designing public policies to reach these goals. Therefore, it is not a surprise that there has not been substantial progress towards sustainability according to the Rio92 (Earth Summit) Principles. In order to achieve sustainability objectives, I think it is necessary to understand the structural reforms needed to promote changes in the public policy decision process, in institutions and in state ministries. The accounting systems for these reforms should include all indirect costs or negative externalities that our patterns of consumption and production generate.

The creation of sectorial environmental institutions, such as ministries of the environment, to resolve environmental problems has not resulted in the reduction of environmental degradation. These environmental institutions were created 10 to 20 years ago with clear objectives and goals to attempt to solve environmental problems. However, environmental degradation has continued and is often a direct result of the development model, which generates indirect costs (environmental and social) that are often undetected by the market. Unfortunately, this market is the decision-making reference for politicians. In order to create substantial and sustained advances towards environmental goals, we must take into account all direct and indirect costs and realize that they are an essential part of development policy.

Our experience indicates that there are two important points that need to be addressed in order to approach environmental problems effectively. First, the environmental institutional setting should create a guided political function so that all institutions with responsibilities in the use and handling of a natural resource (i.e. ministries of agriculture, energy, mines, waters, fisheries, science and technology) respond to environmental problems in a coordinated and coherent manner. Second, our markets should be more efficient at reflecting indirect associated costs for the use of natural resources. Through the inefficient reflection of these indirect costs, profitable development activities have resulted in eroded soils, the disappearance of forests and the exhaustion of fisheries. The responsibility for these indirect costs (environmental and social) is not taken over by economic agents, but handed to society as a whole. These indirect costs are compounded by the preservation of subsidies, such as tax exemptions and financial incentives, including the national fishing fleet fuel subsidy and tree-felling tax relief. These subsidies have resulted in irrational and unsustainable economic activities at both a national and global scale, where international trade sectors and markets protected from free and fair consumer access have generated negative environmental impacts in developing countries.

In a global environment with such a large and developed world economy, indirect environmental costs often exceed the direct benefits of many human activities. Preserving an accountancy that does not recognize these environmental costs creates distortions in the market that do not yield the full picture of our economic behaviour. Erroneous guidance and incorrect signals lead public economic policy-planners and decision-makers in the wrong direction. In order for markets to operate conveniently and for economic actors to make the right decisions, we need accurate and complete information that includes the full costs of products and services. Currently, economic evaluations of a country include all activities for goods production and service provision, but indirect environmental and social costs are not deducted.

Gasoline is a prime example of market failure for total environmental costs for a consumer product. In 2010, gasoline cost close to US$3 per gallon at the pump. This price covers the direct costs related to this product, such as exploration, operation, transfer, refinement, taxes and profit margins of the petrol station. However, the price does not reflect the indirect costs associated with gasoline's contribution to climate change, subsidies to the oil industry, military security costs in unstable geographies, pollution damage, and respiratory disease treatment associated with emissions. According to the International Center for Technology Assessment (1998), these indirect costs would add up to between US$4.6 and $14.4 per gallon, resulting in a consumer cost of between US$5.6 and $15.4 dollars per gallon. Therefore, the unrealistic price of US$1 per gallon encourages unchecked fossil fuel consumption, which results in high environmental impacts such as contributions to climate change or respiratory diseases. These impacts, in turn, generate a huge hidden 'financial' charge' that is being paid by our society and probably will be passed on to future generations.

As long as we do not amend both institutional and market weaknesses, efforts to reverse the serious spoiling of our natural environment will not be successful on the scale and with the impact that the scientific community strongly recommends. More than ever before, we require political leadership that can see the 'big picture' and understand the relationship between economy and environmental services. Since the predominant decision-makers and public policy-makers are economists, it is time that they learned to think like ecologists. We cannot continue relying on the invisible market hand when it has proven to be blind to the real costs of its decisions.

In this context, institutional improvements and changes to the market are urgently needed. We cannot expect payment for ecosystem services (PES) to be successful if it is not accompanied by serious and politically feasible institutional reforms seeking to address the failures and problems just mentioned. Efforts to develop and implement PES mechanisms without understanding and addressing the underlying economic and institutional challenges are of great concern. It is important to understand that PES is only one environmental policy mechanism among many other options for implementing a preset environmental policy. Therefore, if there are no clear policies supported by long-term vision and consensus, it is unrealistic to think that a PES system will yield the expected results or those that were observed in other places. PES is a means rather than a goal. The goal might be to reduce deforestation and preserve biodiversity, but PES will be the instrument to achieve it.

The Institutional Challenge

As mentioned above, effective institutional design that addresses environmental challenges is the pillar for long-term environmental solutions, based on policies and instruments at various scales. Goals must be set at the national, not sub-national, scale, responding to international commitments with a vision of sustainability. Without the proper instruments, institutional design is likely to be of limited success. Therefore, the essential elements for successful economic environmental policy are:

- a strong institutional framework;
- a gradual process of institutional change, with political clarity;
- sustained capacity-building;
- taking stock of lessons learned;
- flexibility to correct mistakes and develop successes;
- political volition and leadership of stakeholders; and
- social support.

With the exception of Venezuela and Costa Rica, environmental institutionalism in the tropical region of the Americas began after the 1992 Earth Summit in Rio de Janeiro. Before the summit, a few countries established offices or agencies within their ministries of agriculture and livestock for forest or protected areas. As a consequence of their institutional objectives and policies, there was little room to grow and develop. The focus of the agricultural sector was on the expansion of the agricultural frontier through increased productivity, incorporation of new production areas, and diversification of production and rural development. Unfortunately, this vision was counter to the protection of forested areas because forests and natural ecosystems were viewed as idle unproductive areas of no value. History shows that these agricultural institutions promoted deforestation through land-use change into high productivity land for agriculture and marginal lands for livestock, aided by the policy framework and economic incentives of rural development.

Thus, when ministries of environment were implemented in the region, there were already some agencies in charge of the management of natural resources, such as forest services, protected areas or wildlife. Consequently, with the creation of the new ministries, the institutional environment was more resource friendly compared to the previous agricultural structures. However, transfer of responsibility to the new ministries in some countries, such as Costa Rica, Mexico and Columbia, utilized the former agencies that specialized in forests or conservation as the backbone of the new ministries. In other countries, the former agencies were transformed into autonomous institutions co-managed by the private sectors concerned. These transformations generated, on the one hand, conflicts of interests (such as the Forestry Institute in Guatemala, the Environmental Authority in Panama and the Fisheries Agency in Costa Rica); on the other hand, due to their political dependence on instances that were institutionally adverse, their management depended on political rather than technical criteria.

For countries to achieve their environmental objectives and PES to have the desired effects, environmental institutions (ministries) need to be designed and structured with operational capabilities that ensure the administration, management, conservation and rational use of renewable natural resources. Ministries of environment should have direct authority for handling, use, development and conservation of forest resources, biodiversity, water resources, protected areas and pollution control, or, at the very least, have political empowerment in the supervision and direction of the autonomous or decentralized agencies in charge of these affairs, ahead of any other ministry. This direct authority must be embedded in a long-term policy that seeks and promotes structural changes to turn our economies into low-carbon or green economies. Moreover, ministries of environment should have a clear legal framework for environmental policy leadership. The minister of environment is the leading public officer directly in charge of generating the political dialogue for the determination and definition of the environment and sustainable development policy; the minister must generate positive coordination dynamics and manage conceptual principles with ministries responsible for public health, transport, energy and mining, agriculture, science and technology, at the very least.

This political and institutional ordering occurs as a result of years of trial and error (not yet observed in the tropical region of the Americas), where countries have learned that the environment is not solely the responsibility of the ministry of environment, but rather a fundamental task of the central government, in the same way that economic and social topics are addressed. Therefore, governments should design complementary public institutions for social, environmental and economic issues. The current political and institutional reality is far from ideal, and in order to move in a sustainable direction, we face a hard task of raising the awareness of political and planning stakeholders.

Consequently, the institutional challenge is to step forward:

- away from the current environmental institutional situation, which is sectorial and often scattered, where ministries of environment do not manage forests, biodiversity, protected areas or territorial planning, thereby generating political weakness;
- towards guiding ministries, which are able to generate inter-sectorial and inter-ministerial political dialogue, and thus successfully address the third pillar of sustainability, assuming that social and economic angles are already being developed and covered by the central government.

The current ministries of environment are often limited by their budget resources, which are allocated by the ministries of finance. These limitations result in a perception that the ministries of environment are focused on moral and ethical issues and are incompetent with regard to progress and prosperity. This perception, in turn, limits their access to resources.

The modern tools designed to resolve environmental problems will have little or no impact unless ministries of environment and environmental policy agendas progress in a guided political direction. In our regions, there are many small-scale PES pilot projects, funded by the Global Environment Facility (GEF) or other agencies, which generate small-scale products with positive results. However, these projects do not generate or improve national capacities or contribute to the structuring of environmental institutionalism in an effective way. Therefore, PES projects must contain strategic political and institutional components based on measurable objectives and long-term monitoring for the improvement and advancement of environmental institutions.

The Economic Challenge

Current consumption and production patterns of modern Western society are rapidly spreading throughout the Eastern marketplace and are based on an economy supported by the consumption of fossil fuels and the production of disposable material goods. It has become blindingly clear that this economic–cultural pattern is unsustainable. Uncontrolled growth in a world of finite natural resources can seriously affect the basic ecological processes that support life if limits are not imposed.

It is imperative that the economy shifts towards a world market that utilizes renewable energy sources, diversified transportation systems, and reusable and recyclable material goods. While global financial resources and technological development expertise are available, we must first understand the economic challenges that this shift entails. A market economic system that ignores indirect costs in the allocation of values and prices is irrational, inefficient and self-destructive. We could avoid environmental bankruptcy if the accounting system would bring forward all costs. In the same way that Marxist societies collapsed because the market hid the economic truth, capitalist societies can collapse if the market does not disclose the ecological truth. This has been mentioned by numerous economists and scientists, but none with so much political impact as Sir Nicholas Stern. In his reference to the fossil-fuel pricing system that does not include the global costs of climate change, he describes this as 'the biggest market failure the world has ever seen' (Stern, 2006).

A structural change in economic policies should identify and modify all perverse incentives, redefine the tax system by reducing burdens on positive and profitable activities, with no impact upon the environment (income, capital gains, transactions, property), and tax all activities associated with environmental burdens or that impair the quality or quantity of environmental services (pollution). Thus, the treasury would not be negatively affected and we would generate suitable conditions in which environmental costs are allocated to the source.

Payment for Environmental Services as a Political Process to Confront Institutional and Economic Challenges

Capitalizing on the experience of PES mechanisms in our region, and taking into account the challenges that we developed in the previous sections, we have drawn the following rules and lessons regarding PES.

The transfer of experiences must include a process of capacity-building, not simply a gathering of individual efforts

Economic valuation and financial instrument topics related to sustainable development agenda resource requirements may be approached from different angles, depending on the objectives pursued. There are several stages in the formulation process of national policies to mobilize resources towards the forest and environmental sectors. Many of these stages are highly conceptual and methodological (e.g. economic valuation of environmental goods and services), while others are more closely related to capacity-building for the design and operation of the final schemes. Internationally, there is more conceptual and methodological progress; however, on-the-ground experiences are relatively few and/or unknown.

Conceptual or methodological models will not solve the problem if they are not clearly framed in national policies, whose objectives require the use of economic instruments to promote compliance with additional or related objectives in environmental or sustainable development. Eventually, and according to the nature of the corresponding legal framework, standards will also be required that encourage the use of such instruments. But what will be of utmost importance is a national institutional capacity for the valuation of services, collection of fees and compensation of providers, or redistribution processes related to the economic instruments or mechanisms.

In this context, the perception that reducing emissions from deforestation and degradation and enhancing carbon stocks (REDD+) as well as projects investments are initiatives of high profit in the short term are erroneous, as these enabling conditions do not exist in most tropical countries. During the recent Copenhagen climate summit, Peru stated that it would attain zero deforestation by 2020. This commitment was made in the same political context in which the government increased oil exploration and mining concessions in forest areas, including areas where concessions for forest management have been granted. Additionally, the Peruvian government is planning three roads projects to access the Pacific Amazon watershed. This is another example where a coherent guiding policy, fully consistent with climate mitigation goals, is lacking.

Costa Rica has several years of experience in public policies linked to different economic tools. Valuation and retribution of services and goods have been adjusted, as well as the collection and redistribution of income. As a consequence, the transfer of experience in this field must be visualized as a national capacity-building process (which does not mean an exclusively public vision) in the design, implementation and evaluation of policies related to the mobilization of financial resources. It is a complex topic, and it should be approached by taking into account this complexity. Due to the complexity of the topic, initiatives to develop PES projects must be observed with a critical eye, especially those led or funded by GEF or United Nations bodies, in terms of adequate components of capacity-building and public policy improvement.

Payment for environmental services is a product of the accumulation of policy and institutional capacity experiences, and still requires improvement

As a result of trial and error, the PES programme in Costa Rica, in combination with other policy instruments, has produced very satisfactory results as a financial tool to mobilize resources. The application of different tools and mechanisms has resulted in the improvement of the legal, economic and operative features of the programme, which in turn resulted in the Certificate of Environmental Services.

The conception of the PES programme was more complex than a single financial mechanism or tool. Besides legal definitions that recognized at least four generic environmental services, different practical modalities were also generated for their recognition and payment, and each of these modalities had its own characteristics. Credit facilitation measures, primarily in forestry, were combined with the programme. Costa Rica's experience and focus on carbon sequestration gradually incorporated water, biodiversity and scenic beauty issues. Identification of new modalities and the development of pilot projects for the 'second generation' of environmental services in Costa Rica have started. These services include services provided by agroforestry and silvopastoral systems, as well as integrated farm management, to name a few. This next generation of services has as its purpose the integral management of natural resources by not only recognizing the value of environmental services, but by promoting rural development and poverty alleviation.

The evolution of the programme has required different stages of legal, conceptual and operational construction, which may be critical in reaching the 'mature' stage required to implement a programme for environmental services. In other words, depending on the characteristics of individual projects, stages of the programme's development should be analysed for convenience and whether certain steps can be bypassed. The assumption behind this analysis refers to the necessity of flexibility in a mechanism in order to adapt to changing environmental, political, cultural and economic circumstances.

The programme must be the product of a collective construction, where the concerns, strengths and interests of a large number of governmental and non-governmental actors are considered

Capacity-building processes require a clear identification of the actors and their roles and responsibilities. It is critical that all relevant actors participate in the model design stage at its inception. This creates the conditions that subsequently facilitate the implementation of concrete actions, as well as enables future 'managers' to take advantage of the learning process.

Sometimes, depending on the national institutional complexity of a country, the call of public- and private-sector actors is required. These actors should participate in the creation of a common, unified and integral vision, with a clear idea of everyone's political, legal and operational responsibilities.

In theory, and based on Costa Rica's national experience, the key process actors are as follows:

- in governance: state forestry administration; management of protected areas; environmental authority; public finance authority; national banking system; water authority; real estate registry; entities responsible for climate change policies and marketing of carbon certificates and for environmental impact assessments, etc.;
- in the private sector: environmental NGOs; business chambers; and groups related to forestry and tourism, etc.;
- in the technical sector: professional associations related to the sector (forest engineers, agronomists, etc.); academics and research scientists, etc.;
- participation of international organizations, both governmental and non-governmental, depending on country dynamics.

The reconciliation of interests from the early stages of model design is central to ensuring the further viability of actions seeking to implement this model. The intervention of consultant teams should be oriented towards:

- the provision of technical and methodological tools; and
- the national definition of process facilitation if contribution of substantive elements is necessary.

National teams should always consider the importance of defining policy priorities and objectives.

The programme should coordinate efforts around a single vision and strategy (state policy)

The identification of economic instruments to promote compliance with environmental, social or economic objectives should correspond to a state policy, which must ensure their sustainability in the long term. Consequently, these objectives must be clearly linked to the national definitions included in the country's strategic and global planning instruments (national and sectorial development plans), which, in most cases, should be articulated with policies from other relevant economic sectors (agricultural policies, competitiveness, social development). Conflicting sectorial interests must be resolved because allocation or reallocation of resources for public investment must be agreed upon. Non-flawed market mechanisms should emerge that permit economic forces to regulate competition for public resources. With the consolidation of these mechanisms, public investments or intervention becomes less necessary.

National sectorial policies must be articulated, although their aim is the same as for national development: improvement of citizens' quality of life. But these policies must also be formulated in two dimensions:

1. between national policies and international agendas and their divisions (conventions, regional policies, international projects); and
2. between national policies and the efforts performed at a lower national scale that are promoted or tolerated by national policies (e.g. studies and pilot projects for the valuation of ecosystem services; collection, payment or retribution; other efforts in previously identified policy areas such as protected areas, forests, biodiversity, hydrological resources, etc.).

The purpose of the articulation is to ensure an efficient use of financial and institutional resources and efforts, which always involve a cost.

In the same way, it is convenient to recognize the limited validity and utility of different exercises in valuation, as well as collection of and payment for environmental services, when their potential to contribute to the generation of policies or mechanisms at a national scale is scarce due to:

- geographic scope;
- market particularities;
- the nature of environmental services; or
- the economic cost of the solution.

Although these exercises will always be an important element of some political decision-making, they are not necessarily sufficient in and of themselves.

Similarly, from a governance perspective, it is essential to recognize the existence of a mutually determining relationship among the three main levels of resource mobilization: politics, instruments and outfitting of the instruments. In other words, none of these has real pragmatic value if taken independently.

Their utility relies on their interdependence. Economic instruments are always a means to achieve greater policy goals, which are mobilized through the respective institutional capacity.

The existence of political will is another essential element to ensure the viability of a proposal to mobilize financial resources. This will is particularly necessary when clear, concrete policies, objectives and goals are lacking. In Panama, early signs indicate that this will not only exists, but has been given a high priority by Panamanian authorities.

The definition of a policy and its related financial system must consider a wide range of elements (legal, political, institutional, cultural, social, economic, etc.) that together yield comprehensive results

Different national policy planning levels must incorporate elements of analysis to integrate interdisciplinary inputs for model-building. These elements are related to legal, political, economic, institutional, cultural and social fields. The identification and clarification of roles, and the expected responsibilities of the different actors must be made during the initial design stages, and must correspond to the specific national reality.

Decision-making should be based on an analysis of the subject in each situation, assessing prior experience and local capacities

Decisions on policy directions and mechanisms for mobilizing financial resources to foster sustainable development should not necessarily integrate all dimensions and issues presented previously. However, in order to enable the required learning process, these decisions should integrate elements to:

- allow on-the-ground actions to utilize legal and political resources and adapt existing instruments to fulfil the desired objectives; and
- supplement the model by focusing on the gaps related to the policies and instruments that are needed for full compliance with the policy objectives.

Conclusions

It is clear that in the case of developing countries, the existence of a series of legal, political and institutional instruments indicates that a great deal has been done. We now need to focus our attention on identifying the difficulties and limitations of these instruments in order to meet the desired objectives. The recognition of existing capacities may be carried out through key elements that have been identified as characteristic of the Costa Rican model, without attempting an automatic transfer of solutions, but by assessing the specific conditions and the relevance of these solutions to those conditions.

Without doubt, the PES scheme and the policy implications required for its establishment have contributed to political dialogue and agreements with civil society, transforming decision-making. This transformation of decision-making is required for our agro-landscape if we want to produce the structural changes needed to achieve our challenging conservation and rural development goals. Nevertheless, this financial mechanism, as innovative as we may perceive it, will not achieve its objectives if we abandon efforts to improve political governance and economic indicators.

References

International Center for Technology Assessment (1998) *The Real Price of Gasoline: An Analysis of the Hidden External Costs Consumers Pay To Fuel Their Automobiles*, Report no 3, www.icta.org/doc/Real%20Price%20of%20Gasoline.pdf, CTA, Washington, DC

Stern, N. (2006) *The Economics of Climate Change: The Stern Review*, Cambridge University Press, Cambridge, UK

18

Measurement and Payment of Ecosystem Services from Agriculture and Agroforestry

New Insights from the Neotropics

Bruno Rapidel, Jean François Le Coq, Fabrice A. J. DeClerck and John Beer

The 17 chapters that form this book give us a good sample of the current developments, successes and challenges of payment for ecosystem services (PES) in Latin America. We reviewed the approaches, methods, tools and applications required to estimate ecosystem service provision related to agricultural land management. We presented methods to assess the value of ecosystem services, and potential means to provide payment for the same. We also compared the PES schemes, including eco-labelling, as alternative means to reward environmentally friendly land husbandry. The case studies provided insights on specific issues at the heart of PES development and sustainability (i.e. access to funding, targeting, monitoring, articulation with other mechanisms such as collective management of watersheds, as well as the institutional and political context required for successful PES development).

Abundant literature has been published on the provision of ecosystem services and the means to promote them. This literature refers principally to forests; agriculture and agroforestry have received much less attention. Thus, in this concluding chapter, we synthesize new knowledge presented in this book, particularly regarding the specific characteristics of Neotropical agricultural and agroforestry systems as providers of ecosystem services, and the means to increase this provision. We then present a brief overview of published information on ecosystem services and payments, expanding this with the new knowledge presented in this volume. The main topics are measurement of the provision of ecosystem services; environmental effectiveness of PES; cost efficiency; effects on social equity and poverty alleviation; and, finally, sustainability and governance. In the discussion of each topic we stress the areas where new research is needed.

Agriculture and Agroforestry as Providers of Ecosystem Services

Literature on ecosystem services and PES refers mainly to forest ecosystems. Moreover, most, if not all, PES experiences in agricultural landscapes have been focused on the inclusion of trees or forest patches. This focus on trees and forestry is natural when biodiversity conservation in the Neotropics is the goal. Many of the species that we need to conserve come from these Neotropical forest ecosystems and have co-evolved with them. Nevertheless, in the Neotropics, as well as globally, about half of the current land surface is under agriculture (Gregory et al, 2002) and it is predicted that the remaining 50 per cent under forest, mainly in the tropics, will be at least partially cleared in the coming decades. It is imperative to devote more efforts to elucidating the significant contributions to the provision of ecosystem services that agricultural landscapes can and will have to provide; in this context, these systems have not received sufficient priority.

Characteristics of PES in agriculture

When dealing with agriculture, the very notion of payment for ecosystem services is blurry for several reasons. The first reason is that, in many instances, the land managers and providers of ecosystem services are also the users of ecosystem services. For example, in the case of pollination services, the services provided by a forest patch within a farm can result in higher yield for the farmers who own the forest patches and the surrounding fields (see Chapter 5 in this volume). However, since bees fly over long distances, this pollination service is also a public good – that is, other farmers with fields within the range of influence of the forest patch will enjoy the pollination service, whether the owner of the forest patch agrees or not (non-excludability), and the provision of this public service will probably not decrease pollination in the fields of the owner of the forest patch (non-rivalry). Therefore, even in this case there is a need to estimate and reward such services. The second reason for this blurriness is more complex. PES in agriculture leads to the possibility that polluters will be paid for not polluting. Many countries' environmental policies rely on the opposite principle: polluters are held responsible for their actions and must bear the corresponding costs – the polluter pays principle. Are land managers potential polluters who are to be charged by society to internalize all of the environmental costs related to their actions? Or have they acquired a right to use the resources on their land, and any reduction in off-site impacts that result from their actions (or the maintenance of a pre-existing environmental service) should be considered as a service to the society that should be rewarded? When we consider eco-labelling, a premium is usually paid for environmentally friendly products (organic, etc.). The argument that the polluter pays suggests that this is wrong. The base price should be the certified product (no externalities). Farmers unable to obtain this certification (polluters) should be paid less (i.e. a negative premium should be applied if we hold to polluter pays). However, from the point of view of the consumer, especially the poor, the lower cost (uncertified) product may be attractive, leading to greater demand for this product and thus promoting undesirable production technologies: a perverse result. The elimination of any incentive, compared to actual market conditions, for good land husbandry is also questionable. These arguments lead us into the tricky terrain of trade-offs between poverty reduction and environmental goals.

The legitimacy of land managers effectively claiming a right to pollute must be analysed in the context of the low price of agricultural goods during previous decades. In most continents, the proportion of household budgets used to buy food has never been so low. Agriculture has made huge efforts to reduce production costs while increasing the quality of products (Tilman et al, 2002). These huge efforts have repeatedly foiled the Malthusianistic prediction of cyclic widespread hunger with population growth (Trewavas, 2002). However, this progress has been made partly at the cost of the environment through the 'mining' of natural resources. Since the environmental costs of production (externalities) are not internalized in production costs, this 'mining' of natural resources to satisfy the growing demand of agricultural products will continue, following the limited but sound rationality of most farmers and consumers. If the environmental costs of food production were internalized, how would global food prices be affected?

The political and human consequences of the recent global food price crisis showed that our societies, particularly in low-income countries, are extremely sensitive to abrupt changes in food prices (Ivanic and Martin, 2008). Therefore, the inclusion of environmental costs in the prices of agricultural commodities seems to be difficult to put into practice. In the context of the global economy, it therefore seems legitimate that farmers who claim to have fulfilled their role of providing cheap food for society should ask for rewards (or compensation) if the environmental impacts of their actions are reduced, especially when doing so increases their production costs and/or reduces their income.

In this context, agricultural PES could be considered as a first limited effort to find a solution to the challenge of maintaining low food prices while protecting the environment. Nevertheless, many of the authors of this book (see Chapters 11 to 13, 16 and 17) claim that a clear political will was required to implement PES successfully in a given country. This political will, in our opinion, is derived from a societal consensus about the fundamental question of respect with regard to the legitimacy of paying farmers for reducing off-site environmental impacts.

This approach is already established in some areas of many developed countries, where farmers' income partially depends on rewards for the ecosystem services that they provide by implementing specific conservation practices on agricultural land (e.g. terraces, set-aside land or organic agriculture in and around drinking-water pumping zones). According to a World Trade Organization (WTO) perspective, these payments can be perceived as disguised subsidies (Krutilla, 1991), responsible, in turn, for artificially low agricultural commodity prices. When implementing PES in agriculture on a large scale, the risk of such objections has to be considered.

PES is not the only solution to foster the provision of environmental services in agriculture. As presented in Chapters 16 and 17 of this volume, PES must go hand in hand with other types of approaches, such as collective management tools (Chapter 16) and national policy frameworks (Chapter 17). One of the most promising approaches is to develop environmentally friendly yet productive agriculture, which is sometimes referred to as 'ecologically intensive agriculture' (Griffon, 2010). PES for agriculture can easily be integrated in this approach – for example, as a means to offset establishment costs of ecologically intensive agriculture practices.

Mutual relationships between ecosystem services in agriculture and agroforestry

As mentioned above, when we deal with agriculture, the protection of the environment is obviously not the only issue at stake. Poverty alleviation and food security are, and will continue to be, of much higher importance for most societies during the coming decades. The challenge of conserving natural resources while feeding the predicted population of 9 billion people in 2050 is a formidable one that agriculture has now to face. According to the ecological view, food provision is regarded as one of the primary services provided by ecosystems (MEA, 2003). Indeed, in the review conducted by the Millennium Ecosystem Assessment (MEA), it was one of the few ecosystem services not in decline. The challenge of producing more food while preserving natural resources in agricultural systems typically results in a trade-off between specific services (i.e. if we want to increase the provision of one service, we have to accept a decrease in the provision of another). The best solution is, then, found by optimizing competing goals. For example, it has been shown in Chapter 1 that nitrogen (N) fertilization of coffee plantations increases soil emissions of nitrous oxide (N_2O), a powerful greenhouse gas. Reduced N fertilization negatively affects coffee production. A coffee grower may have to choose between coffee production and climate regulation. If these trade-offs were immovable in agricultural systems, then protecting the environment would only be possible by decreasing food production: in the present circumstances of most countries, this would not be acceptable. Fortunately, the relationship between different ecosystem services is not always one of trade-offs, and these trade-offs are not immovable (see Figure 18.1).

Some environmental services are facilitated by the provision of another one. Facilitation means that when the provision of one service is increased, the provision of another is increased as well, as sketched in Figure 18.1(a). It is obvious that agricultural ecosystems cannot conserve all of the biodiversity found in the forest that they replace; some species are endemic to forests (e.g. forest-dependent bird species). Nevertheless, the resulting open spaces, when adequately managed, can host abundant and specific biodiversity (Steffan-Dewenter et al, 2007) and can contribute to the movement of forest-dependent species when sufficient patches of natural habitat are retained in the landscape. The importance of considering the landscape scale (e.g. the association of agriculture, forest patches and their transition zones) for the conservation and management of biodiversity is now well recognized (Tscharntke et al, 2002; Shroth et al, 2004). Agroforestry systems are well-known examples of agricultural systems that facilitate the provision of environmental services, particularly the systems that include perennial crops (coffee, cacao, etc.), which are fairly widespread in the Neotropics. On many sites, coffee production is not negatively affected by the presence of trees as long as shade levels are carefully controlled and do not pass a threshold value that depends on site conditions (soil, climate, management intensity). The multi-strata structure of these agroforests offers habitat and food for birds (see Chapter 3), and barriers against the spread of some pests (see Chapter 4); coffee and cacao shade trees can also maintain or even increase fertility by providing soil organic material, recycle nutrients, and some N-fixing species can improve the N nutrition of the perennial crop (Beer, 1987). Similar services have been documented for cocoa and pasture-based agroforestry systems (silvopastoral systems). Some aesthetic services can also be included in this category (e.g. when a watershed covered with coffee plantations offers a breath-taking landscape). Agriculture is also a way of decreasing forest fire risks in dry regions by maintaining bare soil strips in the landscape across which fire cannot cross from one forest patch to another.

The concept of agricultural multi-functionality, well researched in Europe, is based on this particular facilitation between food production and service provision (Le Cotty et al, 2003), where a positive correlation between these two functions is assumed. Successful combinations depend on the relationships between production practices and public goods in particular farms. The identification of such cases where facilitation exists is at the core of the development of multi-functionality.

There may be no interdependence between some ecosystem services (i.e. a neutral relationship) (see Figure 18.1). We define a neutral relationship as a situation where the provision of one service can be increased or decreased without affecting the provision of another. In the case of the joint provision of

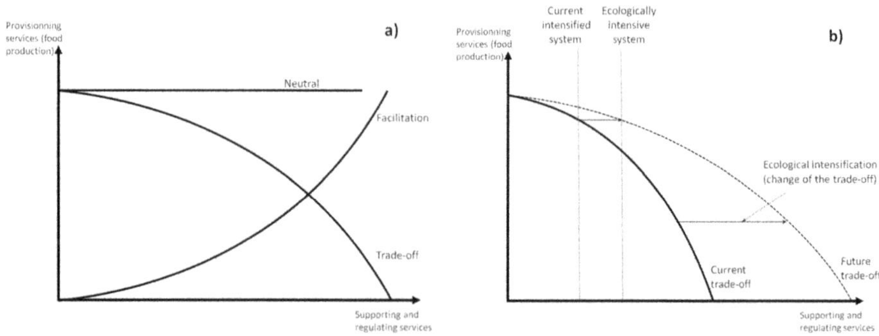

Figure 18.1 *(a) Mutual relationship between ecosystem services provision and (b) ecologically intensive agriculture*

Source: chapter authors

food and other services, a neutral relationship could result according to the following perception by a farmer: 'I could provide this service without additional costs or decrease in the production of goods I can market; but, as this service is not valued, I will not provide it.' Such situations are probably frequent (e.g. see numerous examples described in Part I of this volume). The rotation of annual crops is one example: due to the absence of incentives, this ancient practice has been abandoned in most agricultural areas. It is now common to manage the same crop year after year on the same plot; the negative consequences of this specialization, such as increased pest incidence and reduced soil fertility, are resolved with external inputs. In contrast, crop rotation increases biodiversity, resilience to pests and soil fertility, while decreasing water pollution by pesticides and nitrates, without affecting yield or profitability (e.g. Liebman et al, 2008). Obviously, PES, along with relevant scientific information made available to farmers in appropriate forms, are options to increase the provision of ecosystem services by giving incentives to facilitate the adoption of these practices.

At a landscape scale, the same kind of neutral relationships between services may occur. Pollination and pest control services are clearly favoured, maintaining connected patches of natural and semi-natural habitat in agricultural landscapes (see Chapter 5) (Tscharntke et al, 2005). In order to be effective, this strategy has to be planned at the relevant scale, as different pollinator and pest control agents perceive fragmentation at different scales and not every degree of patchiness works (Steffan-Dewenter et al, 2001, 2002). Although this can be achieved, in some cases, with no decrease in production, it requires a high level of coordination between land managers, and this involves transaction costs to design, monitor and evaluate local arrangements among stakeholders; it also requires solid scientific information to facilitate decision-making processes on land-use distribution.

Finally, there are numerous cases where trade-offs exist: an increase in the provision of one service results in a decrease in the provision of another (see

Figure 18.1(a)). Interventions that permit the conservation of biodiversity, and the provision of biodiversity-related services, often result in a decrease in food production. Here, the problem becomes one of optimization (i.e. identifying the opportunities where the equilibrium can be moved to a more favourable state). A more favourable state is one where, at a marginal cost in the provision of one service, the provision of another service can be increased dramatically. One such example was given in Chapter 10: by targeting the location where carbon sequestration is to be favoured, it is possible to almost double the number of species conserved, while only very slightly decreasing the efficiency of carbon sequestration (–4 per cent). Farm land-use planning strategies can result in large increases in some services while only slightly decreasing production (Smukler et al, 2010). The identification of such opportunities, and of practical ways of moving equilibrium to a more favourable state, requires more research, both at the field and landscape scales. Some preliminary results are presented in this volume; but much more has to be done to convince stakeholders and to establish a sound footing for PES schemes.

Research to be undertaken in this area is mainly the concern of two disciplines: agronomy and ecology. Agronomists work on altered simplified ecosystems, where few species are considered and human interventions are essential. Ecologists traditionally focus on complex natural systems, minimizing human influence. The challenge during the coming decades will be to combine these disciplines in order to design interventions that reduce the trade-offs between services, while focusing on synergies, where the provision of multiple or all services is increased. Plant breeding could play an essential part; but new varieties need to be developed for agroecological diverse landscapes, instead of being dependent on the maintenance of an artificial uniform landscape. On the other hand, the systems to be implemented will be more complex. Knowledge management will play a crucial role, with new ways of sharing information and networking requiring huge innovative efforts.

Main Lessons about Ecosystem Services Provision and PES

This book is primarily intended to synthesize and provide practical information to students, scientists and practitioners about the methods used to estimate the quantity and the value of ecosystem services from agriculture and agroforestry. Most of the chapters dealing with the measurement and value of ecosystem services were based on practical experiences. The key lessons regarding the generation of ecosystem services that can be drawn from these experiences are presented in the following section.

Measurement of the provision of ecosystem services

The prices paid in PES schemes, as we have seen in Part III of this volume, are currently quite low and contrast strikingly with the estimated value of ecosystem services. For example, the Millennium Ecosystem Assessment (MEA, 2005) shows that when the value of all services is taken into account, natural ecosystems have a higher global value than the agricultural systems into which they were transformed. Nevertheless, the amounts paid for the provision of services in the current PES (Parts II and III of this volume) represent a small fraction of farmers' income; the bulk still comes from trading marketable goods. The price of services is frequently high in theoretical studies, based on modelling (e.g. Olschewski et al, 2010); but it rapidly dwindles when it is considered in existing schemes. In reality, this price is based on the buyer's willingness to pay, whether through PES or eco-labelling; unfortunately, this willingness is low.

One of our main assumptions is that the willingness to pay (WTP) for services (the demand side of PES or eco-labelling) is low because, among other factors, the relationships between land management activities and ecosystem services have not been evaluated and demonstrated thoroughly; the existing scientific literature is still insufficient. This assumption is not new; indeed, every paper on PES makes the same claim (see Table 18.1). This gap clearly limits our capability to ensure the final user (demand side) that real provision of ecosystem services will derive from their financial effort, and therefore impairs PES development.

At the field scale, some knowledge has been acquired about the effect of agricultural practices on hydrological services (water regulation, soil sheet erosion) and carbon sequestration in above-ground biomass. However, soil carbon sequestration is poorly assessed, and the local conditions (soil, climate, etc.) that control the provision of climate regulation services are poorly documented, particularly those that exist in the tropics (e.g. in the case of Andosols). This poor documentation reduces our ability to transfer the results to other locations than those where the evidence was produced. Another important specific phenomenon that has been very poorly assessed is the occurrence of landslides in the landscape, and the influence of the combination of key factors such as soil type, vegetation cover and agricultural practices on this occurrence. With regard to the services related to biodiversity, there is still much to be done. Few taxa have been studied (mainly mammals and birds); most insect taxa and soil organisms have not been adequately investigated.

How to scale up from the field to the landscape or watershed is of uttermost importance, particularly regarding two principal services: water and biodiversity. In the first case, some issues related to scaling up to the watershed level have been overlooked, such as the effect of roads in a watershed (Gómez-Delgado et al, in preparation). Data on water fluxes, associated sediment, pesticides or nitrate fluxes at field and watershed scales are clearly needed over the long term in order to capture exceptional events that shape the landscape and provoke most losses and pollution. From long-term series (few are available), models of varying complexities – and, therefore, input data needs – have to be detailed and calibrated in order to fit to a range of applications derived from these pilot watersheds. In the case of biodiversity-supported services,

Table 18.1 *Synthesis from this volume of the main lessons regarding ecosystem services and the means to foster their provision*

Topic	*Key statements from literature**	*New insights from recent experience and research in the Neotropics*
Measurement	Many payment for ecosystem services (PES) programmes are based on shaky scientific foundations; there is an urgent need for better documentation of the relationship between proxies and real environmental benefits. Water ecosystem services (ES) are the least monitored ES. Biodiversity is the most complex and costly ES to measure.	Greenhouse gases (GHGs) are not only comprised of CO_2. Other GHGs must be considered in PES schemes (see Chapter 1). Soil carbon is not properly assessed (see Chapter 1). Spatial scale matching is essential for the estimation of services related to biodiversity and soil and water; but this is a complex issue (see Chapters 2 and 3). ES may not all be measurable on a project basis; but the bulk scientific evidence between land uses and service provision must be stronger (see Chapter 6). Better models and tools are needed to extrapolate local scientific evidence; models and tools should be used more frequently (see Chapters 2 and 6). Landscape beauty requires measurement.
Environmental effectiveness	Additionality is a key concept to ensure effectiveness (payments are given only to foster additional provision). Targeting is required to ensure effectiveness. Leakage due to PES-related price changes must be addressed. Effectiveness of PES schemes must be monitored. Additionality means that when payments stop, land uses may go back to original states.	PES schemes in a clear legal context increase the provision of ES (see Chapter 14). This is a methodologically difficult issue; good databases are needed for correctly assessing effectiveness (see Chapter 14). In agricultural landscapes, additionality enforcement can have perverse effects (see Chapter 6). Regarding eco-labelling, additionality is not accepted as an obvious requirement (see Chapters 11 and 15).
Cost efficiency	Transaction costs are usually prohibitive and undermine PES efficiency. To increase efficiency, payments must be fixed at the opportunity costs (eventual benefits to be given to users, not to providers). There is a trade-off between cost efficiency and poverty alleviation. The more heterogeneous the costs to provide ES, the more efficient PES, compared to regulatory approaches.	Bundling is possible and useful in order to decrease transaction costs (see Chapter 6). In a context where citizen support is secured, there can be mixes between a state-supported framework and a private initiative in order to reduce transaction costs (see Chapters 8, 9, 16 and 17). In the case of agricultural activities that need little impetus to be adopted, PES can provide the incentive to provide permanently inexpensive services (see Chapter 6).

Table 18.1 *continued*

Topic	*Key statements from literature**	*New insights from recent experience and research in the Neotropics*
Social equity and poverty alleviation	Ensure poor household participation is a key feature of successful payment for ecosystem services (PES). Cooperative arrangements are a valuable mechanism for overcoming poor obstacles in accessing ecosystem services (ES) markets. Targeting the poor is not a necessary condition for benefiting them. PES buyers tend to appropriate informational rents due to higher negotiating power than ES providers. Start-up costs are the major problem for poor households in accessing ES markets. The common perception that the poor are interested only in the short term may be misplaced.	It is probably easier to leverage eco-labelling towards poverty alleviation than PES. Targeting the poor is not a sufficient condition for benefiting them. The sharing of benefits of ES provision between providers and buyers must be fostered by more transparent information on costs. PES and eco-labelling can have beneficial side effects on poverty reduction and on changing perceptions about agricultural activities.
Sustainability and governance	PES implementation must begin from the demand side, not the supply side. Government environmental regulations are key to stimulating PES. User-financed programmes probably have higher additionality than government-financed ones. Government environmental regulations are key in stimulating ES market development.	The whole state intervention philosophy must be thought over to increase its coherence in protecting the natural resource base. Social awareness is the key to sustaining PES (and eco-labelling); but we know little about how to increase this awareness. Willingness to accept (WTA) needs more attention. PES must only be thought of as a means, among other complementary methods.

Source: * Landell-Mills and Porras (2002); Pagiola et al (2002); Pearce and Mourato (2004); Engel et al (2008); Jack et al (2008); Wunder et al (2008)

more studies are required on the links between the characteristics of the landscape and species presence and functions, particularly in relation to agricultural production processes. Ecological modelling is required to adapt the findings to other locations; but adequate tools for landscape planning in relation to biodiversity conservation and functions are, in our understanding, still to be developed.

Finally, other services are clearly not adequately measured, such as landscape aesthetics. This can be a fundamental service, particularly as a powerful trigger to increase the flow of financial resources to nature conservation, and therefore to foster the provision of other services less subject to payments. We did not attempt to review methods to measure this service in this volume because experience is scarce; but we feel an urgency to address this gap.

Environmental effectiveness

Are PES and eco-labelling effective means for the provision of ES?

Robalino and colleagues (see Chapter 14) report that the Costa Rican PES programme has been effective; more conservation was achieved where PES was implemented. They also found that the measurement of effectiveness is relatively complex; the isolation of a PES factor in such an analysis is far from an easy task. They could perform this analysis because of access to good databases and knowledge, which probably would have been difficult to find in other developing countries. The comparison of the effectiveness of PES and eco-labelling (see Chapter 11) did not provide definitive answers, but showed the importance of pursuing these analyses.

In the National Forestry Financing Fund (FONAFIFO) experience, additionality has not been explicitly pursued. The overall effects of Costa Rican public policies that were implemented during the 1990s have been, for the most part, successful (see Chapter 12), perhaps more than could be expected from Robalino's calculations. In fact, PES schemes did not come out of a wizard's hat; they were part of a global set of policies, where regulations played an essential part. The Costa Rican public's interest in the environment, interlinked with the economic success of the ecotourism focus that this country has adopted, was also influential. In such a favourable context, environmental effectiveness was achieved to a certain extent, although additionality was not enforced.

In the context of PES in agriculture, additionality is a difficult concept to apply successfully. Agriculture in the Neotropics is characterized by many small and medium-sized landholders, families, co-operatives, local non-governmental organizations (NGOs), etc. As shown by Villanueva and colleagues in Chapter 6, the GEF-Silvopastoral project enforced additionality, and programme participants perceived this as unfair: the 'good' farmers who had already adopted environmentally friendly farming practices were not as well rewarded as the latecomers. Prins and León (Chapter 16) expressed the importance of local dialogue in order to protect water sources effectively. This dialogue could be clearly impaired due to the perceived unfairness of the proposed PES. This may lead to perverse behaviour, such as the artificial lowering of the baseline from which additionality is measured. Although we do not have evidence of this, we feel that when applying PES approaches to agricultural systems, using additionality as a basic requirement and evaluation criteria for PES is prone to generate perverse behaviour and negative dynamics.

Soto and Le Coq (Chapter 15) have observed a great range of strategies for achieving environmental effectiveness and additionality enforcement with regard to eco-labelling. At the extremes are organic seals where additionality is non-existent: the standard is strongly defined, but proof of change in behaviour is not required. At the opposite extreme are 'negotiated' seals, where the standards are adapted to the local situation and the farm must gradually improve its environmental practices. Therefore, not every eco-label enforces additionality as a rule for payments.

Effectiveness comparisons between instruments are relatively rare on the ground. We are still in a phase where experiences are recent and the most knowledgeable stakeholders are those who implement the instruments, and sometimes benefit from them, either because they sell the seal or because their institution depends on the PES. These stakeholders are ill placed to evaluate the effectiveness of the instruments. Other stakeholders without conflict of interest must conduct these comparisons with comparable and thorough methodologies. The challenge is to identify the causes for observed behaviours at different spatial and temporal scales, separating the specific effects of PES and eco-labels from other variables, such as socio-economic conditions and policies. These comparisons are necessary to fill in our knowledge gap.

Cost efficiency

The main drawbacks of PES and eco-labelling are the transaction costs: the costs incurred by the providers to enter into the scheme and the costs to establish the scheme. The cost calculations are somewhat arbitrary. For example, production of scientific evidence on service provision by certain land uses can be included in the project costs. However, universities and research institutes who are funded by the state usually produce the scientific evidence, and their costs are not included in the transaction costs. The transaction costs may be calculated as proposed by the silvopastoral project (see Chapter 6) by including the costs actually incurred to perform the necessary monitoring measurements, but not the costs related to the building of the indicators. The local capabilities of PES research and implementation bodies will be an important facilitating element to increase cost efficiency, as was advocated in Chapter 17. Costa Rica has conducted numerous research projects on natural resources that facilitated the first PES pilot projects, which, in turn, allowed for capacity-building of local stakeholders on PES implementation.

Decreasing transaction costs, however, can have detrimental consequences for other PES features, such as effectiveness, due to poor monitoring. ES bundling, as analysed by Villanueva et al (Chapter 6), is a tool for decreasing transaction costs without detrimental effects: with the same implementation framework (i.e. the same transaction costs), the payments given and the range of the services provided are increased. This bundling is possible when services are actually linked – for example, removing herbicides from coffee plantations' farming practices can reduce the sediment load of rivers downstream, reduce the leaching of herbicide residues in drinking water, and increase biodiversity of soil microorganisms and avifauna. Low transactions costs may also be obtained thanks to adequate simple tools built for this purpose. For example, the GEF-Silvopastoral project (Chapter 6) utilized simple indices to assess both biodiversity conservation and carbon sequestration of farms. Strengthening the indices both scientifically (better representation of actual services provided – that is, adaptation to different environmental conditions) and practically will make them more applicable on a larger scale. In addition, stakeholders must agree upon the representation of services; therefore, development of reliable, simple and shared tools is clearly an area where research is still much needed.

Neotropical agriculture is still in the hands of numerous smallholders, with a high proportion of poor individuals. As we have seen in the first part of this chapter, the agricultural sector offers many opportunities to provide more ES, with little or no reduction in the production of goods. However, lack of incentives, lack of knowledge or lack of resources to take over the upfront costs needed to adopt new practices limits the number of actual opportunities for ecosystem services provision. PES and eco-labelling may provide the means to increase ecosystem services provision by giving smallholders the necessary impetus. These schemes could achieve very high cost effectiveness; but to do so, additional research is clearly required to identify trade-offs and more efficient means to help or induce smallholders to modify their farming practices.

Social equity and poverty alleviation

Poverty is steady in Latin America in spite of the satisfactory global rate of development in this region. Latin America is urbanizing rapidly; nevertheless, approximately half of the poor live in rural areas, where the poor represent a higher percentage of the total rural population. In Mesoamerica, in particular, the share of the global population under the poverty threshold varies between 25 and 58 per cent depending on the country, while the share of the poor among the rural population varies between 29 and 74 per cent, this last figure in Panama, a country with a middle income (Siaens and Wodon, 2003). Furthermore, the rural poor are less linked to markets, and are less sensitive to global development than the urban poor (Fay and Ruggeri Laderchi, 2005). Rural poverty is thus a problem that Latin America has been unable to alleviate. While PES and eco-labelling schemes have been designed in this context, is it realistic to mix objectives as different as natural resources protection and poverty alleviation?

As mentioned by Le Coq et al (see Chapter 11), the fundamental problem of PES and eco-labelling with regard to poverty alleviation is that both are based on assets (especially land) or products. In PES, payments are made on the basis of the land area upon which management practices change. In eco-labelling, payments are made on the basis of the quantity of products sold on the market. Therefore, the poor, who own small farms (or no farms) and produce small quantities, will mathematically receive smaller payments from both schemes. Moreover, the poor usually lack information and are ill organized to prepare the paperwork necessary to enter into PES schemes. There are two clear exceptions to this rule. Some poor, particularly in indigenous areas, can manage large areas of land, including forested land, where PES aimed at forest conservation can bring them relatively high rewards. This is the case of some PES contracts in the Talamanca indigenous region of Costa Rica (O. Sánchez, FONAFIFO, pers comm). Second, the Fair Trade eco-label specifically aims at poverty alleviation and guarantees a price premium on the product. Although the payments are still based on product quantity, the Fair Trade label excludes large farms not organized in a co-operative from certification. The label also includes provisions to strengthen farmers' organizations, relieving the information and organization exclusion motive (see Chapters 11 and 15).

Eco-labelling had a clear effect on farmers' employees, who for coffee, in particular, represent very poor and vulnerable populations. Various eco-labels include provisions for these employees: Rainforest Alliance requires large farms (their main target in Central America) to improve the farm housing provided to employees. By including such poor workers into their certification criteria, eco-labels may have had real effects, even if poorly documented. Eco-labels have another side effect on poverty. Since they require auditing, farmers have to keep record books on their activities and results. This recording, and the importance attached to it, makes enrolled farmers more self-conscious and reflective of their activities, resulting in positive effects on their performance. Furthermore, the fact that their activities are recognized by buyers adds a sense of pride that aids in the relief if not of poverty, at least of certain perceptions of poverty.

PES has not had the same effects. Due to the quest for efficiency, the smallest farmers, if not included in co-operative organizations, will probably be excluded from the schemes; the monitoring costs of small units are too high for PES to be profitable. Even the Mexican PHES, which includes an explicit poverty alleviation goal, targets payments in areas where forests are at risk of deforestation, which excludes the most remote and probably poorest areas where even the means for deforestation are lacking. There are many PES scholars who claim that PES should have clear environmental objectives. Mixing these objectives would endanger the sustainability and final outcome of PES (Bayon, 2002). Nevertheless, national PES initiatives have clearly tried to incorporate poverty alleviation within their outcomes (see Chapters 12 and 13) by targeting PES in areas where rural poverty is a stringent issue. However, as Rolón Sánchez et al concluded in Chapter 13, geographical targeting is not enough; PES could favour rich farmers in areas where there is much poverty.

The solution for PES probably relies on its design and context. The design stage of PES schemes has been identified as a key issue in the future relationship between PES and poverty in Mesoamerica: if the stakeholders, including the poor, are included from the beginning in the negotiation, if they hold a significant place and have some power, then the PES is more likely to alleviate poverty (Corbera et al, 2007). On the other hand, as Rodriguez pointed out in chapter 17, PES must form part of a global policy that includes the reduction of rural poverty to be effective in natural resource management. Whether this goal will be achieved through the inclusion of the poor in PES negotiations, or through geographical targeting, or through other instruments complementary to PES, additional research is necessary. It seems to us that agricultural PES needs to devote greater attention and resources to poverty alleviation, whether for ethical or efficiency reasons. The challenge resides in the way to achieve these joint goals.

PES sustainability and governance

PES and eco-labels face sustainability risks. In Chapter 12, Murillo and colleagues presented the history of Costa Rican PES and its quest for sustainable funding. Soto and Le Coq, in Chapter 15, presented the evidence of coffee eco-labels' market limitations: farmers have more certified products than they are able to sell at the end of the year. Domestic markets for eco-labelled products are nascent or non-existent. Many chapter authors argue that a strong political will is required to push the required reforms in order to make these initiatives sustainable. Since PES and eco-labels are driven by the demand side (Wunder et al, 2008), the issue, then, is how to boost this demand?

Every country has the government it deserves (de Maistre, 1851). Following this old refrain, the grounds for the required political will could be found in society, particularly in the education of its younger generations. Sustainability of these schemes will depend in large part on the efforts of stressing environmental education at an early age.

A good part of the experience related in this book has been drawn from the Costa Rican experience, a middle-income country. This country has been known for its success as a tourist destination because of its green image. However, this green image has been forged over generations and the protection of the environment is an important part of the day-to-day life of this country. Twenty years ago, there were already proposals to include environmental topics in kindergarten education programme (Chacón Mora and Hernández Avila, 1991). Nowadays, most citizens find it normal to pay a tax on gasoline to fund PES and to have 25 per cent of their territory as protected areas. Is this concern towards the environment related to the fact that tourism is the first source of foreign currency? It probably helps, whether it is related to the income generated or to the fact that it generates pride.

Citizen awareness and concern regarding the environment was, and still is, a main priority for many stakeholders in Costa Rica, including the national government. But global awareness (inside and outside Costa Rica) is also important to sustain and broaden the success of eco-labelling. Following the coffee sector's lead, other sectors are developing similar schemes: recently, for example, Rainforest Alliance and CATIE developed a standard for environmentally friendly meat production. Nevertheless, experience shows that these strategies work better in the beginning (the first actors, most dynamic, are those who are best rewarded). Today, coffee producers certify in excess in relation to the market demand. Therefore, the question remains with regard to the future of eco-labels: will this be a race in which the standards will become harder to meet with time so that there always exists a small proportion of dynamic farmers or farmers' organizations who will get the premium? The other possible future is a broad adoption of the current standards; however, it would destroy the niche strategy, premiums would vanish and, eventually, the minority of farmers who did not adopt the label would be excluded from international market access. Obviously, the sustainability of eco-labelling in the long run is also a theme for concern. The progression of the demand side for both eco-labelling and PES is crucial.

The research community should create the necessary conditions to convince society of the importance of natural resources and the means of efficiency to achieve additional provision of ecosystem services. This can be achieved through new evidence, better communication of that evidence and capacity-building in order to better acquaint human resource with the complexity of PES: measurements, contracting, monitoring, etc.

All of these efforts are required for developing the demand side of PES; however, the supply side also requires attention for two reasons. First, not all suppliers agree to enter into contracts, which, in turn, are required for a continuous supply of ecosystem services. We have seen in the case of the Water and Power Utility Company in Heredia, Costa Rica (ESPH) (Chapter 12), that numerous potential suppliers did not agree to commit their land to a long-term contract. Supplier motivation must be further researched to take into account not only financial rationality, but also non-financial rationality and perceptions. This knowledge may help to find ways of adapting PES incentives in a more comprehensive way, and designing strategies to improve willingness to accept and, more generally, environmental motivations. Second, when PES is implemented, there must be a fair negotiation between users and providers or their representatives. All parties must have the relevant information at their disposal in order to be successful, and costs must be known and shared. Therefore, replacement costs and avoided costs for the users, as well as opportunity costs for the providers, must be further investigated on the basis of the scientific evidence discussed in parts I and II.

There is still a long way to go in order to attain a state where citizen and government willingness to pay approaches the 'real' value of ecosystem services. Although the consciousness of the population regarding environmental issues such as climate change, water scarcity and pollution, or biodiversity loss is growing, both in Southern and Northern countries, the willingness to pay substantial sums to reduce these environmental threats is still limited. Socio-economic conditions of households both in Northern countries, struck by the financial crisis, and, more evidently, in Southern countries, affected on a long-term basis by poverty, are weak. Markets for eco-labelled products are still restricted to developed countries. In developing countries, where these eco-markets struggle to emerge, incentives are restricted to exported goods, excluding all agricultural goods consumed locally. Resources are allotted to basic needs such as food, lodging, other consumptions goods, leisure, etc., and the protection of the environment is still considered as a non-basic need. New models of consumption are needed to change the priorities in expenses management both at household and state levels. PES schemes are one of the ways to promote society's learning processes to recognize the value of ecosystems and the importance of the services that they provide.

However, the strong growth of eco-labels (increasingly found on brand names of household goods such as Starbucks, Kraft Foods, McDonald's, Folgers, Nestlé, etc.) suggests that progress is being made; consumers recognize the added values, and corporations are also interested. PES schemes, including eco-labelling, are approaching their adolescent years, and growing pains are clearly evident. The attention given at the birth of these programmes is passing, with increasingly critical attention given to evaluate their effectiveness, and ensuring that the added value being paid for is received. The future, although one of significant challenges, is promising. PES schemes and labels are being developed at all scales (local, national, regional and international), based on a broad spectrum of mechanisms. It is very likely that the next couple of decades will present the second generation of PES schemes, with more direct payments between providers and users, and more cost-effective and rigorous measurements of services provided.

References

Bayon, R. (2002) 'More than hot air: Market solutions to global warming', *World Policy Journal*, vol 19, pp60–68

Beer, J. (1987) 'Advantages, disadvantages and desirable characteristics of shade trees for coffee, cacao and tea', *Agroforestry Systems*, vol 5, pp3–13

Chacón Mora, M. and Hernández Avila, N. (1991) *Propuesta de un manual de educación ambiental para niños y maestros de educación preescolar*, Licence thesis, Universidad de Costa Rica, San José, Costa Rica

Corbera, E., Kosoy, N. and Martinez Tuna, M. (2007) 'Equity implications of marketing ecosystem services in protected areas and rural communities: Case studies from Meso-America', *Global Environmental Change*, vol 17, no 3–4, pp365–380

de Maistre, J. (1851) *Lettres et Opuscules Inedits, Volume I*, Lettre 53, Paris

Engel, S., Pagiola, S. and Wunder, S. (2008) 'Designing payments for environmental services in theory and practice: An overview of the issues', *Ecological Economics*, vol 65, no 4, pp663–674

Fay, M. and Ruggeri Laderchi, C. (2005) 'Urban poverty in Latin America and the Caribbean: Setting the stage', in M. Fay (ed) *The Urban Poor in Latin America*, World Bank, Washington, DC

Gómez-Delgado, F., Roupsard, O. and Moussa, R. (in preparation) 'Water and sediment yield in a coffee agroforestry system at various spatio-temporal scales: From plot to basin and from event to annual scales'

Gregory, P. J., Ingram, J. S. I., Andersson, R., Betts, R. A., Brovkin, V., Chase, T. N., Grace, P. R., Gray, A. J., Hamilton, N., Hardy, T. B., Howden, S. M., Jenkins, A., Meybeck, M., Olsson, M., Ortiz-Monasterio, I., Palm, C. A., Payn, T. W., Rummukainen, M., Schulze, R. E., Thiem, M., Valentin, C. and Wilkinson, M. J.

(2002) 'Environmental consequences of alternative practices for intensifying crop production', *Agriculture, Ecosystems & Environment*, vol 88, no 3, pp279–290

Griffon, M. (2010) *Pour des agricultures écologiquement intensives*, Editions de l'Aube, Paris

Ivanic, M. and Martin, W. (2008) 'Implications of higher global food prices for poverty in low-income countries', *Agricultural Economics*, vol 39, pp405–416

Jack, B. K., Kousky, C. and Sims, K. R. E. (2008) 'Designing payments for ecosystem services: Lessons from previous experience with incentive-based mechanisms', *Proceedings of the National Academy of Sciences of the United States of America*, vol 105, no 28, pp9465–9470

Krutilla, K. (1991) 'Environmental regulation in an open economy', *Journal of Environmental Economics and Management*, vol 20, pp127–142

Landell-Mills, N. and Porras, T. I. (2002) *Silver Bullet or Fools' Gold? A Global Review of Markets for Forest Environmental Services and Their Impact on the Poor*, Instruments for Sustainable Private Sector Forestry Series, International Institute for Environment and Development, London

Le Cotty, T., Aumand, A. and Voituriez, T. (2003) 'Multifonctionnalité et coopération multilatérale: Une analyse du coût de fourniture de biens publics par l'agriculture', *Economie Rurale*, vol 273–274, pp91–102

Liebman, M., Gibson, A. R., Sundberg, D. N., Heggenstaller, A. H., Westerman, P. R., Chase, C. A., Hartzler, R. G., Menalled, F. D., Davis, A. S. and Dixon, P. M. (2008) 'Agronomic and economic performance characteristics of conventional and low-external-input cropping systems in the Central Corn Belt', *Agronomy Journal*, vol 100, pp600–610

MEA (Millennium Ecosystem Assessment) (2003) *Ecosystems and Human Well-Being: A Framework For Assessment*, Island Press, Washington, DC

MEA (2005) *Ecosystems and Human Well-being: Synthesis*, Island Press, Washington, DC

Olschewski, R., Klein, A. M. and Tscharntke, T. (2010) 'Economic trade-offs between carbon sequestration, timber production, and crop pollination in tropical forested landscapes', *Ecological Complexity*, vol 7, no 3, pp314–319

Pagiola, S., Landell-Mills, N. and Bishop, J. (2002) 'Making market based mechanisms work for forests and people', in S. Pagiola, J. Bishop and N. Landell-Mills (eds) *Selling Forest Environmental Services*, Earthscan, London, pp261–289

Pearce, D. and Mourato, S. (2004) 'The economic valuation of agroforestry's environmental services', in G. Schroth et al (eds) *Agroforestry and Biodiversity Conservation in Tropical Landscapes*, Island Press, Washington, DC, and Covelo, London, pp67–86

Shroth, G., Da Fonseca, G. A. B., Harvey, C., Gascon, C., Vasconcelos, H. L. and Izac, A. M. N. (eds) (2004) *Agroforestry and Biodiversity Conservation in Tropical Landscapes*, Island Press, Washington, DC, and Covelo, London

Siaens, C. and Wodon, Q. (2003) *Latest Estimates of National Urban and Rural Poverty in Latin America*, World Bank, Washington, DC

Smukler, S. M., Sánchez-Moreno, S., Fonte, S. J., Ferris, H., Klonsky, K., O'Geen, A. T., Scow, K. M., Steenwerth, K. L. and Jackson, L. E. (2010) 'Biodiversity and multiple ecosystem functions in an organic farmscape', *Agriculture, Ecosystems & Environment*, vol 139, no 1–2, pp80–97

Steffan-Dewenter, I., Munzenberg, U. and Tscharntke, T. (2001) 'Pollination, seed set and seed predation on a landscape scale', *Proceedings of the Royal Society of*

London Series B: Biological Sciences, vol 268, no 1477, pp1685–1690

Steffan-Dewenter, I., Munzenberg, U., Burger, C., Thies, C. and Tscharntke, T. (2002) 'Scale-dependent effects of landscape context on three pollinator guilds', *Ecology*, vol 83, no 5, pp1421–1432

Steffan-Dewenter, I., Kessler, M., Barkmann, J., Bos, M. M., Buchori, D., Erasmi, S., Faust, H., Gerold, G., Glenk, K., Gradstein, S. R., Guhardja, E., Harteveld, M., Hertel, D., Hohn, P., Kappas, M., Kohler, S., Leuschner, C., Maertens, M., Marggraf, R., Migge-Kleian, S., Mogea, J., Pitopang, R., Schaefer, M., Schwarze, S., Sporn, S. G., Steingrebe, A., Tjitrosoedirdjo, S. S., Tjitrosoemito, S., Twele, A., Weber, R., Woltmann, L., Zeller, M. and Tscharntke, T. (2007) 'Tradeoffs between income, biodiversity, and ecosystem functioning during tropical rainforest conversion and agroforestry intensification', *Proceedings of the National Academy of Sciences of the United States of America*, vol 104, no 12, pp4973–4978

Tilman, D., Cassman, K. G., Matson, P. A., Naylor, R. and Polasky, S. (2002) 'Agricultural sustainability and intensive production practices', *Nature*, vol 418, pp671–677

Trewavas, A. (2002) 'Malthus foiled again and again', *Nature*, vol 418, pp668–676

Tscharntke, T., Steffan-Dewenter, I., Kruess, A. and Thies, C. (2002) 'Characteristics of insect populations on habitat fragments: A mini review', *Ecological Research*, vol 17, no 2, pp229–239

Tscharntke, T., Klein, A. M., Kruess, A., Steffan-Dewenter, I. and Thies, C. (2005) 'Landscape perspectives on agricultural intensification and biodiversity: Ecosystem service management', *Ecology Letters*, vol 8, no 8, pp857–874

Wunder, S., Engel, S. and Pagiola, S. (2008) 'Taking stock: A comparative analysis of payments for environmental services programs in developed and developing countries', *Ecological Economics*, vol 65, no 4, pp834–852

Index

For Product Safety Concerns and Information please contact our EU representative GPSR@taylorandfrancis.com
Taylor & Francis Verlag GmbH, Kaufingerstraße 24, 80331 München, Germany

www.ingramcontent.com/pod-product-compliance
Ingram Content Group UK Ltd.
Pitfield, Milton Keynes, MK11 3LW, UK
UKHW022213190926
475702UK00003B/334

* 9 7 8 1 8 4 9 7 1 1 4 7 0 *